BIOSEMIOTICS:

An Examination into the Signs of Life
and the Life of Signs

APPROACHES TO POSTMODERNITY

JOHN DEELY, SERIES EDITOR

VOLUME II

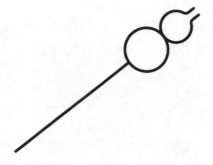

Every movement needs a symbol to grow. If it is the Way
of Signs that leads to postmodernity, how is that path symbolically to be marked. "By the sign of a sign", Brooke
Williams remarked to me in the early days of the Semiotic
Society of America, whereupon she lay down on the table
the symbol above. "And what is that, pray tell?", I queried.
"A sign of a sign", she replied, "the caduceus, The staff of
a messenger bearing a message." Now let it stand also for
this new series as postmodernity advances.

BIOSEMIOTICS:

*An Examination into the Signs of Life
and the Life of Signs*

Jesper Hoffmeyer

Translated by Jesper Hoffmeyer and Donald Favareau

Edited by Donald Favareau

University of Scranton Press
Scranton and London

English translation 2008

Originally published in Danish
as *Biosemiotik. En afhandling om livets tegn og tegnenes liv.*
2005 Ries Forlag, Copenhagen

This translation has been sponsored by the Danish Arts Council Committee for Literature.
It was also supported by University of Copenhagen's research priority area, Religion in the
21st Century; http://www.ku.dk/priority/Religion/index.htm

Library of Congress Cataloging-in-Publication Data

Hoffmeyer, Jesper
   Biosemiotics : signs of life and life of signs / by Jesper Hoffmeyer.
      p. cm.
Includes bibliographical references and index.
ISBN 978-1-58966-169-1 (cloth : alk. paper)
   1. Biology--Semiotics. I. Title.
   QH331.H593 2008
   570.1'4—dc22

2008018733

Distribution:
University of Scranton Press
Chicago Distribution Center
11030 S. Langley Avenue
Chicago, IL 60628

For Thomas Sebeok

Table of Contents

Part 3. Biosemiotics and the Human Being

Acknowledgments

The English language edition of *Biosemiotics* that you now hold in your hands is primarily an updated version of my original book of the same name in Danish, but one that has greatly profited from later critical examinations and new ideas — not the least of these being due to the highly informed and indefatigable editorial work invested in the book by Don Favareau. In addition, many other persons have been creatively contributing to the establishment of this new research area, and almost all of them have their share of insights presented in this work. I have done my best to credit them throughout the text and hope that I haven't made too many oversights.

Most importantly, it should be recognized that this whole work would not have come into existence without the continued support and inspiration of the late Thomas A. Sebeok (1920–2001). Also, my close collaborators through these many years, Claus Emmeche and Kalevi Kull, and my continued meetings with Frederik Stjernfelt, have greatly contributed to my development of the analysis presented here. I am particularly grateful to Emmeche and Stjernfelt for their critical examination of the original Danish version of this book. I also want to thank Peder Voetmann Christiansen, Mogens Kilstrup, Robert Ulanowicz, and Myrdene Anderson for constructive comments on several chapters. It goes without saying that none of these people have any responsibility for the actual text as presented in the book.

Finally, it should be noted here that the late Thure von Uexküll (1908–2004) played an absolutely unique role in the early phases of the establishment of biosemiotics, as did Jörg Hermann, who was the head of the medical section at the ReHa Klinik Glotterbad that hosted our first international meetings on biosemiotics in the early 1990s. Without these meetings, I might never have dared to invest so much effort in continuing my work in biosemiotics. Finally, I should thank the multinational group of scholars that have met and who continue to meet regularly at the annual *Gatherings in Biosemiotics* both for the fruitful and inventive atmosphere that has reigned in these meetings, and for the many important contributions that have influenced and furthered my thoughts throughout the years.

Preface

When a brown hare spots a fox approaching in the open landscape, the hare stands bolt upright and signals its presence instead of fleeing. The explanation for this behavior, according to ethologist Anthony Holley (1993), is that a hare can easily escape a fox simply by running – a fact that the fox seems to "know" (whether by learning or instinct). Apparently, then, what is happening in this behavior is that the hare is telling the fox: "I have seen you" – and as a result, they can both be spared the effort of running.

A modern Darwinist may claim that this communicative exchange is easily explained by the increased fitness conferred upon individuals by the lowering of their energy expenditures. That is an important insight with which I shall not quarrel. And yet, I cannot help feeling that something is missing in this kind of explanation. For how can we be so sure that these animals themselves — as environmentally situated organisms of blood, flesh, and brain — take no creative part in their own behavior? The Neo-Darwinist explanation would require us to delegate to their genetic apparatus the whole burden of anticipating the outcome of all and every future communicative situation these animals may encounter. But why would evolution equip mammals with brains containing billions of extremely energy-costly nerve cells if such brains were then not allowed to make any decisions not already anticipated by the genes?

Charles Darwin got it right in the sense that no behavior is likely to evolve in the organic world that significantly lowers the fitness of phenotypes exhibiting such behavior. This is an extremely important insight — but it should not blind us to the deeper questions surrounding the ubiquitous *intentionality* of communicative behavior. Living creatures are not just senseless units in the survival game; they also *experience* life (and perhaps even "enjoy" it as we say when human animals are concerned).

In other words, there is, as we shall see in this book, an element of *natural play* and not just of *natural selection* discoverable in the natural world. There, organisms never "try to survive" — for the simple reason that they cannot know they are going to die. Rather, they try to escape or to counteract life-threatening events as those events present themselves; more precisely: they just instinctively do what they must to live. In brief: Organisms strive, and this *striving* — a word Darwin did not himself shy away from using — cannot be set aside in any genuine attempt to understand the workings of animate natural systems.

Making scientifically responsible sense of this "striving" is one of the challenges that the emerging scientific field called *biosemiotics* sets out to accept,
and it does so by presenting an understanding that biological communication
is more than just machine-like exchange of information. True communication, biosemiotics argues, is based on *semiosis*, or sign processes. Sign processes
are what this book is all about, and I shall be delving deeply into the question
"What is a sign?" (as well as many questions regarding how signs can be understood scientifically) throughout this text.

For the purposes of this preface, however, it will suffice for now just to say
that a sign is something that refers to something else — with the essential addition that it takes somebody (i.e., a receptive living system) to make the reference. The meaning conferred by a sign is thus acutely dependent on the nature
and the context of its receptive system, the sensing body — and that body's
relations with externality are mediated continually by the active establishment
and disestablishment of such signs. A sign process, then, is more than just a
mechanical transfer of information packets because the sign embraces *a process
of interpretation*. And yet, it is precisely the biological phenomena that comprise this interpretative activity that is neglected — or at least not recognized
as engendering its own measure of causal efficacy in the world — in both traditionally conceived Information Theory and in most contemporary mainstream
Evolutionary Theory.

Yet by making just this slight and empirically well-justified expansion in our
basic view of nature (i.e., to accept that semiosis and interpretative processes
are essential components in the dynamics of natural systems), biosemiotics, as
I hope to show, provides the conceptual tools necessary to explanatorily rein-
tegrate living creatures (including, of course, human beings) into the natural
world from which they came — but from which they have since been effectively
excluded by a scientific ontology that has, at least since the time of Descartes,
consistently encouraged scientists to *de-semiotize* all the naturally communicative and fundamentally interactive processes of living systems.

Caveats

When I set out to write this book in 1999, biosemiotics was still in its initial
phase en route to becoming an autonomous field of its own (I will provide some
further background on the history of the field in the Postscript to this book).
Back then, however, apart from a small group of semioticians who were following the lead of the pioneering linguist and semiotician Thomas A. Sebeok, only
a handful of medical doctors and biologists were actively engaged in investigat-
ing the potential of this newly developing perspective for the study of life. The

whole field was at the time relatively open, but also largely anchored in a shared understanding of semiotic processes that was derived from the naturalistic sign logic of the American logician, scientist and philosopher Charles Sanders Peirce (1839–1914).

Today, only one decade later, the situation has radically changed and biosemiotics has been taken up as a field of study by chemists and physicists, as well as by philosophers and cognitive scientists, anthropologists, psychologists, and scholars from many other disciplines in both the humanities and the natural sciences. This growth spurt in the development of biosemiotics as an international research project is welcome, since it will force every one of us who is working in the field to become more exacting in our arguments, and because it widens the agenda of the biosemiotic perspective to encompass the entire study of life.

But this recent rapid growth also means that the views that I present in this book may not cover all the competing conceptions of biosemiotics now being debated within the field. I have chosen, nevertheless, to keep the title *Biosemiotics* from the original Danish edition of this book — for although there are certainly now other views of the field, this book surely comes close to expressing the original understanding of biosemiotics as envisioned by Thomas A. Sebeok and Thure von Uexküll — as well as, I feel confident, the majority (although certainly not all) of those scholars that are currently engaged in the still ongoing development of biosemiotics.

It is unavoidable too, I suppose, that this text reflects my own original background as a biochemist and scientist. Thus, many readers might have wished for a broader treatment here of areas such as *zoosemiotics* or *medical biosemiotics*, and also perhaps for a more comprehensive analysis of *Peircean philosophy* as a resource base from which to explicate the biosemiotic approach. I can only hope that such readers will show tolerance towards these perceived deficits in the text. No single person can claim competence in all the varied scientific disciplines that must be brought into play in order to reveal biosemiotics in all of its potential and, at the same time, to consistently grapple with all the philosophical implications that it holds.

Similarly, another group of readers may argue that the investigation into sign processes at the "micro level" — i.e., *cytosemiotics* and, in general, *endosemiotics* — have been given undue emphasis in the book. I have done this for a reason, however. For biosemiotics is often dismissed by scholars — not least, those from the humanities — who fundamentally misconceive the project as an attempt to project anthropomorphic features upon an existing world of nature that, as we have all learned in school, can easily be explained without reference to human mental states and constructs. Perhaps no other caricature could be further from the true aims of principles of biosemiotics, however — and, indeed, I too discuss the fallacy of anthropomorphism in some depth later in this book.

Rejecting anthropomorphism, but determined not to fall into the opposite trap of eliminative reductionism, however, one of the main points of the biosemiotic analysis presented here is a rejection of the idea of scientific knowledge as a kind of knowledge obtained by taking the "view from nowhere" (to use Thomas Nagel's famous expression) — a knowledge, in other words, that does not include ourselves as knowing creatures produced by and inside of the very same nature that we are attempting to explain. This conception of scientific knowledge is not only absurd, but it systematically corrupts our understanding of the world because it hides from us the very tools that might help us see how we both emerge from, and still now essentially belong in, nature — i.e., to answer another question that Nagel puts to us: "How can it be the case that one of the 'people in the world' is *me*?" (Nagel 1986, 13).

We may not feel that we can ever answer this question adequately, but a first premise for approaching such an answer must be that we can somehow explain the existence of such *me-ness* in the world. How could evolution create such a phenomenologically odd entity as a "me" or an "I"? An evolutionary theory that does not give us any tools to see how such a question can be meaningfully answered leaves us as object-ified biological robots, or zombies. I must assert upfront that I firmly believe that neither the reader nor I are, in fact, such zombies — and consequently, that a decent biology must search for the evolutionary root forms of *what it is to be* an "I", or a first-person *singularis*.

A key to answering this question, I am going to argue in this work, lies in a sufficiently rich concept of *semiosis*. Biosemiotic analysis takes us back to the questions of how life and semiosis first appeared on Earth, and we shall see that these are, in fact, not two distinct questions, but a single united one: semiosis is an essential aspect of life already at the primitive unicellular level. It is my guess that readers, with no firm background in biology will be greatly surprised to learn how nearly impossible it is to gain an understanding of the interactions and organization constituting the world of living nature without a semiotic terminology — and, perhaps even more fundamentally, a semiotic way of thinking. I do hope, moreover, that this book will show that the attempt to cross the traditional borders between major areas in university life does bring us necessary new insights. Transdisciplinarity will be of no help so long as everybody stays safely inside her own disciplinary borders, politely transgressing no internal disciplinary taboos. Rather, "interdisciplinary scholarship" only becomes fruitful when we collectively take the risk to confront problems *in the ways those problems may be seen within disciplines other* than our own.

Accordingly, the major challenge in writing this book has not so much been to assemble and to present the many and varied kinds of knowledge that must be assimilated in order for the biosemiotic project to take on its real significance. It does indeed appear at first glance to be a far jump from the analysis of *languaging* in the human animal to the analysis of *courtship trembling* in the water mite, or from the *logic of self-organization and emergence* to the *biosemiotics of modern petrochemical agriculture* — and I have tried to make clear all the conceptual links and unique level properties that come into play as we move from one of these phenomena to the other.

A harder problem by far, however, lies in the attempt to surmount the theoretical and philosophical difficulties of putting insights from areas that traditionally have been seen as scientific into play with insights that have traditionally been seen as belonging to the humanities. For, on the one hand, biosemiotics is engaged in developing conceptual tools for *theoretical biology* (and thereby also, indirectly, for *experimental biology*) — while on the other hand, the conceptual understandings made evident by biosemiotics contribute to the development of a *general semiotics* inclusive enough to conceptualize the human being as being not only *in* but also deeply *of* nature. And in doing so, biosemiotics contributes to renewed reflections in *natural philosophy*.

Moreover, because of the ambitious nature of the attempt, I have no doubt that readers coming from backgrounds on each side of the disciplinary Cartesian divide will at times feel their scholarly sensitivities violated by this book. To them, I may only suggest that they attempt to localize their misgivings not to the entire project undertaken here, but only to such particular mistakes and misstatements as they may find herein — and to share my own hope that time will teach us all to do a better job of transdisicplinary scholarship and communication. For there is no way we can allow ourselves to remain stuck, each on our respective sides of the Cartesian dividing line, in our attempts to understand a reality that refuses to so divide itself.

And in that regard, I believe that I need to add one further note of explanation here at the outset of this book. This concerns my belief that if communication across large ranges of scientific disciplines shall succeed in producing new fruitful ideas, it is absolutely necessary to transcend the narrow terminology of each particular discipline in our speaking to one another. Terms are used in very different senses inside different disciplines, and it has been my experience that everyday language should be used as much as it is practically possible to do so when undertaking transdisciplinary work in order to make one's ideas as *explicit* as possible to all participants in the conversation. Thus, for example, throughout this book I have deliberately employed illustrative materials that may seem like simple textbook figures to the expert eye. Such a commitment to baseline clarity and explicitness is often misunderstood as a kind of disciplinary trivialization — but I believe that one should rather see it as a necessary step in the research process itself. We do not get anywhere if we do not understand each other, which means that the mediating process becomes itself a vital part of all transdisciplinary research.

The Structure of This Book

This book consists of three parts and a postscript. Part One contains a general discussion of the biosemiotic perspective as a project in the life sciences. Chapter 1 introduces the basic idea of biosemiotics and addresses some possible initial misunderstandings and misgivings likely to be engendered by the proposal that semiosis is a fundamental aspect of life in general. In Chapter 2, the latter notion is further focused through a discussion of its relation to the basic inside/outside asymmetry of living systems. A biosemiotic model for the origin of life is suggested, and the concept of the generalized membrane is introduced. Chapter 3 discusses the roots of the modern sign concept in the "logic of relations" formulated by the American philosopher Charles S. Peirce, and analyzes the standing of Peirce's sign logic in its relation to modern scientific ideas of causality and thermodynamic irreversibility.

Part Two contains a detailed exposition of biosemiotics as an approach to the understanding of life processes as they are taking place variously and on several different levels in nature. Accordingly, Chapter 4 explores the double-coded (i.e., analog and digital) nature of living systems, arguing that evolution is equally reliant on both kinds of coding, as well as on their interface. Chapter 5 follows the semiotic controls involved along the developmental route from genotype to phenotype. In Chapter 6, the book moves on to an exploration of organismic life as it unfolds within its ecological settings, introducing the concepts of the *semiotic niche* and the *semiosphere*. Chapter 7 deals with *endosemiosis* — the semiotic processes inside the body that control the settings of essential physiological and biochemical states and that assure mind/body integration, i.e. the reciprocal creation of a psychicalized body and of an embodied psyche, a *bodymind*.

In Part Three, the biosemiotic perspective is used in the attempt to throw new light upon the origins of "a speaking animal" — and on the cultural reality shaped and reshaped by this animal through its communal use of signs understood as signs (i.e., words). Chapter 8 examines the problem of human language origins, which — following the lead of American neurobiologist Terrence Deacon — is here seen as the development in the human species of a specialized ability for making symbolic reference. The chapter also briefly introduces Maxine Sheets-Johnstone's theory of bodily movement as the most fundamental basis for linguistic capacity. Chapter 9 addresses the radical consequences that the biosemiotic perspective may have on our thinking in a range of other areas: ethics, aesthetics, biomedicine, environmental understanding, health, and cognitive science. Finally, Chapter 10 considers the prospects of biosemiotic technology, suggesting that such technology might be a necessary tool in overcoming the ecological imbalance that our industrial production systems seem inevitably bound to produce.

In Part 3, too, we move farther and father away from my own primary area of competence, and there I will inevitably enter risky discussions on matters that others might have treated in a more satisfactory way. I can only repeat my request for the reader to show tolerance. It is necessary to follow the biosemiotic perspective into these risky areas, and I see it is an inescapable part of my task to take that risk.

The Postscript gives a brief account of the historical development of the biosemiotic project, as well as a prognosis for its future growth. The late Thomas Sebeok claimed that the idea for a rigorously empirical science of biosemiotics was independently invented at least five times during the last century. The Postscript tells the story of the most recent of these attempts, as this author has experienced it.

Jesper Hoffmeyer
Copenhagen
January 2008

PART I
THE BIOSEMIOTIC APPROACH

I

On Biosemiotics

An Overview

Biosemiotics is the name of an interdisciplinary scientific project that is based on the recognition that life is fundamentally grounded in semiotic processes.[1] This investigation into the semiotic nature of living systems has taken a long time to emerge, since it poses a challenge to many of the prevailing ontological assumptions of both the natural and the human sciences. Yet the biosemiotic perspective that will be argued for in this book has been developed over a considerable period of time in works that, by virtue of their quality and scope, indeed offer path-breaking new ways of understanding both culture and nature. Regrettably, because of their perceived radicalness at the time at which they were written, many of these works were banished to the fringes of scientific study and acceptance. The time for such timidity and denial is surely over, however, and thus the goal of this book is to resurrect these ideas and to scientifically expand on them in light of what scientists have been learning in the last few years.

Most noteworthy among the earlier advocates of biosemiotic investigation are the American natural scientist and philosopher Charles Sanders Peirce (1839–1914), the Estonian-born German biologist Jakob von Uexküll (1864–1944), and, to some extent, the English-American anthropologist Gregory Bateson (1904–80). I will have much to say about each of these men later. However, it should be noted at the outset, that none of these thinkers explicitly used the word *biosemiotics* in their works. Rather, the word apparently was used for the first time by the German medical psychologist Frederich Solomon Rothschild (1962; Kull 1999a). And, in point of fact, the very term *biosemiotics* is thought to have first emerged in Russian literature in the 1970s, where it was used to refer to the study of natural signs — including the study of communication systems in organisms — particularly with reference to Jakob von Uexküll's work (Stepanov 1971, 24–25, 27–32).

---

1   The expression *paradigm* is used by Anderson et al. (1984), Hoffmeyer and Emmeche (1991), Eder and Rembold (1992), and Kull (1993). Here I prefer to employ the less definite term *approach* in an attempt to avoid a premature hardening of the biosemiotic idea into an actual paradigm.

However, it was only in the last decade of the twentieth century that the word began to proliferate in the international literature (Sebeok and Umiker-Sebeok 1992). *According to the biosemiotic perspective, living nature is understood as essentially driven by, or actually consisting of, semiosis, that is to say, processes of sign relations and their signification — or function — in the biological processes of life.* Accordingly, biosemiotician Claus Emmeche (1992) has suggested the following as a succinct definition of the concept of biosemiotics: "*Biosemiotics* proper deals with sign processes in nature in all dimensions, including (1) the emergence of semiosis in nature, which may coincide with or anticipate the emergence of living cells; (2) the natural history of signs; (3) the 'horizontal' aspect of semiosis in the ontogeny of organisms, in plant and animal communication, and in inner sign functions in the immune and nervous systems; and (4) the semiotics of cognition and language."

This idea, then, implies that processes of sign and meaning cannot, as is often assumed, become criteria for distinguishing between the domains of nature and culture. Rather, cultural sign processes must be regarded as special instances of a more general and extensive biosemiosis that continuously unfolds and acts in the biosphere. It is important to emphasize at the outset, however, that the biosemiotic project in no way whatsoever contradicts the conventional scientific understanding of living systems as originating from molecular processes. Biosemiotics accords entirely with this viewpoint but expands upon it by noting that molecular processes cannot be exhaustively described in chemical terms, since such processes, by virtue of their very participation in the constitution of the fundamental processes of life, functionally become distinctive bearers of life's critical semiotic relationships. Thus, rather than making the semiotic aspects of life processes a secondary part of biochemistry, biosemiotics posits that the semiotic dynamic constitutes a vital key to a more complete biochemical understanding, without which the fundamental biochemical organization of life processes cannot be satisfactorily explained.

The biosemiotic understanding, furthermore, is related to *process philosophy*, which considers substance (matter) not as life's fundamental entity but rather as an intermediate stage in an emergent *process*. While process philosophy is normally linked to the work of the English philosopher Alfred North Whitehead,[2] biosemiotics is principally anchored in the evolutionary philosophy of C.S. Peirce. The biosemiotic project also has implications for our understanding of evolution, as it is embedded in a theory about organic evolution as the cornerstone of a cosmic semiotic process that Peirce often

2   Whitehead (1978 (1929)); see also Hartshorne (1970). A current discussion of evolutionary theory in the light of process philosophy is found in Kampis (1998).

described as a tendency of nature to "take [i.e., to form] form habits". The bio-semiotic idea implies that life on Earth manifests itself in a global and evolutionary *semiosphere*, a sphere of sign processes and elements of meaning that constitute a frame of understanding within which biology must work.[3] "The semiosphere is a sphere like the atmosphere, hydrosphere, or biosphere. It permeates these spheres from their innermost to outermost reaches and consists of communication: sound, scent, movement, colors, forms, electrical fields, various waves, chemical signals, touch, and so forth — in short, the signs of life" (Hoffmeyer 1996b).[4]

In the chapters that follow, I will more closely consider various aspects of the biosemiotic approach, but let me deal here with a few warning signals that the critical reader must already have detected. Perhaps the most troublesome alarm alerts those readers with a biological background who think that the biosemiotic approach may be a form of *vitalism*. Empirical researchers may quickly spot a warning sign that suggests *metaphysics*. Humanists, on the other hand, may see a danger marker that points to *reductionism*. Let me confront this last alarm signal first.

3   One could object by saying that not just living beings but also as yet lifeless nature should be included in the semiosphere. From an evolutionary-cosmological standpoint, such an understanding of a continuum, a physiosemiosis, would perhaps be preferable (Deely 1990; Salthe 1999). Here we follow Sebeok (1979) in defining the emergence of life as the threshold for the semiosphere. As we shall see in Chapter 2 (in the section "The Creation of Life"), from a biosemiotic point of view, life's beginning is connected to the creation of encircling membranes, so that self-organizing processes can establish a stable integration of a system of self-reference (via DNA) and a system for other-reference (linked to receptors). The creation of such a self-sustaining feedback between the inner and outer sides in an inner-outer asymmetrical system makes possible, in our opinion, the generation of genuinely triadic sign relationships.

4   The expression *semiosphere* was originally introduced by the Russian-born Estonian semiotician Yuri Lotman, who explicitly used it in comparison to Vernadsky's idea of the biosphere. For Lotman (1990, 125), the semiosphere remained a cultural concept: "The unit of semiosis, the smallest functioning mechanism, is not the separate language but the whole semiotic space in question. This is the space we term the semiosphere. The semiosphere is the result and the condition for the development of culture; we justify our term by analogy with the biosphere, as Vernadsky defined it, namely the totality and the organic whole of living matter and also the condition for the continuation of life (Vernadsky 1926; 1945, 1–12)." One could say that Vernadsky's idea of the biosphere in fact includes what we here would call the semiosphere. But the idea of the biosphere has not retained the connotation that Vernadsky originally gave to it but is used today simply to mean "the ecosystems comprising the entire earth and the living organisms that inhabit it" (*Webster's Encyclopedic Unabridged Dictionary* 1996). For more details about the origin of these terms, see Sebeok (1999). John Deely accepts my use of the word semiosphere and suggests "*signosphere* as a term more appropriate for the narrower designation of semiosphere in Lotman's sense, leaving the broader coinage and usage to Hoffmeyer's credit" (Deely 2001, 629).

Not Reductionist

Is the biosemiotic approach reductionist? The answer is of course yes — if one narrowly defines the words *sign* or *meaning* in terms of human phenomena such as linguistic symbols. In Turkey, those who contemptuously mistreat a certain red cloth decorated with a white half moon and a white star risk a prison sentence, and this is, of course, astonishing to those who do not understand the symbolism of the flag. However, if one understands the meaning of this type of sign, the matter quickly becomes less surprising, even if one may still believe that prison is a too dramatic penalty for this particular offense.

It is unlikely that one will find examples of sign processes of this complexity in the animal kingdom. The biosemiotic approach does not imply that humanity is nothing special but only that the obvious uniqueness of humans is not as users of signs but as creatures who can readily teach themselves to master a special form of sign usage — symbolic reference — that is the basis of linguistic competence. The red cloth with the white half moon and star belongs to the category *flag*, and it is because — and only because — of this that it is legitimately thought that those who trample on the cloth also indirectly stomp on Turkey's good name and reputation.

The American anthropologist and neurobiologist Terence Deacon has made a detailed case for the necessity of distinguishing among iconic, indexical, and symbolic forms of reference, if one will explain the evolutionary emergence of the human brain's special linguistic talent (Deacon 1997). While we share the ability for iconic and indexical reference with all other animals who possess a brain, the capability for symbolic reference is unique to humans, Deacon believes (see Chapter 8 for further discussion). Language's fundamental basis in corporeality is a relatively newfound insight with radical consequences for linguistic and cognitive research (Lakoff 1987; Sheets-Johnstone 1990; Lakoff and Johnson 1999), to which I shall return in Chapters 4 and 8. Here I shall argue, by way of introduction, that humanity's cognitive and emotional characteristics cannot be considered so miraculously great that we can justify setting humans "inside parentheses" in the study of the natural phenomena of this earth. The mental system of humans has grown from nature through an evolutionary process, and we must expect to find phenomena in nature that remind us of humanity in all its forms. Indeed, if this were not the case, we would have to find a theory to explain how we have become such extraordinarily peculiar beings. With all due respect for the distant myth of that theft of the apple from the tree of knowledge, we are still waiting for a plausible theory.

How will arch-humanists (if, indeed, there are any still around) explain that humans could have ended up so totally outside of nature?

There is a perfectly clear answer to this question — or perhaps I should call it an avoidance maneuver — and moreover it is a maneuver that has become quite fashionable of late. One can choose to entirely reject evolution's validity beyond biology itself. According to this view, scientific results cannot be used to critique philosophical positions, because the scientific results themselves have philosophical baggage. Indeed, in the twentieth century, not only historians of science but also philosophers have become acutely aware that all scientific results rest upon tacit assumptions that very often are dependent upon unacknowledged philosophical presuppositions. Natural science, therefore, cannot claim a superior place in our understanding of the world when it is essentially nothing other than a special "story" with no claim to universal validity. This constructivism can have much in its favor, especially as a needed corrective to the almost terrorizing conception of just a generation ago that objective science is the guiding rule for the development of technology and community. However, in its radical version, as outlined here, such arch-social constructivism approaches absurdity. It is true that the fossils found underground could, in theory, have been put there by a capricious god who is amused by teasing scientific eggheads on Planet Earth. However, simply the demonstration that such a "theory" cannot possibly be disproved cannot serve as an argument for taking it seriously — especially in light of the overwhelming quantity of strong and independent evidence we have to support its alternative, the theory of evolution.

Constructivism does, however, contribute to our perceptions as we seek those biases that inevitably conceal themselves in the inner structures of our theories. As we shall see in this book, the neo-Darwinist version of evolutionary theory is lacking in the way that it significantly reflects remnants of ontological ideas that by and large are characteristic of twentieth-century natural science (further on this in chapter 3 and 4. See also Depew and Weber (1995), Wheeler (2006). In this context, constructivism can be seen as a useful counterbalance. But to use constructivism as an excuse to avoid taking serious central, and profoundly well-corroborated natural scientific understandings of this world in which we find ourselves is not a mark of profundity but rather of intellectual complacency. Indeed, in social constructivism's most radical version, where only the cultural and social domains are considered to be real, constructivism itself is reductionistic.

The biosemiotic idea does not maintain that the social, cognitive, and emotional processes of humanity can simply be understood as individual manifestations of traditional biological phenomena. Such an understanding, with varying degrees of radicalness, has been given validity first by sociobiology and later by evolutionary psychology. The biosemiotic idea actually consists in a diametrically opposed understanding, requiring a diametrically radical approach — i.e., if we would build bridges between the *human* and the *natural*, let us assume that nature already possesses semiotic competence. Biosemiotics thereby rejects the idea of the *illusionary* character of human universes of meaning, but it pays for this rejection with its acceptance of *nature's universe of meaning*—or, in other words, the reality of a causally efficacious matrix of biological interaction, the utterly natural product of organisms' interaction that I will refer to throughout this text by use of the term *semiosphere*.

However, this solution is just that which will set the alarm bells ringing for many biologists.

Not Disguised Vitalism

Is the biosemiotic approach simply vitalism in disguise? There is hardly any idea more despised in modern biology than vitalism—perhaps with the exception of Lamarckism, which we will later consider more closely. In biological connections, the word *vitalism* is identified simply with the belief that life consists of the existence of special life forces or *vital forces*, which are not active or known in physical nature. These are unknown and even unknowable forces that biology is claimed to require but which play no role in physics and chemistry. Let me immediately make clear that there is no undertone of this type of mystic power in the biosemiotic idea.

Even so, there is in this book good reason to look more closely at the assessment of vitalism, which in its rational roots was an answer to problems that are still unsolved and that to a large degree nourish the search for a biosemiotic understanding. Thus, the relationship between vitalism and biosemiotics is not based upon theoretical agreement but upon some common conception of what are the central unsolved problems in biology.

Vitalism today is most closely linked to the German biologist and philosopher Hans Driesch (1867–1941), who developed the *Entwicklungsmechanik*, a theory about embryonic development as a kind of mechanical process. The influential German Darwinist and mechanicist Ernst Haeckel (1834–1919) in 1866 proposed his famous (but essentially inaccurate) *biogenetic law*, often summarized in the saying, "ontogeny recapitulates phylogeny" (Coleman 1977, 47).[5] The idea is that every organism in its embryonic development passes through

---

5   Already in 1828, the embryologist Karl Ernst von Baer saw that although the embryo during its development passes through stages which correspond to the forms that one finds in the lower animals, these are the embryonic stages of the lower animals, not—as Haeckel later had it—the mature forms. But this correct view failed to spread outside of disciplinary circles and neither, apparently, was it grasped by biologists such as Haeckel (Coleman 1977, 41–56). It is ironic that von Baer was a type of early vitalist (epigeneticist) and in any case an ardent opponent of mechanicism. Von Baer maintained that it is "the essence (the Idea, according to the new [nature-philosophical] school) of the developing animal form which controls the development of the germ [fertilized egg]" (ibid., 42).

stages, whose sequence follows exactly the stages that the ancestors followed in the historical changes in the species — a process called *phylogenesis*. For example, the human embryo, early on, develops fish-like gill slits, later a three-chambered heart like a reptile, and still later a mammalian tail. Haeckel believed that with this theory he had found a key to a mechanical explanation of the mysteries of embryonic development and the generation of form, and it was to cast light on this question that the German biologist Wilhelm Roux (1850–1924) developed, in the last decades of the nineteenth century, the *Entwicklungsmechanik*, which became the scholarly turning point of the vitalism debate.

Roux believed that the aggregate of the hereditary units (noting that the theoretical concept of *the gene* was itself not yet invented until 1907 by the Danish geneticist W. Johannsen) in the eggs became with each cell division unequally distributed between the daughter cells, so that the potential of individual cells became more and more differentiated and limited throughout the course of embryonic development. Finally, the individual cells of mature organisms formed a kind of mosaic of cells, each with its own genetic content. He therefore called this theory the *mosaic theory*. The beauty of this simple theory was that it led to predictions that could be tested. If the hypothesis was correct, the destruction of an embryonic cell — a blastomere, at the embryo's two- or four-cell stage — would lead to a deformed fetus. If the hypothesis was incorrect, the destruction of the blastema would be expected to have little effect (Allen 1975).

Roux experimented with frog embryos, in which he punctured a single blastomere with a heated, sterile needle, thus killing it while leaving the others (apparently) undisturbed. The experiments seemed to confirm the theory that the embryos developed abnormally, in agreement with the assumption that half of the genes in the blastomere were set out of play because of the intervention by Roux. The egg seemed to contain an inner mechanical device that would ensure that under normal circumstances each cell in a complete organism came to possess the correct characteristics.

In 1891, however, Hans Driesch reached an altogether different conclusion. Driesch's work took place at the zoological station in Naples, where he experimented with eggs from sea urchins. Instead of destroying individual blastomeres, Driesch was content to agitate two-cell embryos in sea water, whereupon the cells separated from each other without becoming damaged. Under these circumstances, it happened that each single embryonic cell developed into a normal, if small, larva (a *Pluteus*). Every cell had, in other words, fulfilled its complete potential, which, one should think, was a powerful argument against the mosaic theory.[6] Driesch, however, was just as eager as Roux to find an expla-

---

6   We know now that the difference between Roux's and Driesch's results is based, among other reasons, upon the fact that the ontogenetic process is dependent upon communication between

nation for the embryonic developmental process of form generation. In other experiments, he left two-cell embryos undisturbed and sought instead to determine which daughter cells would be derived from which of the two original blastomeres. It appeared that when the two-cell stage was allowed to develop normally, each cell became the source of qualitatively different types of tissue in the mature organism. Considered as a whole, Driesch interpreted this research as a sign that blastomeres had an inherent ability to adapt themselves to varying environments, so the process of embryonic differentiation had to be seen as a result of the capability of cells to coordinate internal and external circumstances. Driesch came to regard the embryo as a self-adjusting whole of many cooperating cells that he described as a *harmonic equipotential system* — that is, a system where all parts have equal potential to contribute to the development of the whole organism (Allen 1975).

However, Driesch had to abandon his quest to solve the puzzle that was to give a causal mechanical explanation of the workings of these harmonic equipotential systems. In the beginning of the new twentieth century, he despondently gave up experimental biology and developed in its place (as a philosophy professor in Strasburg and later in Leipzig) that neo-vitalist philosophy for which he would become so notorious. Instead of being able to give a mechanical explanation for the form-generation process, he had to content himself with his idea about a special energy of vitality. Driesch went back to Aristotle's old idea of *entelechy*, the principles or conditions by which things are realized or determined. He emphasized that we must not consider this entelechy to be energy. Entelechy should be understood as a natural effect in its own right, Driesch believed, and he suggested that its efficacy *consists in the changing of energy's direction*. Entelechy thus — in Driesch's conception and as expressed in modern language — belongs to a meta-level; it is not in itself a real force but rather, has a controlling agency upon such forces (Driesch 1908). Even so, *entelechy* became both considered as and named as a vital force, which of course caused the scientific community to strongly disassociate itself from this idea.[7]

---

cells. In Roux's experiments, this communication is disturbed, because the destroyed cells send out false signals. In Driesch's experiments, on the other hand, the cells lack signals from their neighbor cells, which normally will make two-cell embryos act in a certain way, and in the absence of these signals, the process begins anew (but with a reduced supply of metabolic resources).

7   In a recent analysis of the inherent teleology in the theory of natural selection, philosopher T.L. Short (2002) points out that in Driesch's work, vitalism makes the mistake of explaining the *telos* of the evolutionary process, its obvious tendency to favor certain types of organization rather than others (i.e., those selected away) through an efficient causality, namely a vital force. Force is not a common good that one strives for but rather, in the best case, an instrument for such an endeavor. The vital force is therefore not a final cause but rather an efficient cause (a similar point was made by Hans Jonas — see p. 321). This blending together in vitalism of efficient and final causality (in Aristotelian terms) is probably the main reason why this direction of thought cannot

The understanding of *entelechy* as a controlling authority foreshadows a thermodynamically inspired model for the physics of form generation (see Chapter 3), as Driesch had already suspected. It was perhaps unfortunate that thermodynamics in Driesch's time was not yet ready to function as the foundation for such a nonvitalistic solution. Therefore, this story illuminates an important strand of the development of biology. The hardcore reductionism that has characterized many of the most successful biological theories all the way back to Descartes' time has presumably been basically incorrect; nevertheless, it has functioned as a remarkably good foundation for the experimental study of life processes. In contrast, those who persistently deny the oversimplified pictures that are characteristic of reductionist theories have perhaps been on the right path but have not had adequate theoretical means to create a productive alternative to reductionism. The time was not yet right for a scientific confrontation with mechanicism. However, that has changed in the last few decades, with the development of nonequilibrium thermodynamics, chaos theory, nonlinear dynamics, complexity research, and biosemiotics.

Contemporary biology has understood it such that, in the controversy over vitalism, both sides were wrong, but that, while Roux developed a fruitful method, Driesch pursued an unproductive path. Vitalism and Driesch's theory became for many biologists the quintessential example of how badly it can go when philosophical considerations are given credence in connection with internal controversies in the biological disciplines. This assessment, however, is based upon a rather euphoric understanding of molecular biology's potential to solve the problems that Driesch abandoned. The smart idea about genes as information has all too often paralyzed the critical nerves of biologists, who a long time ago should have asked themselves how a *substance* entity, a DNA segment that travels (as a kind of parcel post package) from generation to generation, can at the same time be identical to a *form* entity that can instruct an egg to generate an organism. There is hidden here an unsolved ambiguity that Norbert Wiener (1962, 132), the founder of cybernetics, has already warned against: "Information is information, not matter or energy. No materialism that does not admit this can survive at the present day." The heart of the matter is that it is fairly unclear what we mean with the expression *genetic information*. If the meaning is that the genes contain information that are instructions about how the egg shall create given phenotypic traits (forms), this kind of information is

---

be made fruitful. From a thermodynamic standpoint, however, Driesch's *entelechy* has more to do with the relations between order and entropy than with energy (force). But Driesch was far from unique in this confusion of the entropic and the energetic. We see this, for example, also in Sigmund Freud's idea of *psychic energy* dating from nearly the same period, which in the final analysis has nothing to do with metabolism (the body's production of energy) but instead has a great deal to do with what brings *order* to the psychic house.

of a fairly different mold than the type of information (Shannon information) that is operative in computer science or physics. So it is namely a question of information, which in one or another sense has a meaning, or is meaningful (for the egg?),[8] which again assumes a valuable or preferred direction and it is just the kind of information that we associate with *signs* and *sign processes* or *semiosis* (Hoffmeyer and Emmeche 1991; Santaella-Braga 1999). And thereby we have in a radical sense transcended that molecular genetics, whose results are used to refute the ideas of Driesch. *It thus appears that the key to the problem of form generation will be found neither through the approaches of mechanicist biology nor through the invocation of special life forces, but rather through the elaboration of a semiotics of ontogenesis.*

Specious Metaphysics?

While biosemiotics must reject its connection to vitalism, as narrowly defined by Driesch, it must acknowledge its relationship to the broader strand of nineteenth-century natural science that — outside of biology — is often called vitalism. However, biosemioticians in this instance join company with such renowned scientists — Claude Bernard, Louis Pasteur, and Niels Bohr, among others — that the alarm bells have scant reason to ring. In this more modest understanding of the concept of vitalism, the word refers to the idea that complex phenomena are controlled partially by mechanistic forces and to a certain extent by self-determination. In biology, this conception is regarded not as vitalism but rather as organicism, and this is probably the principal direction among more philosophically inclined biologists of recent times, such as Ernst Mayr and Stephen Jay Gould (El-Hani, Queiroz and Emmeche 2006). An earlier representative of vitalism, as defined in this sense, was the Swedish chemist Jöns Jacob Berzelius (1779–1848), who clearly used the expression *vital forces* but expressly emphasized that he meant to describe a regulative power that could create organization. In 1806, he wrote that the beginning point of life "should be sought

---

8   In sexually reproducing species, the egg is the cell that is prepared to set the ontogenetic process in motion, which begins with the sperm cell's penetration. That the egg cell is *prepared* must be understood here in its phylogenetic dimension. Evolution has developed a finely articulated semiotic interplay between the embryo cells, which already before birth lead to the production of special egg-producing cells, *oocytes*, much as it has created the other numerous components in the play of reproduction. See the section "The Semiotics of Ontogeny" in Chapter 5 for details about the biosemiotic interactions of egg and sperm cells. That this inherited information has *meaning* for the egg refers to the state of "readiness" of the egg for releasing that evolutionarily determined series of processes which is the embryonic development, starting with the penetration by the sperm cell, with which the egg then joins to make a complete genome and a full set of genetic information.

in the basic forces of the elements and it [life] is a necessary consequence of the relationship whereby the fundamental materials are joined. . . . Consequently, there is no special force exclusively the property of living matter which may be called a vital force-for-life; rather, this force arises from the conflict of numerous other [forces] and organic nature possesses no laws other than those of inorganic nature" (Coleman 1977, 147).

That organic evolution on this planet is real and that it is an example of nature's ability to create complex, organized structures from less ordered components is broadly accepted by biologists. Therefore, it is hardly the ability to self-organize that is contested. Instead, at stake is the question of how such self-organizing processes can be understood without seeking explanatory sources other than simple causality. It has become a credo, especially outside of biology's own circles, that Darwinian natural selection constitutes an adequate explanation of historical developmental processes in nature. Therefore, it is most often not considered necessary to introduce further principles into evolutionary theory. The differential destruction of dysfunctional characters is seen by many as the key to understanding the generation of functionality in nature.

As pointed out by the American environmental psychologist Rod Swenson (1999), there is, however, an underlying metaphysical anomaly hidden deep within this apparently so elegant theory. For the most important premise of natural selection is the overproduction of offspring. Only because there are too many mouths to feed can the least fit be eliminated. But this proliferation that leads to overproduction — how can we explain it? For good reasons we cannot here refer to natural selection, for without proliferation, there could be no natural selection. Fertility lurks as some unexplained hypothesis in the heart of Darwin's theory.

Seen from the standpoint of biology, this is perhaps not a serious problem, for the fertility of organisms, their ability to multiply themselves, can fairly easily be explained by our knowledge of the mechanisms of DNA replication and cell division, which are already well described with regard to the simplest organisms. And yet, isn't there a bit of nontrivial ignorance here, that shouldn't be neglected? For even if we know well the mechanisms that cells or organisms use for reproduction, we do not fully understand how these processes are controlled. Strictly speaking, we do not understand the integrated level, which for lack of a better term can be called the cell's *agency*, its tendency to incorporate interactive events into its own project of survival.

That we can explain how living systems behave in order to become prolific does not explain evolution in the deeper sense. How does this fertility come into being? What temporal motive force is at play here? Selection theory also needs a theory about what kind of thing life's agency would be and how that agency comes to be. And so it appears that selection theory is only a surface structure built upon the foundations of a deeper theory of evolution that explains how life relates to the processes that are fundamental to the physical development of the universe.

These are questions that I will discuss more closely in Chapters 2 and 3, but here I will simply mention that modern physics does not leave much ground for dependency on a metaphysics that declares all that cannot be reduced to simple efficient causality (in an Aristotelian sense) to be unreal.

In the book *Evolving Darwinism* (Depew and Weber 1995), the philosopher David Depew and biochemist Bruce Weber point out that Darwinism long ago moved from a Newtonian worldview to a probabilistic mindset. It is just this shift of thinking that we usually label as neo-Darwinism. They maintain that Darwinism is now again transforming itself in a way that adapts to the new worldviews of the physics that have emerged in recent decades. If Depew and Weber are correct in this analysis, the metaphysical divergence between Darwinism and biosemiotics presumably can be transcended.

It is understandable that experimental biologists are suspicious of ideas that imply that nonreducible form determinations underlie life processes. That life processes are embedded in a set of semiotic relationships implies that they cannot be completely described at a single level of complexity, the molecular, since these processes must also always be understood in relation to a significative content that could only be defined when viewed from a superior level. Biosemiotics requires an understanding of living nature that respects discrete levels of complexity as partially autonomous and that therefore is very aware of the hierarchical and emergent character of nature. Such an understanding may seem restrictive in experimental biology, which perhaps not so much as before would be considered a ticket for direct entry into insights concerning nature.[9] However, it is important to understand that the experimental exploration of the processes of life has a central place in biology also from a biosemiotic point of view. The differences lie in the interpretation of observation and experiments, not in the experiments and observations themselves. It may be added that a biosemiotically driven curiosity might be expected to influence those questions it feels relevant to ask and, consequently, the observations and experiments that one will feel tempted to initiate.

To a large degree it is traditional biology's own success which has necessitated and inspired the biosemiotic idea. Throughout modern biology, one encounters expressions and sayings that are in fact essentially meaningless if one attempts to understand them without regard to their semiotic implications. Even in authoritative biochemical texts, these expressions appear quite frequently. One

---

9   Ironically, the success of molecular genetics has to a large degree itself fueled a growing interest in other levels of complexity. See for example Shapiro (1999).

speaks of, for example, lymphocytes that *present* peptide fragments on their surfaces. However, one cannot meaningfully *present* something without reference to communication between some entities. And yet, a biochemist would firmly deny that lymphocytes should possess communicative intentions. Or one speaks about "high fidelity" replication, but reliability is an attribute that presupposes some kind of meaningful content, just as replication — whether it is accurate or loaded with errors — is in itself a sign process. How can something meaningless possibly be reliable?

Let these examples suffice. Claus Emmeche (1997; 1999a) has used the expression *spontaneous semiotics* to describe this well-accepted practice of modern biology. The ordinary assumption is that these are cases of the careless use of language, but this employment of language in each single instance might eventually be transformed into statements that in the end will refer to the attribute's origin through natural selection. As already stated, there could be significant skepticism concerning this explanation.

Considering how widespread spontaneous semiotics has gradually become even in scientific journals (Hoffmeyer 1997a; Yates 1985), a more serious consideration of the scientific status of spontaneous semiotics may be needed. Perhaps this manner of speaking is so prevalent because it reflects an important aspect of the subject matter of biology and biochemistry, i.e., life itself. Rather than getting rid of the semiotic terminology while moving down to living nature's most basic and general level, the biochemical, it appears that precisely at this level an extended use of semiotic terminology has become customary. It's questionable, in fact, if one can at all understand advanced biochemistry and molecular biology without thinking in semiotic terms.

It is my opinion that the answer to this question is no. The biosemiotic approach is already tacitly permitted in the disciplines. It simply remains to become developed as a new integrated paradigm.

We must thus plead guilty to the charges of the experimentalists. The biosemiotic idea certainly implies metaphysical views that have no place in traditional biology. But rather than narrowing down the options — as Driesch's vitalism does — biosemiotics will have a liberating influence on a biology which seriously needs a way to displace the big-brother role that physics has claimed for all too long. Biosemiotics does not turn experimental biology to metaphysics but instead replaces an outdated metaphysics — the thought that life is only chemistry and molecules — with a far better, more contemporary, and more coherent philosophy. Life rather than natural law — and signs rather than atoms — must become natural science's fundamental phenomena. Life is composed of molecules, which manifest themselves as signs (Hoffmeyer 1993; 1996; Kawade 1996).

2

Surfaces within Surfaces

Biological Membranes

Let us in this chapter explore the biosemiotic understanding by applying it to a more concrete topic: biological membranes. An appropriate place to begin this exploration would be to investigate the semiotics of the skin.

From birth we humans are in fact skin more than we are anything else. The skin of the newborn is, as Thure von Uexküll (1999) says, a kind of pre-actual atmosphere (*vorwirkliche Atmosphäre*), and what enters the awareness of the newborn infant is only qualities or differences between qualities—grades of intensities of touch, taste, and smell. In a certain sense, then, the newborn child's skin is a type of brain, in the sense that it is the place where encounters with the world first freeze into the vague structurings of knowledge. And not coincidentally, the skin and the brain both originate from the same germ material, i.e., the embryo's ectoderm layer.

Now traditionally, most of us are taught from childhood that appearances are deceptive, and that we should not judge a book by its cover—which is to say that we should not be superficial but rather go after the heart of truth that lies concealed within the surface statement. The most important and essential aspects cannot be seen or sensed directly, we are repeatedly advised, but must be dug out from their hiding places deep within the depths of things. Deepest within, goes this logic, we will find what we most profoundly feel and know.

Undoubtedly, one could argue that there are good reasons to teach children that things are not necessarily what they seem, and that the world is often full of fraud and deceit (indeed, much of philosophy after Descartes is based on this conviction). But one can also inversely argue that the belief in an innermost being—the idea of the "true I"—is for the most part a cultural bias. And here it might be useful, for a change, to try seeing the world from the skin's perspective (as I will, presently, in this chapter). For the idea that personality has a place, a topological site, is not especially obvious, and by its very nature cannot be scientifically confirmed. Rather, in principle everything that belongs to the domain of reality that grammar calls *first-person singular* is thought to fall outside of the purview of natural science and is therefore considered as imaginary by many philosophers (see Churchland 1986; Dennett 1987; Churchland 1991). Yet the simple fact that our personhood presupposes our brain does not imply that our personhood is *in* the brain — and should we feel so compelled as to finally place our personhood in a definite biological locus, why not place it in the skin? After all, this is where we encounter the world around us and, in so many very obvious ways, the skin is the place where all the fun occurs. And in this chapter, I would like to show you how and why I believe that the skin is an indispensable part of our personality.

Firstly, the skin is the largest and most diversified organ of the human body. If we could stretch a grown person's skin out on the ground, it would cover between one and a half and two square meters. This modest area is crossed by sixty kilometers of nerve fibers and fifteen kilometers of veins and contains millions of sense receptors for pain, temperature, pressure, and touch. Incessantly and whithout our knowing, the skin repairs itself after both physical and psychic traumas, and more than anything else it is the skin that tells us about the person we are, as well as about the people (and the world) that we are facing. Color, texture, tension, and smell provide a constant barrage of information exchange for and among humans, across the interface that is the skin.

The following description from a Norwegian physician who was afflicted with the Guillan-Barrés syndrome, an illness that puts the nervous system out of commission for a time, can illuminate the skin's predominant role in our self-understanding:

> The worst experience was the disappearance and disturbance of the sense of touch. In a way, the borders of my self disappeared. When the hand was placed on the breast, it felt as if it were floating in the air. There was no ending to my breast and no beginning to my hand. *The quilt floated in the air above something that was not me.* A caress couldn't be felt, it was only a fuzzy suggestion of something long gone. I experienced in this situation an intense feeling of being locked inside myself without possibility for physical contact with the surroundings. I saw the responses and heard the words of my loved ones, but I was cut off from being physically present. . . . This confused bodily experience was, I believe, nearly psychotic in its strangeness. The experience, to be without boundaries, that thoughts and feelings were as before, but the body was something different, blurred and long gone and that didn't obey instructions, is difficult to describe. The experience of being physically cut off from physical touch and contact still remains for me an experience of hopeless loneliness (cited in Fyrand 1997, 65; italics added).

It is obvious that the skin protects us against external intrusions, but the Norwegian doctor's account reminds us how indispensable the skin is in semiotic terms as well. *The skin keeps the world away in a physical sense but present in a psychological sense.* It is the skin that gives us the experience of belonging — it allows us to feel the world. But the very fact that the world can be felt is already a complex phenomenon that doesn't just presuppose that there are receptors (sensory cells) in the skin that register touch, pressure, pain, cold, warmth, pH, and various chemical influences, but also that biological *meanings* are assigned to these sensations. It is not enough to sense; organisms must also create functional interpretations of the myriad of sensory stimulations so that these do not become isolated incoming impulses but are integrated into a form that the body understands and can act upon appropriately.

By way of an extremely simple example, consider a mother who sketches the number two on her child's back, and asks the child to guess the number that she just drew. The mothetr's light touch causes small deformations of the child's receptors (the sensory cells that lie close together just under the surface of the outer skin) and these deformations, in turn, cause a depolarization of the nerve endings. If the depolarization exceeds a certain threshold limit, an electrical impulse (i.e., an action potential) is transmitted through the nerve to the central nervous system, where a longer lasting pressure causes a series of impulses whose frequency is a function of the pressure's intensity. And already, even at this level, the organism's own most current contextual situation becomes a relevant factor in the phenomenon, in that the threshold limit is defined by biochemical parameters that reflect the general condition of the organism. Thus the sense of pain, for example, is greatly influenced by prostaglandins that lower the threshold values and thereby increase sensitivity. Prostaglandins are produced in connective tissue, and their production is increased in case of inflammation.[1]

Moreover, when these individual impulses reach the central nervous system, a system-generated integration takes place, so that the impulses from a given nerve ending are compared with impulses from nearby points in the skin (a comparison which constitutes the *discriminative* sense) as well as with impulses from muscles and joints (constituting the *kinesthetic* sense). The meaningful integration of these sense-data constitutes the *stereognostic* sense, and it is this sense that the child in our "simple" example must mobilize to solve the problem — that is, to determine that the mother's drawing is representative of the number two.

As most parents know, it is not easy for a child to solve this puzzle (for reasons having almost nothing to do with numerical knowledge), and the ability **to** gain skill in this game is in large part dependent on deliberate training. Furthermore, the kinesthetic sense is one that must be developed throughout life — it is, for example, highly developed among dancers and football players.

Most importantly, in the job of determining the number two sketched on the back, it is for the most part *the tactile sense in the skin itself* that is trained,

---

1   Such production can be counteracted with acetylsalicylic acid (which is the active ingredient in ordinary pain-relief medicine).

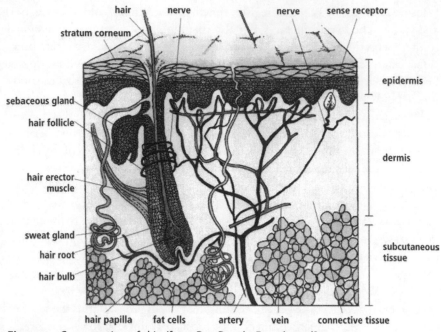

Figure 2.1 Cross-section of skin (from *Den Danske Encyclopædi*).

and the task of linking these tactile signals with signals from the other sense organs is quite negligible. But in more ordinary situations, it is necessary to integrate the stereognostic sense with *all* the remaining senses to create an optimal picture of what is taking place. This involves a long chain of interpretations, or interpretants, on steadily more and more integrated levels. We can sketch this chain as a sequential order of triadic sign processes, as depicted in Figure 2.2.

The Semiotics of a Slap

Figure 2.2, seen above (2a), is a graphic presentation of the elementary sign relations of Peircean semiotics. Here, a sign is "something which stands to somebody for something in some respect or capacity" (*CP* 2:228[2]). We can, for example, think about a slap, which normally is a clear sign that someone is angry. The slap and the anger, however, do not exhaust the implied sign relation here, as the slap itself does not know that it refers to an angry person, so we are lacking a third

2   References to the *The Collected Papers of Charles S. Peirce, Vols 1-8* (Peirce 1931–35, 1958), are given throughout this book as *CP* followed by volume number and the number of the referred paragraph.

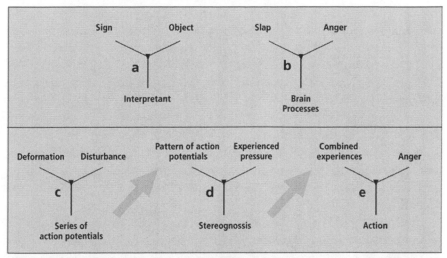

Figure 2.2 The semiotics of a slap. (a) The general sign triad, depicted as a tripod with the interpretant designated as its foot. (b) A slap viewed as a triadic sign. (c)–(e) A slap seen as a chain of sign processes whose interpretant in each articulation emerges as a new sign in a more integrated semiotic relation.

logical instance in the relationship, namely, the interpretation that takes place in the person who is slapped, or perhaps in the mind of the friend who saw it happen. Sign relations are said to be triadic because they always involve three — and only three — logical entities (either real or imaginary). In this case, the slap takes the role of the primary sign (the sign vehicle), and the anger or aggression causing the slapping is the object that the sign refers to. The interpretant is now *the process*, that is to say, the mental connections that take place in the brain and body of the interpreter,[3] *whereby* the slap and the anger *are connected* into a sign relation. The interpretant will often be unconscious, as it probably is if, for example, the observer reacts spontaneously to the slap with a punishing attack on the aggressor (instigating a new slap). Or perhaps the observer (mis)interprets the slap as a sign of malice rather than of anger. The semiotics of the slap can be summarized as in Figure 2.2b.

In the next chapter we will look more deeply into the logical status of sign relations. However, here we must at once (that is, before we turn back to our primary theme, the semiotics of the skin) confront three potential sources of ambiguity. First, it is important to distinguish between what has been called the *primary sign,* or the *sign vehicle*, and the *sign* itself. In Figure 2.2b, the slap

3   I am tempted to use the word *body-brain*. If one simply says *brain*, the reader may not take into account the body in which the brain is located and which is in continuous communication with the brain. This is more closely considered in *Signs of Meaning in the Universe* (Hoffmeyer 1996b, 87–88)

constitutes the primary sign or sign vehicle (Peirce sometimes used the term *representamen* but I will not use this term here), while the whole triad constitutes the accomplished *sign*. The slap, then, *is* a sign only because it is involved in a triadic relationship, as described.[4]

Secondly, it is important to clarify that the triad is an unbreakable whole that cannot be understood as an assembly of three dyadic (two-part) relations. When in Figure 2.2 we chose to link three logical instances in a sign relationship depicted in a tripod rather than a triangle (which is often seen), it is precisely to emphasize this characteristic of wholeness. Sign vehicle, object, and interpretant are of different (but necessarily interdependent) logical types. This will be clarified in the next chapter when we introduce Peirce's categories of Firstness, Secondness, and Thirdness.

Thirdly, it should be noticed that when Peirce, in his designated definition of *sign*, uses the expression *somebody*, he is not necessarily, nor exclusively, referring to a person. And likewise, the Peircean *interpretant* is a much more general category that is directly linked to the triadic logic of sign relations, and should *not* be confused with the very different notion of an *interpreter*. The Brazilian Peirce scholar, Lucia Santaella-Braga (1999) explains, "Since the three elements, sign, object, and interpretant, by themselves, or better, by their existential nature, may belong to various orders of reality as single objects, general classes, fictions, mental representations, physical impulses, human actions, organic activities, or natural laws, what constitutes the sign relation in its logical form is the particular way in which this triad is bound together." In fact, Peirce explicitly referred to this point in the following passage, from a letter to Lady Welby, written in 1908: "It is clearly indispensable to start with an accurate and broad analysis of the nature of a Sign. I define a sign as a thing which is so determined by something else, called its *object*, and so determines an effect upon a person,[5] which effect I call its *interpretant*, that the latter is thereby mediately determined by the former. My insertion of "upon a person" is a sop to Cerberus,

---

4   To avoid this potential confusion, some authors prefer not to use the term *sign* at all when the meaning technically speaking is *sign vehicle*. I prefer to stick to normal language practice and am content to remind the reader that a triadic relation is always knowingly or unknowingly presupposed when something is called a *sign* — as even when in daily speech we simply call the slap a sign, the set of sign relationships that are in play are always implicit.

5   By the word *determine* in this context, Peirce does not suggest that the sign should be involved in a deterministic causality. Peirce sees the sign as determined by its object in a way that reflects the manner in which the interpretant (under the influence of the sign) establishes a relationship between the sign and the object. In another place, Peirce defines the sign (here the representamen) in the following way: "A *representamen* is a subject of a triadic relation *to* a second, called its *object*, *for* a third, called its *interpretant*, this triadic relation being such that the *representamen* determines its interpretant to stand in the same triadic relation to the same object for some interpretant" (*CP* 3:541; Peirce's italics).

because I despair of making my own broader conception understood."[6] (Peirce 1908 (1977), 80–81).

At the bottom of Figure 2.2 is depicted the chain of biosemiotic processes that, taken together, produce the sign relationships of Figure 2.2b. Correspondingly, Figure 2.2c shows semiosis in one of the skin's sensory cells, whose entire architecture and biochemistry create sensitivity to the applications of pressure (a receptivity that, as already mentioned, is influenced by, for example, the concentration of prostaglandins). Here, with the application of pressure, there emerges an interpretant in the form of a context-dependent sequence of action potentials that create a kind of *cellular echo* of the disturbance. This interpretant, the echo, now becomes part of a more complex sign in Figure 2.2d. There, the pattern of action potentials from many different sources, tactile as well as kinesthetic, is processed under the formation of a stereognostic codification of the experience of pressure. Such *stereognosis* constitutes the registering of the disturbance's spatial-temporal characteristics and leads, on one hand, to an even more complex sign that also encompasses other senses (sight, hearing, etc.), as well as to the materialization of contextual memory as seen in Figure 2.2e.[7] The resulting interpretant is here a (conscious or unconscious) awareness of the slap — and it is awareness not of an unintegrated meaningless physical event, but as a sign of anger or aggression from another person. Such knowledge has the potential to create new interpretations in the motor apparatus so that new agent-object interactions might be initiated (whether physical or verbal).

Thus, beneath the conscious (mental) semiosis (here, the perception of a slap as a sign that one has incited anger in another person), a complicated set of altogether unconscious biosemiotic (bodily) processes takes place. The generation of meaning starts in the skin many milliseconds before the brain brings forth a conscious interpretation. Biosemiotics attempts to analyze this sequence of events that traditionally has been considered a simple causal signalling process in no need of interpretive modulation.

Yet there is good reason to hesitate a moment to ponder this difference between a solely biochemical analysis and an analysis that also seeks to determine what is the biosemiotic *function* of the chemical events taking place — i.e., how these events are "understood" by the organism. In both cases the point of departure is a description of the molecular processes that enables the organism to interpret signs of its surroundings. But in traditional biochemistry, the presumption is

6   Peirce had no illusions that his contemporaries would accept his own broader conception whereby nature teemed with beings, for example, bees, that could stand in the place of *persons* as sites for the establishment of interpretants.

7   The chains of sign processes in the body can perhaps be compared with tributary rivers that flow together and integrate (with the generation of interpretants on a steadily higher level) in greater and greater flows.

that such interpretation is either an illusion — a constrained causal process mistakenly experienced as if it possessed a certain freedom — or (and perhaps even worse) a process that only takes place on the psychic level. This last view presumes a dualistic understanding that, strictly speaking, implies that the biochemical machinery is no longer in control when it comes to psychological processes. Considering modern medicine's arsenal of psychopharmaceuticals that can all but put one into whatever mood one wishes, such dualism is greatly at odds with our scientific knowledge. Thus, it cannot easily be denied that our interpretations of the situations that we find ourselves in are strongly influenced by chemical substances (whether naturally produced endogenes such as hormones or neurotransmitters, or externally introduced exogenes such as alcohol or pills). Also, most people seem prepared to accept the reality of an inverse mechanism whereby our health is strongly influenced by our psychological well-being. If then, one (for example, on grounds of religious belief) still wishes to maintain a dualistic view of the psyche-body problem, there still remains the challenge of explaining such well-documented cases of psycho-chemical interaction.

This line of argument is indeed one of the most important reasons to consider the biosemiotic understanding, where psychophysical unity more or less comes for free. Because biosemiotics considers human mental processes not as unique phenomena in the ontological sense, but rather as extremely interesting extensions of a much more general mode of biological organization and interaction that human beings share with all other living creatures. But of course, there exists also a third possible perspective, and this is perhaps the one that most theoreticians would choose — i.e., that one rejects both dualism *and* biosemiotics and simply maintains that all phenomena in the end can be reduced to efficient causality and that our sense of personal causal freedom, or free will, is therefore nothing but an illusion.

And, in fact, this third view is probably one of the most widespread and deeply rooted components in what one could call the *ontology of modern biology*. It has arisen from, and is rooted in, centuries of conflict between church and science, and it therefore carries considerable historical and emotional baggage. And let me here confess that I myself began my studies — and later my professional work — with this exact view as a strong driving force. In the end, however, it proved too simplistic a view to survive my continued investigations into the philosophical problems posed by this approach to natural science.[8] However, even if one denies the concept of a biology-based *semiotic freedom* a place in the ontology of the natural world, it should still be noted that the

---

8   An account of my own history, of course, this is not an argument with which I expect to convince others. Therefore, we shall return to the problem of biological reductionism (eliminativism) in connection with the discussions of sign and causation in the next chapter.

triadic description of a cellular process (such as the sensory cell's production of an impulse that then serves a semiotic function), does offer science a pragmatic advantage because it organizes our knowledge about biochemical processes in a way that reflects their functions in a larger context. Take an analogous example: No urban planner commissioned to analyze subway traffic in Copenhagen would think that such an analysis could be put forth without considering such factors as rush hours, ferry schedules, economic limitations, etc. In the same way, our description of a sensory cell's complete physical activity remains critically impoverished without an accompanying understanding of the exact role that such activity plays in creating the overall function of *sensation*. Accordingly, that a slap is perceived *as* a slap requires that millions of different sensory cells, each in their own specific area of functionality, reacts appropriately regarding the giving and taking of biochemical signs. For even if one believes that the activity of each of these cells obeys the coersive power of natural law down to the most miniscule electronic quiverings — in other words, believes that the cells have absolutely no freedom to interpret and therefore to misinterpret such signs — it could still be useful to describe such activities in the light of their organismic "purposes" — however much we may feel the need to bracket the term *purposes* at this point.8

The Self

The semiotics of the skin encompasses numerous other elements beyond those associated with the senses of pressure and pain. Generally, the skin might be considered a user interface that couples us to the outer world. On one hand, the skin thus serves us as a kind of topological boundary; while, on the other hand, its semiotic capacity opens up the world to us — so that the question of where our self begins and ends is not at all an easy question to answer scientifically. Are not the impulses generated by the blind man's stick really a part of his self? Similarly, as I can just now see houses more than four kilometers away on the other side of the fjord, it seems as if a part of my self reaches out over such a large area. And if, for example, a lightning bolt strikes on the other side, "I" will see it in an instant, even before "I" hear the thunderclap. "I" exists, so to speak, in places over there.[9] The problem here, of course, is that we lack a clear understanding of what our self really is. Does the self have a mass or dimension, or is it a purely mental entity? And do purely mental entities *exist*? In the biosemiotic analysis, the problem of the self is closely associated with the problem of biological *reference*. The skin has both an inner side and an outer side and an

9   Henri Bergson even said, "My self reaches all the way up to the stars" (cited in Kemp, Lebech, and Rendtorff 1997, 64).

asymmetry is therefore established by the skin between that which is inside and that which is outside. The self exists only insofar as that which is inside contains an intentionality toward or reference to that which is outside — an *aboutness*, as it is often called. But this outward reference rests upon a corresponding inward reference, such that one could say that other-reference presupposes self-reference. The French philosopher Maurice Merleau-Ponty (2002 (1945)) has expressed it this way: "L'évidence d'autrui est possible parce que je ne suis pas transparent pour moi-même et que ma subjectivité traîne après elle son corps." (The evidence of the other is possible because I am not transparent to myself and because my subjectivity pulls its body behind it", cited in Zahavi 1999.[10] The key point here is precisely corporeality. For Merleau-Ponty, subjectivity is bodily, and to exist as a body is to exist neither as a pure subject nor as a pure object but rather to exist in a manner that overcomes this opposition. When I experience my self, and when I experience an other, corporeality is the common denominator; we are similar because my experience of both my self and of an other is incarnated. And because my experience of the self is necessarily an experience of a kind of corporeality, it can not be separated from an experience of the other — "I am always a stranger to myself, and therefore open to others," as the Danish phenomenologist Dan Zahavi (1999) has subtlly expressed it.

Yet even if we let ourselves be inspired by the insights of phenomenology, we cannot let our curiosity be paralyzed by the conception of phenomenology as transcendental and as in any way eliminative of scientific knowledge. Thus, while bearing in mind Merleau Ponty's understanding of the self, we shall therefore now pursue the semiotics of the skin that will show us this self in its evolutionary ancestry — i.e., by generalizing it from the particular human self we think we know so well to the self as it occurs in other living organisms.

Under the skin, we come upon even more cellular layers, layers that envelop tissues or organs — or, in other words, beneath the surface we encounter even more surfaces. And if we go further below these surfaces, we again find more surfaces, i.e., the membranes that surround single cells. Now, the number of cells in a mature human's body may be estimated as fifty trillion; and if all of these cells are thought of as being spheres with an average radius of 2.3×10^{-3} centimeters, then the combined area of the cell membranes can be calculated as some 300,000 square meters, or almost a third of one square kilometer. Yet even this figure is probably set too low — for the assumption that all the cells must be spherical is a gross idealization. The vast majority of the cells are, in fact,

---

10   In Søren Kierkegaard's religious philosophy (as expressed in *The Sickness unto Death*) there is the following view of the self: "The human is spirit. But what is spirit? Spirit is the Self. But what is the Self? The Self is a relationship that acts upon its self, or it is in the relationship by which the relationship relates itself to its self; the Self is not the relationship but the relationship that relates itself to its self" (Kierkegaard 1944 (1849)).

oblong, and thus their surface area is somewhat greater than if they were spherical in form.

Moreover, this is not yet the end of the "surfaces inside of surfaces" principle in biological organization. For when we next move *into* the cell, we again encounter a plethora of biologically important surfaces. The cell's interior is packed with bodies inside bodies, e.g., the organelles with names such as mitochondria, lysosomes, Golgi apparati, cell nuclei, and the endoplasmic reticulum (see Figure 2.3). It is difficult to estimate the cumulative area of these additional surfaces, but we would hardly exaggerate if we would say that it is tens and perhaps hundreds of times larger than the total area of the cell membranes alone. A human body, then, consists of perhaps as much as thirty square kilometers of membrane structure. And across all of these membranes there occurs constant biosemiotic activity whereby molecular messages are exchanged in order to bring the biochemical functions on the inside and the outside of these interior membranes into concordance. Thus, the meta-membrane that is the human skin is indeed a highly specialized manifestation of the very same interior interface principle whereby life processes are most generally built up.

But here we must immediately address an essential reservation about this abstract generalization. For an organism's outermost layer (whether consisting of skin, fur, plumage, chitin layer, or something else) is indeed special, in spite of everything, in that it encloses a system whose uniqueness as a particular *kind* of interface cannot or should not be overlooked. And in this regard, we would like to point out a general difficulty that newcomers to the biosemiotic perspective sometimes find themselves struggling with — i.e, the nature of *the interpreter's role*. For semiotics is too often presumed to be a science that attempts to explain

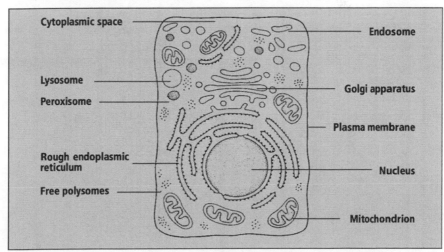

Figure 2.3 Schematic cross-section through eukaryotic cells with organelles.

away the person or organism, whose accustomed principal role is displaced by semiotic processes that go on behind his, her, or its back. Indeed, from Adam Smith's *invisible hand* and Karl Marx's *capital*, through Sigmund Freud's *oedipal repression* and the behaviorists' *conditioning*, to the sociobiologists' *selfish genes*, Western science has been beset with attempts to postulate *virtual agents* whose forces are efficiently causal and yet unseen (in the same way as that highly valued ideal, gravity?), and are considered to be the essential dynamic behind the surface functioning of things.

Biosemiotics, however, is capable of transcending this tradition of disregard for the autonomy of individuals through a theory of semiotic emergence that I will attempt to unfold in the course of this book (Chapter 7). In the Peircean schema of sign relationships sketched above, there are interpretants but not interpreters — but this does not mean that the category of *interpreter* is an empty one. For as we shall see, the idea of *semiotic emergence* implies that while there is no centralized director "behind" the person or organism, the organism or person as an entity is continuously regenerated as an active, creative authority. *The person is thus not a stable being but rather a constant becoming.* The critical point is to recognize the emergent autonomy of various levels of organization.

A single example must suffice here: Some billions of years ago it happened that a number of daughter cells from a single monocellular organism developed a series of symbiotic relationships with one another and, in time, even entered into a process of shared ontogenetic differentiation — so that there emerged a small multicellular organism, consisting of cells with connected life histories but with differentiated roles in their newly collaborative mode of being. These cells having reached this state, it would no longer be sufficient for us as scientists to describe the activity of each single constituent cell in isolation. Rather, one would hereafter have to consider the presence of a new holisticly autonomous actor, an interpreter — or a system of interpretants as Stanley Salthe (1993) has formulated it — that is able to organize the semiotic life processes of negotiating an external environment for the benefit of the collective, and at the expense of the interests of single cells.

Thus, in a certain sense, the appearance of a multicellular organism might be seen as the appearance of a new kind of causality in the natural world, i.e., a formal causality, as suggested in the Aristotelian scheme. In the next chapter, we shall consider this type of formal causality in more detail.

The Lipid Bilayer

The human skin is indeed something special; it serves as the outermost semiotic interface for an organic community of some fifty trillion individual living cells. And the human skin is of course perfectly suited to this role as it is itself

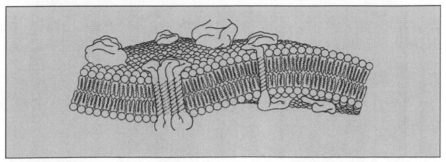

Figure 2.4 The lipid bilayer is the most prominent feature in this depiction of a segment of a cell membrane. The single lipid entity (a phospholipid) is pictured as a circle with two ends, where the circle represents the hydrophilic end and the tails depict the hydrophobic ends. When such bipolar lipid units are placed in an aqueous solution, they spontaneously assemble themselves, as shown here, into a bilayer in which the water-repellent lipid ends turn against each other. This creates a barrier that water or water-soluble compounds cannot pass through without help from proteins, which are also seen in the figure and which play a variety of roles in the semiotic functions of biomembranes.

an evolutionarily advanced example of a fundamental principle of life — i.e., the organizing principle of the membrane or interface, the *semiotic bridge* driven by the persistent asymmetry between the inside and the outside relations of biological organization and interaction. The operation of this semiotic bridge, and the inside-outside asymmetry that drives it, is the theme I shall develop in the following pages.

In its most basic manifestation, a biological membrane is composed of a simple lipid bilayer mixed with numerous protein molecules, and such membranes encapsulate each and every cell on Earth. (Hoffmeyer 1998b; Hoffmeyer 1999a).

(A schematic presentation of the lipid bilayer is given in Figure 2.4). In this system, the protein molecules are responsible for a large range of membrane functions that the lipid layer alone cannot support — first and foremost, the ability to transport material through the membrane and to translate the molecular signals that hit the outside of the membrane into meaningful signals for the inner side of the membrane. When a protein receptor molecule detects such a signal (such as a hormone molecule or a message substance given off by a neighboring cell), what happens is a binding of the signal molecule to the receptor. This then gives rise to a steric change of the receptor molecule that extends across the membrane. On the inside of the membrane, the sterically changed receptor molecule may now, depending on the contextual situation in which the cell finds itself, initiate a variety of specific cascades of biochemical activities that can influence the processes that control the reading of the cell's genetic material.

One should take care not to underestimate the complexity that these cell membranes exhibit. Thus, for instance, the number of surface receptors, i.e., the protein molecules that specialize in translating specific molecular signals in the cell's surroundings, has to be counted in the millions. Accordingly, more than a third of the human organism's total basal metabolic energy production is consumed just to maintain the intense activities taking place across the body's widely distributed membrane interfaces.

And this is because membranes, or proteins topologically linked to membranes' interior and exterior surfaces, control by far most of what goes on in a cell. In the prokaryote (primitive single-cell organisms without cell nuclei, such as bacteria), the plasma membrane attends to the transport of molecules and ions, the biosynthetic translocation of proteins and glycosides; the building up of fat material; communication via receptors; electron transport and the related phosphorylation processes; photo-reducing phosphorylation; and the anchoring of the chromosome, and thereby to cell continuity and replication. In the larger and more advanced eukaryotic cells (cells with nuclei, such as those found in plants, fungi, and animals, as well as in many single-celled organisms such as yeast and amoeba), these same functions are taken over by subcellular organelles — the individually membraned mitochondria, chloroplasts, Golgi apparati, ribosomes, lysosomes, etc. Some of these membranes (perhaps, even all of them) are almost certainly descendants from once free-living prokaryotic organisms, which at one time in the remote past probably were engulfed by some other kind of prokaryotes and which by luck managed to survive, reproduce, and after some time enter into a symbiotic relation with the host organism — what Margulis (1970) called the process of *endosymbiosis*.

Membranes also play a leading role in the functioning and survival of multicellular organisms; the coordination of growth and development in multicellular organisms requires the constituent cells to engage in activities that reflect their exact spatial position in body and tissue. These specifications cannot be obtained from the DNA or gene because the DNA cannot "know" where in the organism any given constituent cell is located. Such knowledge must be obtained solely through the cells' communicative surfaces. The form generation process is mainly the result of local cell-to-cell interactions in which signal molecules from a cell act upon the neighboring cells' receptors. For example, animal cells constantly search out their surroundings with the help of small feelers called *filopodia* that stretch out from the cell. "These cytoplasmic extensions that drive cell movement and exploration are expressions of the dynamic activity of the cytoskeleton with its microfilaments and microtubules that are constantly forming and collapsing (polymerizing and depolymerizing), contracting and expanding under the action of calcium and stress," writes Brian Goodwin (1995, 36).

Membrane structure thus exists relatively independently of the DNA. For while it is true that all membranes are dependent on the *availability* of enzymes that catalyze the synthesis of new lipid units — and that the proteins necessary for the function of the membrane structures must of course also be present — it is critical to note that the growth of the membrane itself takes place by the insertion of new elements into the already existent membrane, and that its undertaking of this process alone thereby increases its area, while using itself as a template. In other words, membranes can only be created using themselves as the prototype, and each cell must therefore inherit fragments of the cell-specific types of membranes that will be involved in the cell's future life. In particular, the egg cell must always emerge with a complete layer of pre-existing cellular membrane types. And this, in turn, should alert us to the fact that the common myth that is broadcast on a nearly daily basis in newspapers and textbooks (and even in some scientific articles), telling us that DNA is the crucial controlling element in the cell's life — and even in the organism's life — is fairly misleading.

For as we shall see in Chapter 5, DNA is a hermetically closed and, for the most part, completely passive molecule, which, as Richard Lewontin (1992) has put it, makes nothing: "First, DNA is not self-reproducing, second, it makes nothing, and third, organisms are not determined by it." Rather, The DNA molecules contain in their nucleotide-based code an array of vital instructions mainly aimed at the growing embryo, enabling tissues and cells to produce whatever protein resources that may be needed in different contextual situations as advertised by the reception of molecular signs at the cell surface. Therefore, If any agency in the body deserves to be called directive or controlling, it would not be DNA but instead the membranes that permeate the body. Yet it would be more accurate to say that the biological orchestration of an organism is created by a well-tuned symphony of biosemiotic relationships across the membranes, which in each instant of an organism's life controls and coordinates the biochemical, physiological, and even cognitive processes that together constitute life (Chapter 7).[11]

The Creation of Life

The shift of emphasis from DNA to the membrane interface solves a problem that has long been a mystery to biology — i.e., that of how a one-dimensional and fairly static molecule such as DNA could be able to specify the generation of a three-dimensional embryo in time and space. There is, in fact, no reason to expect

11   On a still higher level, we can also talk about an ecosemiotic regulatory operationality that is crucial for the ecological and even for the evolutionary dynamic (Hoffmeyer 1996b; Kull 1998; Hoffmeyer and Kull 2003).

that the lifeless DNA molecule would be capable of such a counterintuitive feat. Yet where there is DNA, there is always also the cell with its membranes,[12] whose patterned organization is autonomously determined by the continuity of cellular life through cell division.[13] Living cells, through their membranes, use DNA to construct the organism, not vice versa.[14] It is the active functioning of these membranes as well as the membrane-connected proteins that direct life's activity, not the passive and inanimate DNA. It is, in other words, in the semiotic functioning of the cellular membranes that we shall seek what can be called life's *agency*, its inherent future-directedness, its survival project.[15]

It follows from these considerations that the generation of membranes has been a principal milestone along the path that led to the emergence of life on our planet some four billion years ago. How exactly the first living event happened we can, of course, never know, but it helps that modern nonequilibrium thermodynamics and complexity research support the viewpoint that organized structures *can* arise from less ordered structures under the conditions that prevailed on our planet at the time of life's beginning (Prigogine and Stengers

12   This applies also to the virus, which is certainly a kind of membraneless DNA, but which cannot reproduce itself without the participation of the host cell's mass of membranes.

13   Because of the strong contemporary focus upon the genes, one can easily forget that (with the exception of the chromosomes that suddenly double in number right before the division, after which they are scrupulously partitioned into two of each type to each daughter cell), the cell's remaining parts grow in a smooth and continuous way. The cell itself does not die during cell division — it just continues its life under "two different flags."

14   Kenneth Schaffner (1998) has used the expression *causal parity* to describe such processes. In this case, however, the problem rather is one of incommensurable kinds of causality. The causality of membranes would be an *efficient causality* in the classic Aristotelian schema, while the causality of the genome is instead a *formal causality* (see Chapter 3).

15   The astute reader may at this point ask the question: Why not use Jean-Baptiste Lamarck's and Henri Bergson's concept of *impulse* or *strive* here? I would reply: Because the risk of being misunderstood is far too great. The problem is that the term *impulse* can be understood as an *aspiration* in an unspecified sense (such as a plant's diffuse quest for light) or as a *pre-planned end* in a specified sense (such as a predator's hunt for a given prey). In the latter case, it is a matter of a striving after a fairly defined result (a *transcendental end*), whereas in the former case there is only a situation of a preferred direction (a purely *local teleology*). If one attributes an *impulse* to life in the first sense — which was actually what Lamarck and Bergson did — I do not see a great problem with that (see Chapter 9). But if one attributes to life in general a working, or a striving toward a *definite* aim or end, this invites the justifiable accusation of preordaining from the beginning a *fixed purpose* for life — something for which there is no scientific justification and which, furthermore, traditionally has been characteristic of ideological or religious encirclings of science. Lamarck's and Bergson's unjust destiny was to become misunderstood as spokesmen for the idea of a natural impulse linked to a *transcendental* purpose, which was in fact not their view. And it has been my experience that an insufficiently informed notion of what biosemiotics is and is not asserting is also subject to — and will have to guard itself as much as possible against — the same kinds of facile misunderstandings.

1984; Kauffman 1993). "Order comes for free," as complexity researcher Stuart Kauffman has succinctly expressed it.[16] And Kauffman's theory of life's emergence almost by necessity, through a growth of complexity in autocatalytic systems, offers a good starting point for a biosemiotic theory about life's origin.

Autocatalysis refers to situations in which the chemical reactants that take part in a system accelerate the processes by which they themselves are produced. Now, the standard understanding of life's origin has long been that the emergent chemistry of life had RNA as its central element. Stuart Kauffman (1995) has pointed to several problems with this scenario, however. One objection is that the belief in a "magic molecule" (be it RNA, DNA, or some as yet unspecified protein) holding the key to the very *origin* of life rests on a too simplistic idea of what life is about in a chemical sense. For as Kauffman points out, the simplest independently living cells that are known, the pleuromona,[17] are greatly stripped-down examples of bacteria, yet they themselves still contain between five hundred and one thousand genes.[18] "All free-living cells have at least the minimum diversity of pleuromona" writes Kauffman (1995, 42), alerting us that "your antenna should quiver a bit here: Why is there this minimal complexity? Why can't a system simpler than pleuromona be *alive*?"

The answer that Kauffman himself offers to his rhetorical question is the result of his work with the mathematical modeling of combinatorial chemistry. This research has shown that when the number of individually *random* reactions that are catalyzed in a system of chemicals becomes great enough, there suddenly emerges an orderly and complete *network* of catalyzed reactions. "Such a web, it turns out, is almost certainly autocatalytic — almost certainly self-sustaining, alive" (ibid., 58). Kauffman refers to this phenomenon as *autocatalytic closure*,[19] and his schema for the origin of life then can be summarized as fol-

16 Thermodynamically speaking, Kauffman's aphorism is not meant to be taken *literally*, for to produce order, energy must necessarily flow through the system, thereby inevitably leading to entropy. With the expression *for free*, Kauffman simply means that ordered structures emerge without selective processes, and that self-organizing-structure generation in this context is the means or condition by which selective forces can have something with which to work (rather than structure being a result of selection). I thank Claus Emmeche for having pointed out this important detail to me.

17 The simplest organism is now considered to be *Mycoplasma genitalium*, which includes over four hundred genes.

18 The virus is, of course, much simpler, but it dos not live its own life. Vira are parasites that invade other cells, where they overcome the metabolic workings in order to reproduce themselves. By all accounts, the vira did not exist at the beginning of life but, on the contrary, depend upon the existence of more advanced forms of life.

19 A logically closed system fulfills the following conditions: If one has a quantity of components, the system will be closed in relation to operation M if it happens that when x belongs to the quantity and x is linked with another component, y, by way of operation M, so y will also be part

lows: Complexity leads to autocatalytic closure, and autocatalytic closure leads to life. "The secret of life, the wellspring of reproduction, is not to be found in the beauty of Watson-Crick pairing [DNA nucleotide base pairing], but in the achievement of collective catalytic closure. The roots are deeper than the double helix and are based in chemistry itself. So, in another sense, life — complex, whole, emergent — is simple after all, a natural outgrowth of the world in which we live" (ibid., 48).

While an important advance in our understanding of the natural world in its own right, however, relative to the discussions of this chapter, Kauffman's scenario has a quite significant insufficiency. This is that it all but totally ignores the issue of organismic *semiosis*, which is a vital element in all processes of life. Thomas Sebeok (1985 (1976), 69) expressed this when he stated that "a full understanding of the dynamics of semiosis may in the last analysis turn out to be no less than the definition of life." From the biosemiotic viewpoint, Kauffman's idea about the generation of autocatalytic self-sufficiency is only a necessary *first step* on the way from a chemical system to a living system (see figure 2.5).[20] Beyond this, we must add a *second step*, which is the establishment of the very conditions that could make semiosis possible in the first place — i.e., the generation of a closed membrane around such an autocatalytically closed system of chemical components and thereby *the creation of a basic asymmetry between an inside and an outside, making the membrane a potential interface structure through wich the autocatalytic mix on the inside might learn to adjust cleverly to conditions outside.*

Closed lipid bilayers are, as already mentioned, indeed spontaneously generated when simple components (the phospholipids) are present together in an aqueous solution. But, as we also noticed, a simple lipid bilayer alone will not allow the transport of water-soluble compounds, and membranes in actual living systems always contain numerous specialized protein molecules that attend

---

of the quantity. In logic, the components are propositions, and the operations are logical conclusions; in biochemistry, the components are the molecules in metabolism, and the operations are individual metabolic processes.

20   In a later work, *Investigations*, Kauffman takes a big step toward accepting semiosis as a fundamental element in the creation of life (Kauffman 2000). The proliferation of organization and diversification that Kauffman's later analysis here implies, is caused by the necessary creation of *autonomous agents*, that is, "autocatalytic system(s) able to reproduce and able to perform one or more thermodynamic work cycles." Kauffman very explicitly emphasizes that this definition leads to inconvenient questions about measurement and recognition. For if work is defined as "the constrained release of energy," from where do these constraints come? To this he answers, among other things, that "autonomous agents also do often detect and measure and record displacements of external systems from equilibrium that can be used to extract work, then do extract work, propagating work and constraint construction, from their environment" (ibid., 110). And since a measurement is always an interpretation, we are here in the middle of biosemiotics.

to this task. The spontaneous generation of such a more realistic proto-cell membrane, then, is actually a fairly demanding challenge.[21] Yet even if we cannot here point to a solution to this explanatory difficulty, we can yet maintain that the generation of a kind of proto-cellular membrane with the ability to canalize a selective flow of chemical reactions across itself was at some point a critical initial step on the way *towards* the evolution of genuinely living systems (Hoffmeyer 1998b; 1999a), see also (Weber et al. 1989; Morowitz 1992; Weber 1998a).

But such a closed autocatalytic system, locked within a proto-membrane, still exhibits no capacity for autonomous biological *agency*, and thus we hesitate to call it a genuinely *living* system. There is not yet in the system, as we have described it so far, any proper mechanism for registration of the outer world. There may occur a flow of chemical compounds across the membrane, but this flow is not yet incorporated in the project of sustaining a self. As we reach back to Merleau-Ponty's definition of the self, we may now pose a new question: What *would* it take to attribute a self to a closed membrane system? It seems to me that we should be content if we could point out a mechanism that supplied the system with a genuine other-reference — which again, if it were a case of genuine reference, would necessarily assume a corresponding basis in a network of self-referential activity. The question then becomes this: How could such a thing emerge within the membrane system?

Figure 2.5 suggests a conceptual five step model for the origin of agency and semiosis or, in other words, for prebiotic evolution. The figure does not pretend to detail the kinds of chemical processes that most models of prebiotic evolution have focused on. Instead it departs from a semiotic understanding of the internal logic of the origin process. If - as suggested above - we take Kauffman's concept of the generation of autocatalytic self-sufficiency as a necessary *first step* on the way from a chemical system to a living system, and the *second step*, also depicted, to be the establishment of a closed membrane as a basis for creating the *asymmetry* that makes life-sustaining semiosis with its environment both possible and necessary in the first place, we then propose the following three additional steps as necessary for satisfying our criteria for a living system, exhibiting true biological *agency*, in every instance:[22]

Third step: Once the conditions for autocatalysis within a closed membrane system are in place, there must be established not only one closed membrane

21   Although we acknowledge this difficulty, we must note that, thus far, all alternative scenarios of life's origin equally run into great chemical difficulties. This suggests that there is still (at least) one major part of the puzzle we still are not yet even seeing.

22   Although we acknowledge this difficulty, we must note that, thus far, all alternative scenarios of life's origin equally run into great chemical difficulties. This suggests that there is still (at least) one major part of the puzzle we still are not yet even seeing.

BEGINNING OF LIFE

1.   Formation of autocatalytically closed system.

2.   Membrane formation: Establishment of an asymmetry between insides and outsides.

3.   Emergence of higher order autocatalysis in swarms of membrane units.

4.   Digital redescription of protein components in DNA or RNA (self-reference system).

5.   The membrane becomes an interface (other-reference system). Integration of a system for self-reference and a system for other-reference.

Figure 2.5 Five-step model of pre-biotic evolution. See text for explanation.

system, but a whole swarm of closed membranes interacting chemically and reciprocally through the flows across their membranes. One might understand this as the establishment of a kind of autocatalytic closure on a higher level. By virtue of this step, there would be created the capacity for a kind of proto-communication between membrane entities.

Fourth step: Membranes and their inner autocatalytic reactive combinations must be stabilized through a digital redescription (such as is found today in the RNA or DNA molecule arrangement common to all true cells). Such reproducible molecular arrangements serve as the structural specifications, which (at least in modern cells, where a sophisticated system for protein synthesis has developed) can be translated into defined proteins. Thus the idea here is that the proteins contained in the autocatalytic soup that we believe surrounded the primitive structures of early life, over time became matched with a sequential or digital coded description (presumably first in RNA but later in DNA). In cells as we know them today such a matching of protein strings with nucleotide strings occurs via the tRNA code (see Chapter 5), but this complicated machinery was probably a later invention. At first the matching mechanism may have been much more partial, perhaps just connecting essential domains of the protein surfaces to distinct nucleotide sequences. The advantage of this step in the beginning was to function as a buffer against disturbances of the system, because the components that were lost in this process could then be recreated from the corresponding digital descriptions. Moreover, as we shall discuss in more detail in Chapter 4, this step would have accelerated the evolutionary process by the establishment of that dynamic we have referred to elsewhere as the principle of *code-duality* (Hoffmeyer and Emmeche 1991). The creation of *information-carrying molecules*[23] is often seen as an absolutely central step in pre-biotic evolution (Dawkins 1976; 1989). In the biosemiotic scenario, outlined here, the matching of the proteins' structure to a digital code is an important, but in the end, a secondary step.

Fifth step: To complete the evolution into a true living system, the autocata-lytic system's proto-membrane would need to develop into an actual interface. And to do this, the chemical conditions on the membrane's system-external side must be registered on the system-internal side by a semiotic mechanism that can activate the synthesis of the exact protein species that are needed at any given moment in order to cope with an ever changing external chemical situa-tion. There must, in other words, be established a *feedback loop* between the sys-tem of other-reference (the membrane system) and the system of self-reference (RNA, DNA).

The psychological concept of *individuation* is nowadays most often linked to the analytical psychology of C. G. Jung, where it refers to the processes through which a person comes to understand her own individuality and con-ceive her self as a "blended other." However, the outline that we have given in this chapter of life's beginnings also basically describes a process of individu-ation, especially if the word is understood in its broader philosophical mean-ing as a process through which an individual form of existence develops itself from the basis of the general continuum. The emergence of life as outlined here defines an act of *biological internalization*, whereby a semipermeable mem-brane closes itself upon itself and thenceforth connects its continuing exis-tence to a partially trapped, autocatalytic, and agentive self-existing system (Hoffmeyer 2001a).

Finally, one should not forget that life's emergence was inseparable from that of its environment, much in the same way that the pattern in a carpet is depen-dent upon the existence of the rug. Superficial understandings, which identify the origin of life with the inexplicable creation of information-bearing (some-times derisively called *magic*) molecules, far too often neglect the primacy of this deep connection between organism and environment. But by positing the cellular membrane as the essential locus around which life originated, biose-miotics from the beginning sees the inside-outside asymmetry of agents in the

---

23 It is unfortunate that information, in the context of this all too common use of language, is identified with sequential information. The use of the word *information* is on the whole fairly unclear, a situation that we shall examine more closely in Chapter 3. In the context of biology, the information that is contained in the very architectonic structure of the complete and fully func-tioning cell is of great significance, which is easily overlooked, especially when RNA and DNA are designated as *information-bearing molecules*.

world — and the semiotic bridge that joins the two — as the turning point of life's ongoing evolution.[24]

This is why we can speak about *our* skin. Some four billion years after the first swarming membrane vesicles discovered how to make the components of their inner spaces into helpmates for their own survival project, people now walk around thinking that the skin is there to protect them. But in a certain sense the inverse is true. It was the skin that gave us life.[25]

24   Augustin Berque (2004) proposes a similar view from the standpoint of philosophical geography.

25   One can note in this connection that we know as little about what is inside as we do about what is outside, which of course adds an edge to the Cartesian problem.

3

Sign and Cause

Final Causes

In Western cultures, there is a strong streak of *voluntarism*, an intense belief that human will — good or evil — is responsible for whatever goes on in the cultural world. Accompanying this (perhaps slightly arrogant) belief in and worship of human freedom, however, there is also a distinct *determinism* with regard to nonhuman nature. Here, the law rules to such an extent that many adhere to the belief that the ways of the world could not have been any other than what they actually are. In nature, absolutely no freedom is allowed.

These two beliefs can coexist with little opposition, so long as one thinks that Humanity and Nature are God's separate creations, whose fundamental relationship is therefore determined by God. The moment that one accepts the theory of evolution, however — and with it the view that humans are produced from nature and are, indeed, a part of nature — then humanity's freedom and nature's total unfreedom are logically set on a collision course. For how can anything that is itself unfree ever generate something free?

Such an intransigent contradiction undoubtedly accounts for the continuing power of *creationism* in North American thought[1] and, more generally, behind the mistrust of natural science that has grown so marked in recent decades across the entire Western world. For there are few who can fully accept the opposite solution — i.e., that humans in a profound sense are as unfree as physical nature — and this presumably is the belief many natural scientists would express, if one would closely question them.[2]

In this metaphysical landscape, the philosophy of Charles S. Peirce offers an original solution, a comprehensive and coherent *cosmogonic philosophy*, as he called it, though this is a solution that few know much about, and that

---

1 Recently a new and refined version of natural theology has seen the light of day: the intelligent design (ID) theory, whose main supporting argument is again nature's manifest quality of design — which does not fit well with natural science's traditional view of nature's essential unfreedom (Behe 1996). And while ID theory is itself scientifically unsubstantiated, it must be admitted that natural science's ontological biases provide legitimate targets for such religious revisionism.

2 As I will argue later, such materialism is in reality a disguised dualism (see Searle 1992).

fewer still have fully understood. The essence of this solution, as we shall see, is that on the one hand, natural laws are not absolute, because they themselves are products of an ongoing evolution, and on the other hand, humans' cognitive processes in the deepest sense are of the same kind as all other processes in nature.

The extraordinary difficulty that philosophers and natural scientists have had in following Peirce along this path is attributable, to a large extent, to the fact that the Cartesian body-mind dualism — from which the nature-culture split is born — entirely lacks a concept of purpose that can free itself from its uniquely human connotations. For if one were to follow the underlying logic of Peirce's cosmology, one would then have to accept the existence in nature of so-called end causes or final causes. But among those who have not freed themselves from Cartesianism, such an idea will quickly ring alarm bells. Must there be a *purpose* in nature? Are we not thus returning to religion by asserting this?

Now it is a fact that I, in writing these words, have a purpose in so writing. And it is also as nearly a fact as anything can be — in any case for the nonreligious — that I as a human being am, in the end, a creation of nature. It is, in other words, a proven fact that there *are* at least *some* purposes manifesting within nature, even if they are, as here, only connected with humanity. And yet the scientifically oriented mind, especially, should feel challenged by this nearly incomprehensible, and in fact unfounded, privileging of just one creation among all others endowed with the talent to operate with purposes. This talent must come from somewhere. If it is not from God, and human beings are themselves fully part of nature and nothing supernatural, then a scientific paradox obtains. And it is just here that Peirce's solution comes into play: "It is a widespread error to think that a *final cause* is necessarily a *purpose*. A purpose is merely that *form* of final cause which is most familiar to experience" (*CP* 1:211; italics added), or in other words, "purpose is the *conscious* modification of final causation" (*CP* 7:366; italics added).

That which in the Cartesian tradition is integrated into one concept, the concept of purposive, consciously conceived end causes (and which in a strict sense has only validity in the human world) is thus in Peirce's philosophy two things — one specifically human, and the other a general principle of emergent organization — that should not be confounded. *Psychological* end causes, such as the distinct purposes I might have in writing this text, are in Peirce's thinking just a special subcategory of the much broader category of final causes — and these, according to Peirce, are at play in any sort of goal-oriented activity in nature, as well as in culture. For, to Peirce, a final cause is simply the general form of any process that tends toward an end state (a finale). Peirce expresses it in this way:

[A final cause is] . . . that mode of *bringing facts about* according to which a *general description of result* [read *lawlike system of regularity*] is made to come about, quite irrespective of any compulsion for it to come about in this or that particular way; although the means may be adapted to the end. The general result may be brought about at one time in one way, and at another time in another way. Final causation does not determine in what particular way it is to be brought about, but only that the result shall have a certain general character (*CP* 1:121).[3]

When one is socialized into the Newtonian world view — as are virtually all people today, whether they are conscious of it or not — then the idea of final causation is very difficult to accept. For the Newtonian view of causes is temporalized in a very thin one-dimensional thread. That event A is the cause of event B means that first comes A and then comes B, that is to say, event A happens at time t_1 and event B at time t_2, where $t_2 > t_1$. Yet according to this scheme of thinking, final causes imply that events that happens at time t_1 are caused by an event that first takes place at time t_2, which must be said to profoundly contradict both scientific and everyday experience.[4]

In order to understand the idea of final causes, one must broaden one's conception of time so that it contains more than just one dimension. Event B is not *just* caused by the preceding event A, for the very fact that A takes place is already part of a pattern of events — a pattern that also includes occurrences of type B — which in the end obeys general laws of a more compelling kind. *Final causes* in this sense have the character of *natural law*, as pointed out by Lucia Santaella-Braga (1999) with the important distinction that Peirce did not think that laws took precedence over the dance of randomness, but were the products of it.[5]

For example, the natural law that goes under the name of the *second law of thermodynamics* expresses an entirely universal condition pertaining to all known physical existence, namely that the net amount of entropy will always increase with each spontaneous process that takes place in this universe. What this law determines, as Peirce observed, is not "in what particular way a given end is to be brought about, but only that the result shall have a certain general character" (*CP* 1.211).[6] The second law of thermodynamics can be understood as

3   Thanks to Lucia Santaella (e.g., see 1999) for pointing out this and many of the following points concerning final causes in Peirce's thought.

4   Perhaps more than anyone else, Robert Rosen has, from a mathematical foundation, seen that the Newtonian tradition rests upon a needlessly narrow concept of causality (Rosen 1991). Rosen defended a return to Aristotle's broader understanding of causality, but suggested the term *functional entailment* in place of final cause, using *functional* in its biological sense (see below).

5   I will discuss this in some detail below.

6   It is worth remarking that this understanding of a final cause corresponds to the original Aristotelian view of *telos* as "that 'for the sake of which' something exists or occurs or is done." T.L.

one of the final causes that must be part of any satisfactory explanation of evolutionary phenomena in general, including Darwinian selection.

Relevance of the Second Law of Thermodynamics

The second law of thermodynamics, often called the entropy law, has a fairly ambiguous status in physics, perhaps precisely because it can be interpreted as a kind of final cause — as done here. Many physicists will probably object to this characterization and claim that we are led to this misinterpretation because we focus on the macro level (effects) and fail to follow the law back to its roots in microdynamics (interactions). But even doing so does not solve the problem, it turns out. For, as the Danish physicist Peder Voetmann Christiansen has pointed out, it is simply not possible to define an underlying microplan of any macroscopic system without proceeding from (and within) the preexistence of some boundary conditions — conditions which themselves are "given" thanks only to the macroplan that is the experimenter himself and, in the end, the ecosystem in which he or she exists and whose present state must necessarily be far from equilibrium (Christiansen 1999).

To get a grasp on all this, we must ask this first of all: What, in fact, should be understood by the phrase *the micro level*? Molecules can be divided into atoms, which can be divided into elementary particles. Elementary particles appear to consist of parts, and these parts again to consist of parts, and so forth. What is the bottom line in all of this? The traditional answer to this concern has been to say that the contribution from the submicroscopic parts is locked by virtue of the quantum character of energy, and that the smaller the parts are, the greater will be the separation between their energy levels.

Therefore, as tradition would have it, we are fully entitled to consider a gas at room temperature as a classic system consisting of *stiff* molecular particles, whose dynamics we can describe. In other words, we need not (and in fact *must* not) take the elastic properties of molecules into consideration. The irreversible growth of entropy thus does not reflect a true finality in the universe, but simply our human ignorance about the energy and movements of individual molecules — an ignorance that by its very nature increases every time something happens in the system.

Short (2002, 326) has expressed it this way: "The stock example is familiar: It is the nature of an acorn to grow into an oak, not into a spruce tree or a butterfly. The final cause, then, is a potentiality whether or not actualized." This, however, is not the view of *finality* that one normally finds in the literature. Most often, final causes are there linked with human (or at least human-like) intentions or desires about the future. But for Aristotle, such causes are in fact *efficient causes*, that is, forces or energies that concretely make things happen in our familiar sense. Final causes, on the contrary, are general goods, desirables, that are the deeper causes of such actions of intentions (ibid., 325–26).

But notice that a strange subjectivity has unwittingly entered into the argument here. For had we, as human experimenters, chosen a system at a very high temperature instead of a system at room temperature, then we *would* have had to consider the elastic qualities of the molecules — and therefore the quantum mechanical laws — and at that point we *would* have to deal with another microdynamic. Thus, Christiansen (1999, 53) observes, "The very notion of [what constitutes a] 'microscopic state' depends crucially on our ability to heat and isolate systems, and this ability is not reducible to microscopic laws but depends on technology and intention. The physicists do not just *isolate* a natural system for closer study, but with their methods of preparation *create* the system, including the notion of microscopic parts and the laws that govern them."

The fact that the Nobel Prize in physics in 1996 was awarded for the discovery of the He^3 isotope's superfluidity corroborates this point. Superfluidity can only exist at temperatures less than one thousandth of a degree above absolute zero. The universe's background temperature is more than a thousand times higher than this, i.e., several full degrees above absolute zero. Thus we can, in Christiansen's words, "be pretty sure that superfluid He^3 only exists where there are physicists to study it" (ibid., 54).

Accordingly, then, the entropy law cannot be interpreted in purely epistemological terms — i.e., as an expression of a fundamental ignorance on our part — but must be interpreted ontologically as an expression of the universe's intrinsic irreversibility. The entropy law thus implies that the universe finds itself in a state *very* far from equilibrium — which is why a flow of events continue to happen in it. Among other things, there is an irreversible flow of low-entropy (that is to say, well- organized) wave energy that goes out from the sun into cold space. The Earth receives a small fraction of this energy flow, which allows it to warm itself and set in motion winds and ocean streams and the dynamic geothermal cycle of water from ocean to ocean by way of fog, rain, snow, glaciers, floods, and so forth.

An even smaller part of this irreversible flow is captured by the chloroplasts of green plants — where photosynthesis sets another cycle in play by nourishing the cyclical web of food chains. The whole time, highly entropic energy is given off in the form of thermal radiation that exudes off into the universe. On Earth, the instreaming solar energy and the outstreaming thermal energy (it appears) are more or less equally great, otherwise the Earth's temperature would be subject to radical changes, which apparently does not happen — or has not yet. But even if all the energy leaves the Earth again, its flow through the system has perpetually changed the microstates of the Earth system, and gradually it also changed the Earth's macrostates, as witnessed by organic evolution and by the gradual emergence of a self-regulating biosphere.

Most interesting in this connection are the insights that were first formulated by the physicist Erwin Schrödinger (1944) in his famous book, *What is Life?* Schrödinger saw that life does not so much draw upon energy for energy's sake — but rather that the energy flow *through* organisms allows organisms to get rid of the entropy that they themselves produce through the processes of life. The trick is to consume low-entropy energy and to transform it into high-entropy energy. In this way, life sidesteps the entropy law while it creates organized (low-entropy) structures — living organisms — without violating the thermodynamic requirement that entropy must increase with each spontaneous process. Entropy *does* increase, certainly, but all of the surplus is exported to the surroundings and, in the end, to outer space.[7]

Christiansen's point is that physicists are subject to these thermodynamic conditions since they, too, like everybody else, must live forward in time. Rather than a paradox, then, this shows that the entropy law is actually a tautology: "For the prerequisite of being able to say anything is that the entropy of the universe is higher after the saying than it was before. The same entropic condition applies to any significant event, to every difference that makes a difference, i.e., rises appreciably above the noise level of fluctuations" (ibid., 58).

The expression *downward causation* is often used in this type of causal relationship, where a macrostate acts upon the very microstates of which it consists (Campbell 1974).[8] The expression is probably an attempt to say the same thing as is meant by the expression *final causes* without linking oneself to the cultural inheritance of Aristotelianism, where the term carries so much baggage. But to the Newtonian mind, the idea of downward causation isn't especially popular, either. Perhaps this is because many scientists are unwilling to let systems possess properties that do not, upon closer analysis, show themselves simply to be reducible to the properties of their component parts.[9] Water can clearly be seen to be wet, which neither oxygen nor hydrogen are (at room temperature in any case). But those *physical characteristics of the*

---

7   This last point was applicable, at least, until human-induced flows of energy began to create a significant contribution to the energy economy of Earth (Hoffmeyer 2001b). I consider this idea further in Chapter 9.

8   Valentine Turchin's work on *meta-systems transition theory* also examines the resulting top-down control of bottom-up processes, and Vefa Karatay and Yagmur Denizhan (2002) have shown the relevance of this work for biosemiotics.

9   Frederik Stjernfelt has noted that a second reason for natural science's general skepticism about the idea of downward causation could be that this form of causality, in contrast to efficient causality, does not function in the linear dimension of progressively sequential time. It is thus connected instead with *formal* — rather than final — causes in the Aristotelian scheme. (See more on this below.) And yet, to explain the kind of order we see in naturally occurring systems, modern-day physics has had to introduce the idea of the *attractor state* — which, of course, must be regarded as a genuine form of downward causation, if anything is.

water molecule that makes a person who puts his finger into a glass of water experience the sensory quality of wetness, it can be argued, could be fully accounted for in terms of the well-known properties of hydrogen and oxygen, and it is therefore not reasonable to describe water as a *cause* of the experience of wetness. Wetness, in this view, is not really a property at all, but an epiphenomenon.[10]

The resistance of many physicists against ideas of final or downward causation may derive from a deeply held reductionistic intuition that phenomena are always caused or at least explained best by the dynamics that play out on the underlying levels (and thus, in the end, among elementary particles). But this intuition seems more and more forced — the more it becomes clear that there is no bottom line to such reductionism. As we have seen in the discussion above (Christiansen 1999), an ultimately deepest level cannot be identified *even in principle.*

What is worse, even the idea of some kind of *finally elementary* particles at the bottom level, from which all causality ultimately must derive, is at odds with other fundamental physical theories. Thus, according to quantum field theory, reality consists exclusively of fields, and elementary particles are to be seen as nodal points in such systems, that is to say, as purely topologic occurrences that cannot be ascribed the property of substance (Bickhard and Campbell 1999). That particles are topologic occurrences means that they belong to definite form categories that cannot gradually transform themselves into each other. A surface with a hole in it can, for example, gradually be reshaped so that it becomes a teacup but it cannot be gradually turned into a surface with two holes in it. In order for that to happen, the surface must be punctured, and that would create another topological category.

One could perhaps now think that the reductionistic intuition can live as easily by now simply positing *quantum fields* as its deepest layer as it has attempted to do by positing *elementary particles.* But fields don't have the same independence of the patterns by which they exist as particles have. Mark Bickhard writes,

---

10   Epiphenomenality is often associated with the idea of supervenience, which implies that there cannot be changes of the higher level without there being related changes on the lower level. That consciousness is supervening on the brain implies that changes in consciousness cannot take place without concomitant changes in brain processes. Sometimes a distinction is made between *weak* and *strong* supervenience (Kim 1990). We will not pursue this discussion here, as we associate ourselves with Mark Bickhard's view that systems that find themselves in the thermodynamic state called *far-from-equilibrium* (such as organisms and ecosystems) are woven into unbreakable relationships with their surroundings, so that they cannot be characterized as merely supervenient. A flame, for example, is not simply dependent upon its constituents but also upon oxygen supply, temperature, and other factors (Bickhard and Campbell 1999).

The critical point is that quantum field processes have no existence independent of configuration of process: quantum fields are process and can only exist in various patterns. Those patterns will be of many different physical and temporal scales, but they are all equally patterns of quantum field process. Therefore, there is no *bottoming out* level in quantum field theory — it is patterns of process all the way down, and all the way up. Consequently, there is no rationale for delegitimating larger scale, hierarchical patterns of process — such as will constitute living things, minds, and so on. . . . The recognition that everything is organization of process — just at differing scales and with differing hierarchical organizations — makes the choice to see pattern and organization as of lower level, and thus to render properties of those patterns and organizations as epiphenomenal, a choice that renders everything epiphenomenal because there is no level at which anything is other than an organization of quantum field process, including even the smallest scale quantum fluctuations (Bickhard and Campbell 1999, 331–32).

But is the status of quantum field theory such that we could confidently base a rejection of the reductive particle ontology upon it? It is acknowledged, at least, that particles can both be created as well as disappear during observable physical processes. And this result is precisely what relativity theory brings about when it is used in the context of quantum mechanics. The merit of quantum field theory is that it allows us to handle these remarkable phenomena at the theoretical level. So even if one takes the modern and pluralistic standpoint that fields and particles are equally fundamental, one has already thereby subscribed to the idea that final causes (or their explanatory equivalent) may eventually have to be considered. For the mere fact that fields are fundamental (even though for the moment they must share this honor with particles) will imply that one cannot ignore the holistic configuration of those fields in the attempt to understand our physical universe.[11]

We thus conclude that the rejection of finalistic interpretations of the second law of thermodynamics is based upon (largely unexamined) ideological positions as much as or more than it is upon compelling scientific arguments. And accordingly, we will not therefore let such hidebound ideology deter us from seeking out a broader idea of causation than has been current in natural science since Newton.

For in ontological terms, reality does not consist in fundamental particles whose dynamic interactions cause the whole world. Instead, what I will advocate in these pages is an ontology whereby everything fundamentally consists of *process organization*, in which causality is a characteristic of such process organization.

11   The possibility of a particle interpretation of quantum field theory is dismissed by Halvorson and Clifton (2002).

Final Causes in Biology: *Utricularia floridiana*

Let us now examine the problem of causality by studying a concrete biological example. With inspiration from the ecologist Robert Ulanowicz's analysis in the (1997) book *Ecology: The Ascendent Perspective*, I choose here to look at the sophisticated feeding strategy of the water plant *bladderwort*, of the genus *Utricularia*.

These humble plants are found in freshwater lakes over much of the world where they feed themselves by eating small animals. The bladderwort's leaves (Figure 3.2) are covered with a film of bacteria, diatoms, and blue-green algae called *periphyton*. Species of *Utricularia* in the wild are always covered with such periphyton, and some types of bladderworts excrete complex sugar compounds in order to bind algae to the leaf surface and to attract bacteria. Furthermore, the growth of bladderworts support the growth of the periphyton simply by providing an area substrate — which is necessary because the types of algae making up the periphyton are not adapted to a pelagic (free-floating) life.

A new trophic level now comes into the picture in the form of a number of small, nearly microscopic (about 0.1 millimeter) animals, collectively known as *zooplankton*, that graze the periphyton. The zooplankton include a range of various species, such as *Daphnia* and *Cyclops* (both are types of water fleas), *Rotatoria* (rotifers) and *Ciliata* (multicelled animals with hairy cilia which are used to gather food).

Occasionally the animals will hit a hair on the small bladders (the *utrica*, which gives the bladderwort its genus name). When this happens, an opening

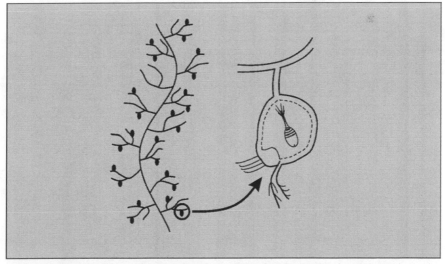

Figure 3.1 Sketch of a typical "leaf" from *Utricularia floridiana*. The detail shows how the interior of the bladder contains an undigested prey (from Ulanowicz 1997).

will emerge at the end of the bladder — and because osmotic pressure is maintained inside the bladder, the small animals are then sucked up into it. The opening quickly closes again, and a slow digestion of the animal begins, whereby nourishment is released and taken up by the plant through the bladder's walls (Figure 3.2).

This kind of interaction between different species is technically called *indirect mutualism,* because all the involved species benefit from the arrangement. For even if the bladderwort nourishes itself by eating the zooplankton, so do the zooplankton, despite all, derive great benefit from the opportunity to graze upon the phytoplankton layer — and the phytoplankton layer, for its part, gains by broadening its *lebensraum* when the bladderwort's surface area increases. Thus, by ensuring the bladderwort's growth, the periphyton increases its own biomass, even in the very process in which it is itself eaten.

The bladderwort system is an example of a biological autocatalytic system. Ulanowicz points out that the generation of an autocatalytic cycle at the species level distinguishes itself from chemical autocatalysis, in that the entities that are involved in the cycle are themselves able to adapt. The result is a self-organizing autonomy that cannot arise in the corresponding chemical system. The systems possess a type of *centripetality*, says Ulanowicz: Each arbitrary change that

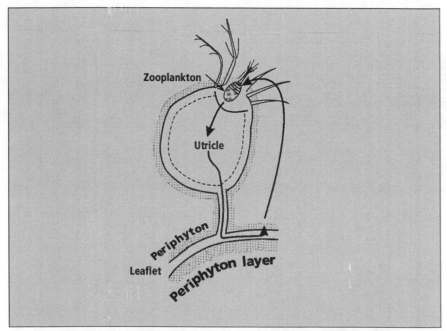

Figure 3.2 The *Utricularia* system's self-catalyzing cycle. The bladderwort plant provides the necessary surface for the growth of phytoplankton. Zooplanktons graze on the phytoplanktons and become themselves caught in a bladder (from Ulanowicz 1997).

randomly increases the speed by which material and free energy are carried to a given element in the system (for example, periphyton) will increase this element's ability to catalyze the flow of nourishment and energy to the next element (for example, zooplankton), and the change will eventually be rewarded (by increasing the growth of the whole system).

Thus, the self-catalyzing *ensemble* becomes the organizational locus of the growing flows of matter and energy, which are drawn into the system: "Taken as a unit, the autocatalytic cycle is not acting simply at the behest of its environment: it actively creates its own domain of influence. Such creative behavior imparts a separate identity and ontological status to the configuration, above and beyond the passive elements that surround it. We see in centripetality the most primitive hint of entification, of selfhood, and id" (Ulanowicz 1997, 47–48). Ulanowicz then goes on to cite Karl Popper, who opined, "Heraclitus was right: We are not things, but flames. Or a little more prosaically, we are, like all cells, processes of metabolism; nets of chemical pathways." I stated this in another way in *Signs of Meaning in the Universe*: We are "infinite swarms of swarming swarms" (Hoffmeyer 1996a, 125).

Very few people will deny that a human individual is *real* in an ontological sense — and yet all of us consist of nothing but cells that, with the exception of our nerve cells, did not exist at all even seven years ago. As humans, we do not identify our self with the constituent units of matter that exactly at this moment make up our body. In the same way, we would ordinarily consider Beethoven's *Grosse Fuge* as an autonomously existing phenomenon of our world, even in those cases (should they ever happen) where this particular piece of music were not being played anywhere on Earth, the whole has a genuine *identity* that is not exhausted by its immediate configuration. Nor is it a mere nominalism, as I will discuss below.

Likewise, we must attribute real ontological status to such self-catalyzing cycles as the bladderworts. Self-catalyzing systems generally outlive their constituent elements, which most often come and go as illustrated in Figure 3.3. Yet by their very existence, they have genuinely *causal* effects, because they constrain the number of possible sustainable adaptations and innovations available for the evolutionary process.

If we look at Figure 3.3d and ask *how it has come to be* that a certain species — for example, species D — is located in this ecological configuration, the answer must necessarily reflect the self-catalyzing system as a whole, for in the absence of the self-catalyzing system, species D most likely would not have displaced species B. It seems difficult to avoid saying that it is the *relationships between species* that causes these replacements to occur. Or, in other words, relationships and not individual species are carriers of causality. What takes place in the bladderwort's system is, of course, fully determined by the material processes

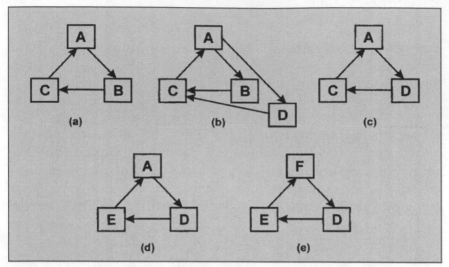

Figure 3.3 Step-by-step shifting of components in a self-catalyzing cycle. (a) The original cycle. (b) Component D appears and is more effective in the feedback loop. (c) D eventually replaces B. (d) In the same way, C is replaced by E. (e) Finally A is replaced by F, so that none of the original components remain. The system endures; the components perish (from Ulanowicz 1997).

that happen within it — but these processes are grounded in a *dynamic* whose inner logic is dependant upon the way the system is structured, and upon the relationships among the populations in the ecosystem.

John Deely (1994, 15) calls this, interestingly enough, *intersubjectivity*: "This contrast between the intrinsic constitution of a thing and its relation to other things provides us with the fundamental contrast between the *subjective* and the *intersubjective,* between what things are in themselves and what they are as parts of a system, that is, relative to other things." He notes that in the Middle Ages (and in most of the modern age), such intersubjectivity has not been considered to have independent existence. And here, I suppose, do we find the essence of what is called *nominalism*. This oft-contested doctrine holds that the relation between existing things is not itself part of what exists, but should instead be regarded as a mere construction in the mind of the observer. That a mammalian shoulder bone fits into the shoulder joint of the same animal is, according to this view, not part of reality, for *fits into* is a purely *relational concept*, and relations are not real things in themselves, only things like shoulder bones and shoulder joints are. The fitting-in *relation*, correspondingly, does not — to a strict nominalist — refer to any independently existing reality.

But as Deely (1994, 22) carefully explains, "What is at stake in this peculiarity of *relation as an intersubjective mode of being* is the very possibility of semiosis — or, perhaps I should better say, the opening in nature to an action of signs not reducible to the more typical, or at least more widespread, brute force interactions of Secondness constitutive of [purely] subjective being and the physical environment of material objects."[12]

The tendency of natural science to deny the ontological reality of relations is probably the reason that some biologists in recent years have looked back to Aristotle's more articulated understanding of causation (Rosen 1991; Salthe 1993; Riedl 1997; Ulanowicz 1997; see also Juarrero 1998).[13] Autocatalytic organizations such as the bladderwort can then be considered as an Aristotelian *formal cause* for species D's existence in the bladderwort's periphyton. And this overall cycle relates to its individual component parts in the same way that an organism relates to its cells, namely by establishing a set of organizational and interactional boundary conditions for the activity of its subunits — activities that persist far beyond the lifetime of the subunits

Indirect mutualism of the kind seen in the bladderwort demonstrates the development of *asymmetry* in the ecosystem, as the flows of energy and nourishment are displaced from what is called equiprobability — i.e., the state where they are most equally distributed among different components and channels. Instead, the flows of matter and energy are drawn into those parts of the ecosystem where self-catalyzing cycles have established themselves. In this way, new structural organization emerges in the ecosystem demonstrating the self-organizing dynamics of the system. Ulanowicz (1997) shows, furthermore, that this "propensity to move away from a state of equiprobabilility" is a characteristic property of living systems in general. Thus behind the kind of formal causation that we have seen exemplified here, a more general propensity is at play — one which could be described perhaps as *a general propensity of complex systems to self-organize.*

As even Kant long ago observed, the self-organizing property of living systems obviously is in opposition to the physicalist conception of nature shared by most scientists. Kant, as a consequence, claimed that living creatures must always be understood teleologically.[14] And it was for this reason that Kant maintained that biology was not a natural science. But rather than accepting Kant's radical renunciation of our possibilities for achieving a naturalistic understanding of life, biosemiotics follows Peirce in seeing the principle of final causation as a natural property of the world at large.

12   In using the term *Secondness*, Deely is referring to the Peircean *categories of relation* that I shall be discussing in more detail shortly.

13   I discuss this problem more thoroughly in Hoffmeyer 2008a. See also the thorough analysis of the ontology of relations by Paul Bains (2007).

14   This is discussed in Chapter 6 in the section, "Self-Organization, Semiosis, and Experience."

At this point, however, we must distinguish the semiotic conception of *causation* from the Aristotelian conception. Aristotle distinguished, as is well known, four kinds of causes: the material, the efficient, the formal, and the final. To illustrate this, let us again borrow an example from Ulanowicz's excellent treatment of the subject (Ulanowicz 1997). The example is a military conflict. Here the *weapons* constitute the *material* cause to the battle, the *soldiers* comprise the *efficient* cause, the armies' *placement* with respect to the particularities of the landscape becomes the *formal* cause, and the *final* cause will consist of *the actual intersection of particular social, political, and economic conflicts.*[15]

This example at once reveals the hierarchical dimension to the Aristotelian idea of causation. As Ulanowicz states, it is the officers who represent the battle's most *central* level. Ulanowicz calls this level the *focal* level, which is precisely the level that is concerned with the *formal* causes to the course of the battle — e.g., the location of the armies in relation to the landscape. In comparison, the soldiers can only influence limited elements of the entire activity, whereas the heads of state for their part can mostly influence the events that take place before and after the actual battle. It is precisely this inherent hierarchical dimension of the Aristotelian concept of causality that was incompatible with the reductionistic and universalistic perspective that has been urged upon us by Newtonian science.

And ever since the Newtonian revolution, we have had to live with a natural science that almost exclusively concerns itself with efficient causality. Material causality, for example, is now accepted to a certain degree in biology, but most often only in the somewhat pre-causal role of *initial conditions*. Formal causes were difficult to avoid in practice since organizational structures obviously always influence the particularities of natural processes, but the predominant credo has been that formal causes can always be reduced to efficient causes. With regard to final causes, it must be said that an unquestionable taboo has ruled — a taboo that has been especially troublesome for biology, where it is difficult to avoid teleologic (often called functional) explanations of the following type: The heart beats *in order to* pump oxygen into the body. But this is an explanation that, strictly speaking, would imply that the heart would have to possess an intention of helping the body — an unacceptable thought, of course, if only for the simple reason that the heart has no brain, and cannot therefore possibly know that it is located in a body in need of oxygen.

As we shall see below, most biologists nevertheless consider themselves fully justified in using functional explanations of this kind — probably not even

---

15   In a deeper sense, however, the *final cause* in this case is the desirability of circumstances (e.g., wealth and power) that both of the involved powers cannot simultaneously realize; thus, they have to fight over it (see Short 2002).

seeing it as contradicting the taboo against final causation. This is because it is assumed that such teleologic explanations could ultimately be justified reductionistically through the application of the principle of natural selection. For this principle is quite generally supposed to be responsible for any functional optimization of the gene pool that might arise, and for this reason, explanations of the kind discussed above are said to be *teleonomic* rather than *teleologic*.

Before we look more closely at the functionalist ideas of biology, there may be reason to warn against an uncritical adoption of the Aristotelian causal categories. For even if we accept the necessity to operate with formal and final causes in biology, this does not imply that we can adopt Aristotle's understanding of *matter* as a passive victim of the form-giving principles of nature. This latent dualism of Aristotle's, on the other hand, was easily incorporated into the compromise between Aristotelian philosophy and Christianity that was worked out by the medieval scholastics. It is perhaps not a coincidence that the alternative to this way of thinking has perhaps never been expressed more elegantly than by Giordano Bruno in the work *Cause, Principle, and Unity*, a book that more than anything else led to his execution by burning (by papal decree) in the year 1600. Here Bruno allows his hero, Theofilus, to get into a discussion with the orthodox Aristotelian Discono, who in dismay accuses Theofilus of implying that "matter is *propre nihil* [almost nothing], pure potency, bare, without act, without virtue and perfection." Theofilus answers to this: "So it is. I claim that it is deprived of forms and without them, not in the way ice lacks warmth or the abyss is without light,[16] but as a pregnant woman lacks the offspring which she collects and expels forth from herself, and as this hemisphere of the Earth is without light at night, which it manages to reacquire by its turning around" (Bruno 1584 (2000)); translated by the author from the Danish version).

The Aristotelian matter-form dualism was itself originally a way of solving a philosophical problem that seriously occupied the minds of the philosophers in antiquity, the question of change: How can anything change without ceasing to be the thing it was? By positing the idea of a "form-giving principle that rules over matter," it became possible to think of change as change in form only, thereby conserving the identity of matter in spite of the change. Change then, in the Aristotelian conception, was change in form, and as such, was brought about by the processes of formal causation.

But Aristotle saw formal causation itself as guided by final causation. And at work here is a somewhat questionable metaphysics in which nature's purpose is posited as harmony, perfection, or "the good." In short, Aristotelian formal and final causation depends very much on a whole set of ideas that one should

---

16   See *Genesis* 1:2. "And the earth was without form, and void; and darkness was upon the face of the deep" (King James Version).

take care not to import into the contemporary scientific discussion when using these terms. In the Peircean conception of final causes, we find none of this. Quite to the contrary, Peirce believes on the most fundamental level that both *randomness* and *irreversibility* characterize final causes. Peircean final causation therefore is a very different thing from Aristotelian final causation, and the two should not be confused.[17]

The Functionalist Idea in Biology

The identification of final causes with *purposeful causes* — where purpose is understood in the psychological sense — probably explains why the rejection of final causes has been, and remains, all but total among biologists. In a book entitled *Nature's Purposes*, Colin Allen, Marc Bekoff, and George Lauder (1998) collected a group of key articles from the previous thirty years' debate about teleology, functionality, and purpose in biology.[18] Yet once one has already overcome the fear of, and defensive taboo against, the very possibility of final causes, as we have here, the subtle discussions in the perceptive and well-argued articles assembled in this volume feel a little strenuous to say the least. In fact, what these articles illustrate more than anything else is the absurdity of explaining away final causation in the workings of living systems. Like the more and more subtle play of epicycles invented by astronomers to uphold the geocentric universe, thoughtful biologists have had to invent more and more subtle arguments in order to uphold a view of organic evolution as obeying the irreproachable scheme of efficient causation as the *only* basic explanatory tool in the natural world. Rather than dwelling upon the many subtle details of this descussion, I'll limit myself here to a short presentation of the main arguments.

Should we speak about a modern consensus among biologists concerning the questioin of functionality, this consensus would be best framed in the terms of Karen Neander's *etiological theory*, where the point is that a trait or characteristic of an organism is *functional* if it derives from parents whose offspring survived through natural selection by virtue of exactly the function that this characteristic served for the parents. Or, in Neander's (1991) own words, "It is a/ the proper function of an item (X) of an organism (O) to do that which items of X's type did to contribute to the inclusive fitness of O's ancestors and which caused the genotype, of which X is the phenotypic expression (or which may be X itself where X is the genotype), to increase proportionally in the gene pool."

17   This is one reason why I prefer to use the term *semiotic causation* for this phenomenon in my own work. For although *semiotic causation* does, of course, imply some sort of final causation as an operative principle, it does not import the whole baggage of Aristotelian metaphysics.

18   For a very different discussion of this problem, consult Havel and Markos (2002)

There are divergent opinions about this, though, especially as concerns the very selectionistic conception of evolution expressed in the etiologic theory. For example, Ammundsen and Lauder (1994) note that the idea of *homology* is a pre-Darwinian concept that originally was based upon comparative anatomy and which therefore cannot be defined through a theory of functionality. For example, the forelimbs of humans, dogs, moles, whales, and bats are all homologous, but their functionality is totally different.

Against the etiological theory, they point to the wings of insects — which, according to all accounts, originally had no aerodynamic function, but only served the purpose of thermoregulation (a role they still can have, for instance, for bees). "If we identify the function of insect wings as the effect for which they were *first* selected, then we would say that the function of wings is thermoregulation" (Allen, Bekoff, and Lauder 1998, 357). Ammundsen and Lauder (1994) instead suggest that we base the definition of function upon the relevant trait's *causal role* — as originally suggested by Robert Cummins (1975). This is to say that functionality must always be understood *in relation to a task*, and the function of the trait or characteristic must thus be understood as its contribution to the fulfillment of this task. A dove wing's causal role is to contribute to the flying ability of the dove, which will thus be the function of the wing even in the hypothetical case where conditions change so much that flying skill is no longer of use to the bird in its quest for survival and reproductive success.

Cummins's attempt to define biological functionalism in relation to causal roles was dismissed by philosopher Ruth Millikan who referred to the absurdity of maintaining that clouds have the causal role (i.e., function) of raining so that streams and rivers are filled with water, which — as part of the hydrologic cycle on earth — work to keep the soil moist so that plants can grow (Millikan 1989). For this is obviously a different idea of *function* than we are using when we say that it is the heart's function to pump blood throughout the body, even if its placement in a system of interdependent causal relations looks, at a first glance, somewhat the same.

Seen from the biosemiotic standpoint, Ammundsen's and Lauder's suggestion has the advantage that functionality is ascribed to the *lineage* as an autonomous historical system. It cannot therefore be reduced to the bare rational reflection of external selective forces, but rather, follows its own inner (semiotic, I would say) dynamic of development. But since the authors do not explicitly set themselves apart from the commonly accepted mechanical model of causality in biology, it is difficult to see how they can defend themselves against objections of the kind that Millikan and others have put forward (Millikan 1989; Neander 1991; Sober 1993).

At this point, however, the etiological theory is better only in appearance. By linking *functionality* to an already selected *effect*, this "effect" may seem to assume an aspect of value that in a crucial sense separates it from the raining that Millikan facetiously attributes as a "function" of a cloud. Insect wings with effective aerodynamic potential were selected (which is to say that they were in some sense *favored*) by nature, in preference to those insect wings that could only be used for thermoregulation. But if one looks closer into this favoring, called *selection*, then it is emphasized again and again that in the end, it too is all just a result of blind coincidence, and *not* purpose (see Andrade, 1999). Thus, functionality of *this* kind is scarcely different from the cloud's raining; both are *coincidental*, rather than *functional,* in the normal sense of this word.

In this debate, it seems that only philosopher Ruth Millikan really avoids this vulgarization of the idea of *causality*, perhaps because her source of inspiration is not primarily biology, but rather problems in language philosophy and cognitive research concerning the eventual possibility (or impossibility) of intentionality in computers. Millikan (1998) thus has no fear in approaching the idea of *purpose* — on the contrary, she succinctly declares that "the definition of 'proper function' is intended as a theoretical definition of function or *purpose*." And just as function is inextricably linked to purpose, so also, in her view, is *purpose* unbreakably linked to historical *becoming*: "The mouse may have a disposition to take measures that will remove it from the vicinity of the cat, but not if under water, at 1000 [degrees] Fahrenheit, in the absence of oxygen, while being sprayed with mace, or just after ingesting cyanide. Indeed, there are thousands of stressful conditions under which the mouse might be placed, which would extinguish its escaping behavior" (Allen, Bekoff, and Lauder 1998, 308). The point is that the escaping behavior of mice only works under normal circumstances. Functionality thus presupposes an idea of *normal circumstances*. But where, in turn, does *that* idea come from? "At the very least, we must make a reference to something like the conditions in which mice have *historically* found themselves, or better, found themselves when their dispositions actually aided survival. . . . Mice must be born of mice. Consider if a seeming mouse were born of a fish, what would set the 'normal conditions' for manifestation of its relevant disposition?" (ibid., 309).

Millikan seems in no doubt that when a mouse flees, it does so with a purpose — but that this purpose is not *planted in* the mouse's genes or in its brain in any kind of naïve, simplistic sense. Rather, *purpose* here constitutes an aspect of the historic process in which the mouse is only a (short-lived) current participant or exemplar. And as with mice, so with humans. For us, too, our conscious and intentional actions stand under the influence of the inheritance from evolutionary history. We are, so to speak, historically formed by the world — and this is basically why we can learn to know it.[19] Accordingly, our "conscious inten-

---

19 We can only learn anything about the world because our inherited talent for forming intuitions (literally, from the Latin, "*in*-sight") has been evolutionarily molded by the conditions

tions" are not something that develop de novo in an isolated brain; instead, they always emerge from historical relationships between the head and the world (which includes, of course, the actions of other similarly situated agents).

And although Millikan apparently is comfortable with *natural selection* as an explanation for the functionality exhibited by animals, her project nevertheless collapses the ontological barrier between "free" humans and "unfree" animals. In so doing, the door opens a crack to admit actual, nonpsychological, *final causation* as an operative force in nature. And, as we saw in Chapter 1, natural selection itself depends upon the existence of a deeper dynamic in the physics of the universe that can explain the emergence of *agency* in life:

> That we can explain how living systems behave in order to become prolific does not explain evolution in the deeper sense. How does this fertility come into being? What temporal motive force is at play here? Selection theory also needs a theory about what kind of thing life's agency would be and how that agency comes to be. And so it appears that selection theory is only a surface structure built upon the foundations of a deeper theory of evolution that explains how life relates to the processes that are fundamental to the physical development of the universe.

Thus, biological functionality has its roots in the universe's fundamental *directedness*.

The Central Dogma

Natural selection could hardly maintain its credibility as an explanation of biological functionality if it hadn't coincided so well with another one of biology's authorized stories about the innermost nature of life: the *Central Dogma*. It was Francis Crick, one of the discoverers of the double-helix structure of DNA, who jokingly codified the dogma whereby information can only be transferred from DNA to RNA and further to protein, but never the other way around (i.e., from protein to DNA).[20]

---

prevailing in our world. Peirce puts it this way: "Thus it is that, our minds having been formed under the influence of phenomena governed by the laws of mechanics, certain conceptions entering into those laws become implanted in our minds, so that we readily guess at what the laws are. Without such a natural prompting, having to search blindfold for a law which would suit the phenomena, our chance of finding it would be as one to infinity. The further physical studies depart from phenomena which have directly influenced the growth of the mind, the less we can expect to find the laws which govern them 'simple,' that is, composed of a few conceptions natural to our minds" (*CP* 6:10).

20   It has since been shown that there is a reverse-transcription enzyme that uses RNA as a template for the synthesis of DNA, and therefore transmits information the "wrong" way. This, though, still leaves intact the fundamental principle of the Central Dogma, which is that genetic material cannot be created from, nor altered by, proteins.

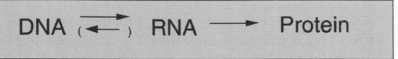

Figure 3.4 According to the Central Dogma, the information flow in a cell is one-way. Information is passed from DNA to RNA and further to protein, but never in the reverse direction.

In Crick's own original formulation (1988, 109), the Central Dogma was simply this: "Once 'information' has passed into protein it cannot get out again." But this glib formulation contains great ambiguity. First of all, of course, there is the question of what information actually *is*. Crick is himself fairly explicit in this regard. "Information means here the precise determination of sequence, either of bases in the nucleic acid or of amino acid residues in the protein" (ibid.). Information is to be understood, then, as a kind of *specification*. It is important to note here that this conception of information, which we may call the *biochemical information concept*, is decisively different from the concept of information that underlies Information Theory, or Shannon information. Shannon's information concept is a purely probabilistic measure that in principle can be objectively determined, and is deprived of any internal relationship to relevance or meaning.[21] A manuscript created by a monkey, for example, may contain more Shannon information than a manuscript of the same length created by a normal user of language. For a language user writes according to rules, which make his letters more predictable than a monkey's.

The problem with the biochemical information concept is that a specification has no unequivocal or measurable reality. Even in the simplest cases, such as when we say that the RNA sequence AAA specifies the insertion of the amino acid phenylalanine into the growing protein chain, such specifications are hugely dependent on the prevailing present circumstances. A mutation might perhaps substitute a different amino acid in place of phenylalanine in the same location in the protein chain, and then suddenly the same information (AAA) would specify something entirely different. The question of the ambiguity of genetic information will be thoroughly examined in Chapter 5. Anticipating that discussion a little bit, we can say right now that generally, the problem is that a *specificational information concept* is one that is necessarily context-dependent — and thus, in the end, is necessarily dependent upon

---

21   One objection to this holds that the objectivistic interpretation of the information idea is not, as such, implicit in Shannon's original idea, but rather is a consequence of its frequent use within closed systems that have well-defined realms of possibility. But Shannon himself clearly instituted this practice, stating that the semantic aspects of communication are irrelevant to the engineering problem.

what effect (or, in Peircean terms, *interpretant*) the information will elicit in the receiving system.

To see how this is so, we can, for example, following Sterelny and Griffiths (1999, 102), look at a simple system for information transmission such as a doorbell. By pressing the button, one specifies the message (out of a theoretical infinity of possible other messages) that there is someone seeking contact. The difference between the ringing sound and the contrastive quiet here depends not simply upon the capability of the ears of the resident — as, for example, he or she may be at that moment playing loud music or making some other kind of noise such as vacuuming or cutting grass — but also depends crucially upon such nonreceiver elements as whether or not the power is off, there is a loose connection in the circuit, or if the bell is in order. Here we naïvely presume that the *source* of the information is the pressure on the button, while all the other conditions fall under the subservient classification normally called *channel conditions*. But as Sterelny and Griffiths remind us, our thinking this so may not so much accurately reflect a fact about the world, but reflect, rather, "a fact about us. The sender/channel distinction is a fact about our interests, not a fact about the physical world" (ibid.). "What is noise and what is signal depends on what you are interested in. When you see a white dot passing across your television screen, it may be a tennis ball (signal) or it may be atmospheric interference or the cat sharpening its claws on the aerial (noise). But nothing in nature dictates that one dot is signal and the other is noise" (ibid., 103).

In Crick's formulation, *information* stands forth as an indubitable category, virtually as a thing or a postal package that one can send anywhere in the world, without any change in its contents. And in the aftermath of Crick, biochemists have repeated this rather peculiar use of the word *information* such that it has now become so seemingly self-evident that no further explanation is needed. I quote as an example from a nearby biochemistry textbook on my bookshelf: "*Of course*, all metabolic reactions are controlled ultimately by genetic information, which specifies the structures and properties of enzymes" (Mathews, van Holde, and Ahern 1999, 876; italics added).

Typically enough, the next chapter in this textbook has the title "Information Readout: Transcription." And so we have become so accustomed to this type of language use that now it barely surprises us. But, notoriously, it is somewhat other than normal to maintain that cells can *read* or that *information* can control. Both of these worthy activities usually presume cognitive abilities that biochemists will be the first to deny could occur in cells. And if we consider the word *information* in its everyday meaning, then we can, for example, say that there is some information contained on this page. But I hope that the reader doesn't believe that she becomes *controlled* by this information. I hope the situation is rather that the reader feels *inspired* by the act of interpreting these words *as* information and interprets them in relation to her own interests and needs. Exactly the same is what is happening in cells and in organisms, I will argue, when they act upon their inherent genetic material *as* information with which to survive and develop.

The idea that information can stream, flow, or let itself be transported through a channel derives from classical information theory as developed by Claude Shannon and Warren Weaver (1949), as mentioned briefly above. The starting point of the work of Shannon and Weaver was an engineering-related interest in developing telecommunications media, and consequently, in optimizing the ratio between signal and noise. What might be the meaning of the signals transmitted through channels was for good reason not a question that interested the fathers of information theory. The information that flows through channels is only defined in relationship to its external characteristic of *non-noise*.

Information transmission, of course, always depends on material processes. The words on this page can be read because of the patterning of black ink on the white of the paper, which is reflected onto the retina of the eye, as light bounces off it. (In writing for the blind, this optical technique is replaced by a tactile approach.) But as soon as we switch from an external to an internal approach to information, the explanatory viability of the *transport* metaphor ceases to be obvious. Shannon information can flow, stream, or pass, but in what sense can a *specification* flow?

The tacit transfer of the transport (or flow) metaphor of Shannon information to genetic information is not only meaningless (since genetic information cannot be understood independently of the particular context in which it is situated), it is also very misleading. If the expression *genetic information* shall be maintained at all, then what is at stake cannot be equated with such context-free (and semantics-free) information. Genetic information is deeply linked to the survival project of organisms and must be understood as a set of inherited instructions for the production and delivering of protein resources that can help one to function in the world.

The transport metaphor has here the effect of objectifying genetic information, giving it a thingish character and a consequent autonomy that it does not possess. And this reification of information, for its part, opens the way to the misunderstanding that natural selection can optimize genetic information bit by bit. For when *information* is understood as bound to the gene and not to the context, the idea easily emerges that the population genetic combinatorics at the DNA level correspond with an equally unambiguous combinatorics at the semiotic level. In short, by considering information as genes, one eliminates the idea of genes — and life — as a semiotic phenomenon.

Such misplaced metaphor and oversimplification can hardly be expected to survive the ever deepening scientific developments that, more and more, demand a refined understanding of the massively complex and interdependent processes of life. A little maliciously, perhaps, one might compare the Central Dogma with the chemical theory of phlogiston of an earlier time (Hoffmeyer 2002a). For in a somewhat similar fashion, chemists of the eighteenth century reasoned that when substances burned, one was witnessing the liberation of a special internal substance called *phlogiston*. Flammable materials such as oil or coal, for instance, were thus considered to contain large amounts of phlogiston, and this accounted for their propensity to readily burn.

In 1794, however, the French chemist Antoine Lavoisier pointed out that, what really happened when something burned, was that oxidation was taking place — that is, burning (and rusting) material was not emitting some mystical phlogiston, but instead consuming (binding) oxygen. And the exact same process, he also helped to show, was soon proven to take place as well in the metabolism of animals. And just as the phlogiston theory, in its time, stood in the way of the further development of chemistry — we will soon see, I believe, that the idea of the transport of information within and between living systems today stands in the way of the development of a biology in keeping with the progress of the times.

For an up-to-date biology must acknowledge that the biochemical concept of information is just too impoverished to be of any explanatory use. Information isn't just something that is transported or that streams through a channel; information is something that must be understood — and which therefore can be misunderstood. Recently, Eva Jablonka (2002, 580) has made a resolute attempt to save the concept of information through a semantic definition, by placing the concept in a functional evolutionary perspective where "the focus should be neither on the evolution of the signal 'carrying' the information, nor on the evolution of the final specific response. Rather it should be on the evolution of the system mediating between the two — on the interpreting system of the receiver." However, even if Jablonka's understanding of the information concept comes close to the biosemiotic concept of a sign, she strangely enough still thinks that it is important to preserve the notion of information as something that can be moved from one place to another.

But if information is to be understood as a relationship that involves an interpreting system, it is difficult to see what it could mean to say that information is transported. It is well known that if one looks, for example, at a little white disk on a red background, it seems to be green. Make a copy of the disk, place it on a blue background and the disk will appear to be orange. The disk's color information *travels* as little with the disk itself, as the gene's information travels with the DNA. Information, understood as a relationship between a cue and an interpreting system, is not itself an entity that can be transported. Still, the increased depth in Jablonka's idea of information consequently parts ways with a reductionistic approach whose ideas are still considered useable in the theoretical understanding of evolution that Jablonka is challenging.

Semiosis and Finality

Biosemiosis, the never-ending stream of sign processes that regulate and coordinate the behavior of living systems, depends upon the special receptivity that evolutionary systems over time have developed towards selected features of their environment. Peirce considered "the tendency [of things] to take habits" (*CP* 1:409) — or, in more modern parlance, *self-organization* (the tendency for ever new regularities to arise in natural systems) — as the most fundamental characteristic of nature. And living systems are examples par excellence of the tendency, to *take notice*, that is to say, to detect regularities in their surroundings that, if properly interpreted, might guide them to perform well.

The tendency to generate new habits is, of course, fundamental for all true learning, as well as for all cognitive activity. A bird must generate the habit of flying before it can find food, and a doctor must be in the habit of deciphering X-ray pictures before he can decide that a given shadow is a tumor. For Peirce, this law of habit formation, which he also called the *law of mind*, was more fundamental even than the universe's current physical laws — for these latter are themselves the product of the former.[22] And here we have come upon an element in Peircean cosmology that at first might seem somewhat counterintuitive. And yet, upon a closer look, the idea is strikingly modern.

In opposition to the natural science of his time, Peirce considered *indeterminacy* and *chance* as the primordial condition of the world. Given this starting point, the real "mystery to be unravelled" about our world is not so much that it contains an unbelievable mass of ungovernable activity, but that there is any stable order in it, at all. For very good reasons, science has set itself the task of explaining the events and structures we find in our world as being in all cases the outcome of natural lawfulness. And, although scientists are generally confident that aberations from the determinacy of natural laws will ultimately be

---

22   "The law of habit exhibits a striking contrast to all physical laws in the character of its commands. A physical law is *absolute*. What it requires is an exact relation. Thus, a physical force introduces into a motion a component motion to be combined with the rest by the parallelogram of forces; but the component motion *must actually take place exactly as required* by the law of force. On the other hand, no exact conformity is required by the mental law. Nay, exact conformity would be in downright conflict with the law; since it would instantly crystallize thought and prevent all further formation of habit. The law of mind only makes a given feeling *more likely* to arise. It thus resembles the 'non-conservative' forces of physics, such as viscosity and the like, which are due to statistical uniformities in the chance encounters of trillions of molecules" (*CP* 6:23; italics added).

explained in good manner, it remains a striking fact that natural systems are — as they are actually found in nature, not as they may be reductively described in the classroom or textbook — are "an unruly mess," as ecologist Peter Taylor (2005) has succinctly put it.

Yet if indeterminacy and chance are the basic properties of our world, rather than determinacy and order, such, in fact, is exactly what might be predicted. And if so, the mystery that has to be explained is that in this swarming jumble, there are nevertheless domains where one can meaningfully talk about *lawfulness*. The task for natural science is thus not so much to corral the slimy and messy diversity of life into the straightjacketed uniformity of natural law. The task is rather to explain how ordered structures emerge out of unordered, chaotic diversity. Peirce (*CP* 6:12–13) writes,

> Uniformities are precisely the kind of fact that need to be accounted for. That a pitched coin should sometimes turn up heads and sometimes tails calls for no particular explanation; but if it shows heads every time, we wish to know how this result has been brought about. *Law is par excellence the thing which wants a reason*. Now the only possible way of accounting for the laws of nature and the uniformity in general is to suppose them results of evolution. This supposes them not to be absolute, not to be obeyed precisely. It makes an element of indeterminacy, spontaneity, or absolute chance in nature (italics added).

The italicized sentence — *Law is par excellence the thing which wants a reason* — is the key to Peirce's critique of natural science. He asks us to consider why natural science is not interested in the most central question: How is it that there can be *natural law* in the world at all? It is precisely in his quest for the answer to this question that Peirce comes to formulate his admittedly provocatively titled *law of mind* as the most general (and even as the only possible) source for the emergence of lawfulness. In his presentation of this "law" Peirce used a terminology that is suitable for provoking all alarm bells to ring in the mind of a modern scientific reader. For, read through contemporary lenses, the implicit anthropomorphism in the Peircean ascription of *feelings* or *habits* to natural systems will shock many who are accustomed to an absolutely rigid wall between humanity and freedom on one side, and nature and necessity on the other side. But it is exactly this dichotomy that Peirce would have us transcend in his deliberately provocative use of terms. When Peirce says that nature possesses *mind*, it is not a cheap anthropomorphism, for Peirce's idea of mind is not linked to the human mind. Rather he sees the human mind as a special instantiation of a general principle in the universe, the *law of mind*: "The psychologists say that consciousness is the essential attribute of mind; and that purpose is only a special modification. I hold that purpose, or rather, *final causation, of which purpose is the conscious modification*, is the essential subject of psychologists' own studies; and that consciousness is a special, and not a universal, accompaniment of mind" (Peirce 1902; *CP* 7:366, italics added).

Peirce's *law of mind*, his cosmogonic philosophy, is described as follows: "In the beginning — infinitely remote — there was a chaos of unpersonalized feeling, which being without connection or regularity would properly be without existence. This feeling, sporting here and there in pure arbitrariness, would have started the germ of a generalizing tendency. Its other sportings would be evanescent, but this would have a growing virtue. Thus, the tendency to habit would be started; and from this, with the other principles of evolution, all the regularities of the universe would be evolved" (CP 6:33).

Thus, Peirce sees natural laws as habits that emerge from prior — and then serve to structure subsequent — interactions in nature. These habits then self-reinforce and can ultimately freeze into more or less fixed patterns, a process that will gradually broaden the domain of *necessity* and *natural law*. And Peirce notes that,

seen from the standpoint of semiotics, such habits can be considered to be *interpretants*, and here we see how the determination of final causation lies at the root of Peircean philosophy:

By thus admitting pure spontaneity of life as a character of the universe, acting always and everywhere, though restrained within narrow bounds by law, producing infinitesimal departures from law continually, and great ones with infinite infrequency, I account for all the variety and diversity of the universe, in the only sense in which the really *sui generis* and new can be said to be accounted for. . . . At the same time, by thus loosening the bond of necessity, it gives room for the influence of another kind of causation, such as seems to be [also] operative in the mind, in the formation of associations, and enables us to understand how the uniformity of nature could have been brought about (*CP* 6:59–60).

This other kind of causality alluded to by Peirce is exactly what I have suggested we call *semiotic causality*, i.e., bringing about things under guidance of interpretation in a local context (Hoffmeyer 2007. Semiotic causality cannot be reduced to efficient causality but, on the contrary, is dependent upon the workings of efficient causality, since interpretation, even in its most primitive modes, is of no need if not followed by habit formation, or, in other words, by anticipatory action — and action unquestionably depends on efficient causality. Semiotic causality thus gives direction to efficient causality, while efficient causality gives power to semiotic causality. Their reciprocal relation is one of interdependence, not one of exclusion.

But the exclusion of semiotic causality as an explanatory tool is exactly what the traditional scientific ontology induces in the scientist. The discussion in this chapter has served to clear the passage for the workings of this other kind of causality to be explored. By undermining the belief in natural selection as a sufficient explanation for the existence of agency and functionality in the world, we have observed the necessity to consider the more deeply buried directedness of our universe as stated by an ontological reading of the second law of thermodynamics. We have shown then, that the banishment of final causation from the accepted ontology of science is not a necessary consequence of physical knowledge such as we have it, but is derived rather from old ideological bindings of science such as the past wars between science and church.

Peirce's definition of *sign*, writes the German semiotician Helmuth Pape, actually constitutes a general description of the *inner structure* of final causality. Pape (1993) elucidates, "There cannot be a sign process without there being a final cause involved in it" (cited in Santaella-Braga, 1999). In the biological world, certainly, signs incite the generation of interpretants in the form of actions which are future-oriented, inasmuch as living beings always seek signs for survival and for reproduction. That organisms react to signs necessarily implies that these signs are meaningful, and that they are directed toward latent activities, whether now or later. Even a historian doesn't study his history for *history's* sake, but because he *hopes* that it can help him to earn money, to build a career, or perhaps even to become wiser about the world.

Firstness, Secondness, and Thirdness

At the same time that semiotics always involves finality, it necessarily also makes use of ordinary efficient causality. To more precisely define this relationship, we must look more closely at Peirce's idea that the world contains three and only three fundamental categories of phenomena, which he laconically called Firstness, Secondness, and Thirdness. This categorical system is deeply original and can be seen as a realistic turning rightside up of the human-centric Kantian categorical system. For while Kant's classification is based upon the formal characteristics of human thought, Peirce's classifications analogically rediscover natural categories and their cognizable patterns.

Most profoundly, Peirce's realism is thus based on an evolutionary intuition: "Thus it is that, our minds having been formed under the influence of phenomena governed by the laws of mechanics, certain conceptions entering into those laws become implanted in our minds, so that we can readily guess what the laws are. Without such a natural prompting, having to search blindfold for a law which would suit the phenomena, our chance of finding it would be as one to infinity" (Peirce 1955, 317).

This, of course, does not mean that there is any simple isomorphism between our scientific theories and that part of reality that they seek to describe. Rather, Peirce was a conceptual realist — but not in the kind of naïve way that is so often seen in natural science circles in the form of an unlettered arrogance about philosophical "speculation." About *truth*, Peirce (*CP* 5:565) writes this: "*Truth* is that concordance of an abstract statement with the ideal limit towards which endless investigation would tend to bring scientific belief, which concordance the abstract statement may possess by virtue of the confession of its inaccuracy and one-sidedness, and this confession is an essential ingredient of truth." In keeping with this pragmatic criterion of truth, Peirce's idea of *realism* consists in the claim that ideas shall be known by their consequences (but not just consequences as they are judged here and now, but such as the common efforts of the scientific community will, with time, come to approach closer and closer).[23] In this way, neither individuals nor scientist should attempt to substantiate ideas in any other way than by an evaluation of their effects: "In order to ascertain the meaning of an intellectual conception one should consider what practical consequences might conceivably result by necessity from the truth of that conception; and the sum of these consequences will constitute the entire meaning of the conception" (Peirce 1905; *CP* 5:9)

Accordingly, Peirce (1955, 322–23) describes the three fundamental categories in his system in the following way:

> First is the conception of *being or existing independent* of anything else. Second
> is the conception of *being relative to*, the conception of reaction with, something
> else. Third is the conception of *mediation* whereby a *first* and a *second* are brought
> into relation. To illustrate these ideas, I will show how they enter into those we
> have been considering. The origin of things, considered not as leading to anything,
> but in itself, contains the idea of First, the end of things that of second, the pro-
> cess mediating between them that of Third. . . . In psychology feeling is First, sense
> of reaction is Second, general conception Third, or mediation. In biology, the idea
> of arbitrary sporting is First, heredity is Second, the process whereby the acciden-
> tal characters become fixed is Third. Chance is First, law is Second, the tendency to
> form habits is Third. Mind is First, matter is Second, evolution is Third.

Relations belonging to the category of *Firstness* are generally characterized as including potentialities — possibilities, simple and uncompounded (for could they be broken into components, they would already be relations of Secondness) (Dinesen and Stjernfelt 1994, 14). In perception, for example, Firstness thus also denotes pure quality, such as redness or the stench of rotten cabbage — provided that these expressions are not understood as being also attached to a concrete existence.

---

23   The "deplorable degeneration" of the pragmatist's truth criterion to a sort of "practical validity criterion as judged in the here and now" caused Peirce to later insist upon distinguishing his own ideas from James's and Dewey's *pragmatism* by re-christening his own conception *pragmaticism*.

Relations of Secondness manifest as separateness, rupture, opposition, manifestation, existence, and quantity, and appear as soon as quality is imposed upon, or adheres to, something else. "Secondness encompasses Firstness and can be said to be its realization, which is a 'brute fact,' and where the First is only the possible, the Second is the necessary," write Dinesen and Stjernfelt (1994, 15).

Relations of *Thirdness* occur in and with the establishment of a connection between the universe of possibilities that is Firstness and the plethora of events that is Secondness. Thus, relationships of Thirdness emerge as *habits*, whereby quality and quantity are themselves put together in a relationship. Cultural, linguistic, and biological habits are all thus relationships of Thirdness.

The significance of these reflections for our discussion of *causality* follows from the recognition that actions or processes can either play out within relations of Secondness or within relations of Thirdness (but not, however, within relations of Firstness, which are as yet potentialities). The processes that take place at the level of Secondness are dyadic and constrained, and these are the types of processes that are controlled by efficient causality. Peirce characterized such actions as "brute, unintelligent, and unconcerned with the result" (*CP* 6.332). As an instance of mere Secondness, a dyad is a discontinuous fact occurring "here and now," writes Santaella-Braga: "there is nothing general about it. As an efficient causation, it is brute force or compulsion, an effective action, blind, nonrational, singular in occasion, just a factual compulsion and the *hic et nunc* of an event" (Santaella-Braga 1999, 501). Consequently, one should not confuse *efficient* causality with *deterministic* causality, which presupposes a lawfulness that is not yet present on the level of brute Secondness.

When we speak about lawful processes, we are already dealing with phenomena at the level of Thirdness, and it is only on this level of relations that we will, according to Peirce, encounter the organizational principles of final causation. But it must be remembered that even this, for Peirce, does not entail any sort of fixed and unchanging determinism — since such natural law itself arises only within the processes of evolution, and as such is necessarily incomplete. The oft-conflated relationship between final and efficient causality in contemporary science now becomes distinguishable and clear: Efficient causality is in effect the raw power that final causality needs to impose itself upon events. Lucia Santaella-Braga (1999, 502), paraphrasing Peirce, explains it this way:

> A law is something general and for that reason it is not a force. "For force is compulsion; and compulsion is *hic et nunc*. It is either that or it is no compulsion. Law, without force to carry it out, would be a court without a sheriff; and all its dicta would be vaporings." Thus the relation of law, as a cause, to the action of cause, as its effect, is final, or ideal, causation, not efficient causation (*CP* 1:213). . . . Final causation without efficient causation is helpless, but efficient without final is worse than

helpless "by far, it is mere chaos; and chaos is not even so much as chaos, without final causation; it is blank nothing" (*CP* 1:200).

Semiosis, sign action, is necessarily embedded in sensory material processes, and therefore has both a dynamic side, which allows a process of communications to take place, and a complementary *logical*, or *mediating*, side. The first of these sides stands under the force of efficient causality, and the second expresses the controlling agency of final causation.[24] Natural and medical science's systematic reduction of the domain of Thirdness to the pure and simple indexicality or Secondness — as it is subject to the coersive raw power of efficient causation — unquestionably has won many victories for these sciences.[25] But uncountable numbers of plants, animals, and humans have had to pay the price for the the failed visions of these sciences, that for centuries saw it as the highest wisdom to treat all living beings as asemiotic devices. Lately, and in an attempt to counter the inherent unworkability of this short-sightedness, science has added the *information* metaphor to its theoretical repertoire. But as we have seen throughout this chapter, since even the information metaphor, whenever it is operationalized, is again consistently deprived of every trace of the dimensionality of the embodied relations of Thirdness that characterize living organisms, I believe that in the end this futile metaphor will only serve to pull the wool over the eyes of ourselves and others.

24   The term *agency* is being used here as a general concept (*see* chapter 2, note 22), for I explicitly reject the common contention of the humanistic tradition, since Brentano, that humans alone are privileged with regard to the possession of intentionality and agency. At most, they may be unique only insofar as they are able to use the tools of symbolic reference to be able to conceptualize this fact (see Deacon 1997).

25   It is necessary here to specify that a law, a natural law, clearly possesses relations of Thirdness in terms of Peirce's categories, but that Peirce was an implacable critic of modern science's way of viewing natural laws as purely mechanical laws whose essence is to be absolutely unbreakable, or a "grand principle of causation which is generally held to be the most certain of all truths" (*CP* 6:68). Santaella-Braga elaborates on this: "In clear opposition to this deterministic notion of causation based on a conception of law as absolute, Peirce conceived of law as a living power founded on a peculiar conception of habit giving room for chance, growth, and evolution" (Santaella-Braga 1999, 505). In Peirce's thought, natural laws belong to the category of Thirdness, but it is precisely this dimension of natural law that was cut from modern science's concept of a law per se.

PART 2

THE SEMIOTICS OF NATURE

4

Code-Duality

Weismannism

In the 1890s, the German naturalist August Weismann famously showed that somatic cells and germ cells lead totally separate lives in the organism. A cell line consists of a clone of cells that can be traced directly back to a single ancestor cell, and Weismann argued that the separation between the cell lines that will become sex cells and those that will differentiate into all of the other cells of the body takes place early in embryogenesis. This finding was significant, for it seemed to imply that properties acquired by the organism during the course of its own life cannot in any way affect the sex cells, and thus, the offspring. In other words, *acquired properties are not heritable* — and in showing this, Weismann hoped he had safely refuted the mechanism for evolution proposed nearly a century earlier by the French biologist Jean Baptiste Lamarck (1809) — which is discussed in some detail in Chapter 6.

Yet as we shall see, Weismann's new doctrine has turned out to be far from as certain as it was for a long time thought to be. For by and large, it is difficult to understand the great and enduring significance ascribed to Weismannism if one does not also reflect upon the ideological and religious aspects of the scientific controversy between Lamarckism and Darwinism that this doctrine, more than anything else, would put an end to. That the cruel and purposeless process of natural selection should be the real cause behind the wonderful arrangement of life forms in animate nature — and thus also the explanation for our own human nature — was not a message that most people liked to believe. Many, therefore, took refuge in various versions of neo-Lamarckism, doctrines that not only implied the possibility that traits that one had cultivated in one's own life could be inherited by one's children, but also, and more importantly, implied the belief that there was in nature an inherent striving for perfection — *la marche de la nature* in Lamarck's terms.

It should be noted here that at the end of the nineteenth century, Darwinism was nothing like the uncontested theory in biology that it is today — and there was indeed no really compelling biological evidence for its truth (Ruse 1979. Religious forces in the United States, in particular, were capable of successfully fighting Darwinism for a considerable portion of the early twentieth century — and, especially in paleontology, the eventual conversion to Darwinism became a long, drawn-out affair. Similarly, the animosity against religion was a frequent component in Darwinian rhetoric and Weismann himself was a dedicated materialist with an aversion to the spiritualist and vitalist overtones present in much neo-Lamarckian theorizing of the time (Depew and Weber 1995).

Dusty as these conflicts may seem today, there is no doubt that fires are still smoldering within their ashes. For to the modern mind, Darwinism stands for rationality, materialism, and clear separation of science from ideology or religion — whereas Lamarckism brings associations of irrationalism and religious interference in scientific matters. Thus, still today, critical reconsiderations of Darwinism automatically call forth a suspicion that it is, in fact, not just the specific postulates of Darwinism, but scientific rationality as such, that is being rejected. The old conflicts between science and the Church are not so easily forgotten.

Here, however, we must nevertheless venture to have a closer look at this whole complex of ideas. For we are going to argue that Weismann's doctrine is based on a specific semiotic trick that is characteristic of living systems in general — a trick which Weismann himself could not have uncovered, because it presupposes a knowledge about the molecular semiotic dynamics of life processes that had not yet been discovered in his time. And yet, while the inner core of Weismannism will be in a certain sense confirmed by this reformulated understanding, it is one which in no way supports the aggressive transmission-genetic[1] reduction of evolutionary biology that Weismannism otherwise seems to justify. On the contrary, it appears that the Weismann doctrine depends on a duality between analogue and digital representations that opens the door to a sophisticated interaction between the domains of the genetic and the somatic.

Life as a Dead End

In textbook versions of Weismann's doctrine, the reader usually is shown a figure with a sequence of sex cells connected by arrows from left to right. From each sex cell, a diagonal arrow branches off to point to a picture of an adult organism.

---

1   Genetics, properly speaking, comprises both a transmission aspect and an expression aspect. The transmission aspect is concerned with the transmission of genes from generation to generation, whereas the expression aspect deals with the question of how genes actually do produce the traits ascribed to them. For most of its history, however, the science of genetics has restricted itself to the first of these two aspects, the transmission phenomena, while more or less leaving the expression problem for the future. It is one thing, for instance, to claim that the taboo against incest is genetically anchored, quite another to explain how a gene might manage to cause such an effect. See more on this in Chapter 5, especially Lenny Moss's distinction between gene-P and gene-D (Moss 2001). This emphasis on the transmission aspects, of course, has made the reductionist strategy so much easier to perpetuate.

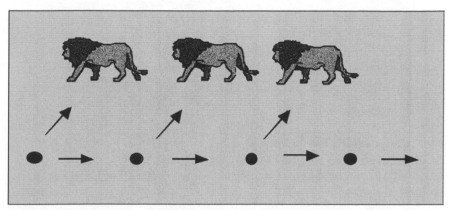

Figure 4.1 Weismann's doctrine. The filled circles represent sex cells (see text).

This last icon is normally depicted as a dead end with no further connections to its future sex cells or the progeny derived from these sex cells, as shown in Figure 4. 1. The adult lions in Figure 4.1 do of course contribute their sex cells to their offspring, but since, according to Weismann's doctrine, the sex cells are unaffected by the actual life of the lions themselves, this contribution plays no autonomous role in the scheme and should be left out. This figure is well suited to illustrate why Weismann's doctrine blocks the way for the Lamarckian belief in the inheritance of acquired properties.

But if the arrows are meant to symbolize straightforward causal connections, then the figure is less satisfactory. Parents certainly influence the life and survival chances of their offspring in many other ways than through delivering their sex cells. Many species exhibit prolonged periods of parental care — and even in species with no distinct parental care, offspring are typically left to hatch in suitable places where their chances of survival are higher than they would be elsewhere. Organisms, to a large extent, choose their own habitats and their own mating partners. They select and consume resources, generate detritus, and construct important components of their own environments such as nests, holes, burrows, paths, webs, dams, and chemical environments (Lewontin 1983).

The British biologist John Odling-Smee has suggested the term *niche construction* for the organism's own contribution (positive or negative) to the establishment of suitable ecological conditions for its offspring: "Organisms, through their metabolism, their activities, and their choices, define, create and partly destroy their own niches. We refer to these phenomena as niche construction" (Odling-Smee 1988; Odling-Smee, Laland, and Feldman 1996, 641). That organisms often outright destroy their own living conditions at a locality is well known — as, for example, can be seen by the many documented instances of *ecological succession* wherein one species typically changes the ecology of its local environment such that a rival species can then flourish there and, in flourishing, wipes out the original species.

The Darwinian consensus has long been that we can, by and large, neglect these kinds of influences, because such activities are seen as genetically determined. According to this understanding, the genotype determines not only a morphological phenotype, but also a behavioral phenotype. Thus, in a statistical analysis, the sex cell and its genome can represent the individual and, for all intents and purposes, its characteristic activity. It is, of course, admitted that the determinism is not complete, and thus the phenotypic variety that may be observed in one and the same genotype (or gene pool, if the analysis concerns variations inside a population) under different environmental conditions is then referred to with (and supposedly explained by) concepts such as *reaction norm* or *phenotypic plasticity*. However, since natural selection plays out over a huge number of generations and among many individual organisms, it is routinely assumed that the effect of this type of variation will ultimately be averaged out, so that one may safely bypass niche construction effects when accounting for the mechanisms driving evolution.

Odling-Smee (2001, 118), not surprisingly, considers this an explaining away of the problem:

> To us the idea that niche construction can be dismissed because it is the product of natural selection makes no more sense than the counterproposal that natural selection can be disregarded because it is a product of niche construction. From the beginning of life, all organisms have, in part, modified their selective environments, and their ability to do so is, in part, a consequence of their naturally selected genes. Niche construction and natural selection are two processes, operating in parallel, but also interacting.

The intricate interplay between natural selection and niche construction may be nicely illustrated by the survival strategy that has evolved in one of Darwin's finches, *Cactospiza pallida*, from the Galapagos Islands. This finch has constructed a niche identical to the niche that elsewhere in the world is occupied by woodpeckers. But instead of using its beak directly to drill a hole in the wood, *C. pallida* has developed a technique of using a cactus spine held in its beak to dig the insect out. In this case, Odling-Smee explains, it was not first the anatomy of the bird that was formed by natural selection to solve a given task — but rather, this finch, like so many other species, utilized a much more general and flexible adaptation: the capacity for learning. The ability of the finch to use a cactus spine is not guaranteed by the presence of some relevant gene (the *gene* for thorn digging?) — and yet it develops quite reliably as a consequence of the general ability of the bird to profit from its experiences in interacting with its milieu (Odling-Smee 2001).

In the context of this book, it would be tempting to say that the bird makes use of its own semiotic competence to *create an interpretant* (a habit) upon seeing a cactus spine in the relevant situation. And, as Odling-Smee remarks, this creation of a new way of living (or *niche*) in itself implies that natural selection is now driven to reinforce the selection pressure on the bird's ability to learn.

In the traditional scheme, then, evolutionary processes may in principle be described by just two equations. The first describes the change of organisms (O) as a function of organism and environment (E): $d(O)/dT = f(O, E)$, and the second describes the change in the environment as an autonomous process: $dE/dT = g(E)$. Already in his influential 1983 paper, Richard Lewontin showed that if one was to acknowledge the fact that organisms themselves may often influence their niches, this scheme would have to be changed to a set of coupled differential equations reflecting the reality of the interaction between organisms and environment — and that, in fact, each of them is a function of the other: $dO/dT = f(O, E)$ and $dE/dT = g(O, E)$. While the process in the first of these scenarios figuratively might be depicted as species that climb mountains, then, in the niche-construction scenario, species would jump trampolines in the sense that in this situation there are no fixed equilibrium points (Lewontin 1983).

We shall return to the difficulties of upholding an efficient Weismann barrier, in the sense of a noninterventionist view of individuals in the evolutionary game of natural selection. But before we pursue this problem any further, it is necessary to take a closer look at a trend that I will call *the genocentric turn* in biology. This is the tendency of biologists to shift the emphasis of their analyses and investigations from the level of the organism to the level of the gene — seeing genes, rather than organisms, as the operational level for selection.

Gene Selectionism

With Crick's articulation of the Central Dogma (see Figure 3.4), Weismannism obtained a sort of molecular confirmation. For since proteins may, roughly speaking, be identified with the machinery of the body — and are as such directly responsible for its activity and behavior — the implication of the Central Dogma is that the body machinery cannot feed back into the genome. The body thereby ended as a strange blind alley in evolutionary theory landscape. It is little surprise that a theory that sees organisms as causal dead ends in the evolutionary game would sooner or later give birth to the idea that organisms, rightly considered, are just instruments for the strategic interests of genes. It was the Oxford biologist Richard Dawkins who, back in 1976, disseminated this idea in his well-written book *The Selfish Gene*. And this was an idea that met with considerable sympathy both inside and (particularly) outside of biology.

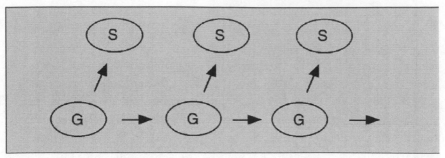

Figure 4.2 Molecular Weismannism. G = germ cells, S = somatic cells.

According to this idea, often called *gene selectionism*, the organisms that classical Darwinism focused on as the substrate for natural selection should be seen rather as kinds of "survival machines" or "vehicles" that are merely *used by the genes* in *their* competition for a ticket to the next generation.

According to this scenario, genes that have instructed their organisms well increase their chances of becoming multiplied via their offspring. Thus, while the organisms themselves are short-lived, their genes will continue to survive as copies or replicas in (statistically speaking) half of the offspring. Genes in Dawkins's terminology thus become *replicators*: "What was to be the fate of the ancient replicators? . . . Now they swarm in huge colonies, safe inside gigantic lumbering robots, sealed off from the outside world, communicating with it by remote control. They are in you and me; they created us, body and mind; and their preservation is the ultimate rationale for our existence. Now they go by the name of genes, and we are their survival machines" (Dawkins 1976, 21).

But something is jarring here, isn't it? Perhaps it's an over-smart semiotic loop: "They created us"? After all, on the face of it, it is *we* who created the *gene*. This is not a joke, for it highlights the essential trouble with Dawkins's concept of the *gene* — i.e., its misleading ambiguity as to what kind of a thing the gene really is. For in fact, it was the Danish botanist Wilhelm Johannsen who, in 1909, created the terms *gene*, *genotype*, and *phenotype*. But Johannsen's gene was very different from the modern idea of genes, since Johannsen certainly did not imagine that the gene existed as a material thing.[2] For him, the gene was simply a tool for calculations (similar to the measure of IQ that had been introduced at about the same time by the French psychologist Alfred Binet). However, as is well known, the concept of the gene was soon taken over by fruit-fly geneticists who did not have Johannsen's reservations about a straightforward material interpretation. The famous *Drosophila* Group, led by Thomas Hunt Morgan, came to see the

---

2   Johannsen actually compared the concept of *the gene* as a material structure that resides within the cells with the naive belief — of the peasants of his own time — that a team of horses was hidden inside the locomotive.

genes as "pearls on a string" (Morgan et al. 1915) — and after Watson and Crick's discovery of the DNA's double-helix structure in 1953, we finally got the modern biochemical concept of the gene. It is this biochemical concept of the gene that is implied when we are told that the human genome contains some twenty-five thousand genes.[3]

But when Dawkins, writing about genes, claims that "they created us," a closer inspection reveals that this *they* does not at all refer to the biochemical genes but to a much less well-defined entity called an *evolutionary gene*. The evolutionary gene refers to any short piece of DNA on a chromosome that segregates and recombines with appreciable frequency. Such pieces must be reasonably short, since it is essential for Dawkins's hypothesis that the gene *itself* is a replicator, meaning that it can multiply through replication. And this again presupposes that this *gene* is not often broken up; and the longer a gene is, the higher is the risk that it may be broken up by the processes of genetic recombination (e.g., by the crossing-over effect of meiotic division, the process whereby new germ cells are formed.)

Thus, one strange thing about Dawkins's gene concept is that nothing guarantees that the gene itself need be a functional unit. Sterelny and Griffiths (1999, 79) have put it this way: "If genes are just arbitrary DNA sequences, then most of them will have no more systematic relation to the phenotype than an arbitrary string of letters has to the meaning of a book." We shall not pursue the implications of this observation any further here, but content ourselves by noting that a clear understanding of what a *gene* in the above sense really *is* recedes even further when the question is studied in more detail (Neumann-Held 1998). So much more odd, then, is the claim that these ambiguous and mysteriously *intentional* entities should have "created us body and mind."

What we will note here, however, and what is especially problematic to a Dawkinsian understanding, is the well-known chemical fact that DNA molecules are essentially passive under normal physiological conditions. As I have argued elsewhere (Hoffmeyer 1997a), it is the very inertness of the molecule under normal conditions that makes DNA so adequate as a *memory* carrier. One cannot well store memory in a substrate that also takes an active part in the everyday business of the cell.

But DNA molecules, as Evelyn Fox Keller reminds us (Keller 1995) do nothing and remain inert until the arrival of a varied and highly specific set of proteins that (as we shall see in Chapter 5) have the manifold tasks of opening up

---

3   I had better be careful here. A few years ago, the estimated number of genes in the human species was something like 80,000. After the publication of the map of the human genome in 2001, there were reported to be only 30,000–40,000 genes. And the latest estimate I saw had only 20,000–25,000 genes (*Nature* October 21, 2004).

the double helix, copying the signifying strand onto mRNA at a chosen site, modifying this mRNA in a diversity of ways, transporting the mRNA out of the nucleus to the "protein factories" (endoplasmatic reticulum) in the cytoplasm, and much, much more.

In eukaryotes, the passive DNA is protected by specific proteins inside the cell nucleus, where it is continuously controlled and repaired when it has become damaged. In short, the *active* components of the cell are the proteins that work together in subcellular complexes or membranes — and, as we will see in Chapter 5, these complexes are the *real* "doers" in cellular life that constitute the agency of the cell. To ascribe such agency to DNA is highly contraindicated from the chemical point of view.

Digitalism

The widespread tendency to see DNA as the main actor in the unfolding life process is probably due to the fact that the genetic code, as carried in the DNA molecules, is a digital code. And, in fact, Western culture is profuse with *digitalism* — by which I mean an exaggerated trust in the performative power of digital codes (Hoffmeyer 2002a). Let me take a moment here to explain what I mean.

The word *digital* comes from the Latin word for finger, *digitus*, and a digital code is quite generally a code based on symbols that, like the fingers, are discontinuous — for instance, the numbers 1, 2, 3, 4, 5, or the letters of any writing system or alphabet. The lengthy strings of zeroes and ones that make up the algorithms in binary computation language have recently come to stand as the quintessence of digitality, but the prior invention of the book was already based on a digital code (of letters) — so the modern distinction between digital electronic media and old-fashioned books is dangerously misleading when it comes to a clear understanding of codes.

The alternative to a digital code is an analog code, and this kind of code is based on the principle of analogy. In an old watch, for instance, the pointers circle the watch dial in a kind of analogy of the (perceived) rotation of the sun around the Earth. In the same way, the height of the mercury column in a thermometer can be said to analogously reflect the magnitude of the actual temperature. One might also call a glove an analog coding of a hand, and in a certain sense one might even see the wings of the bird as an analogue codification of the aerodynamic properties of air currents.

Note, however, that the coding principle in digital codes is necessarily *arbitrary*, relative to the content. That the word *horse* in English refers to a big grazing mammal with a certain kind of mane cannot be recovered from the form of the letters but must be known through another process, as when speaking to a Frenchman, that same animal in French has the equally arbitary designation *cheval*. Digital codes are necessarily based on symbolic signs (in the Peircean sign trichotomy), whereas analog codes are based on icons and/or indices (as discussed in Chapter 8).

Since the genetic code is based on a sequence of discrete signs that are grouped together in sequences of triplets (e.g., the trinucleotide sequence UUU that is translated to the amino acid *phenylalanine*), and whose relation to the amino acids they are coding for is mediated through a complex historically established interplay between protein and RNA molecules (see further discussion in Chapter 5), the genetic code is a clear case of digital coding.

Let's return to my earlier point. In Western culture we are accustomed to think of digital codes as superior to analog codes. He who masters numbers and letters (or computers!) clearly has a higher status than he who relies on the less formalized, analog ways of understanding reality. That science accords high authority to digital coding schemes may of course be traced back to the Galilean idea (now so widely adopted) that the language of God, or now reality, is mathematics. But it is hard to escape the feeling that this privileging of the digital representation also somehow reflects the separation between plan and execution that has been such an important principle behind the success of the industrial society.

Since the execution of work in industrial production is supposed to follow deterministically from the prescribed plan, there is left little or no space for *interpretation* at the level of execution.[4] In traditional industrial production, all creativity was delegated to the planners, and workers were not supposed to add anything innovative to the process. The element of creativity was not part of work on the floor.

If we take this model and project it upon the theory of evolution, a digitalist reading of genetic expression follows naturally. For just as the overwhelming productivity of industrial society was obtained through a separation of planning from execution in a master-servant relation, evolutionary theorists have argued that evolutionary creativity should be explained by a separation of the genetic master plan (the DNA) from the mundane operations of the cytoplasm — with the cytoplasmic processes themselves, of course, obeying the dictums of the genome.

Seen in this way, digitalism may be thought of as yet another example of the well-known tendency of science to transform the dominating structures of social reality into natural reality. But this account makes digitalism as a historical bias

---

4   Planning itself, of course, may contain elements of analog coding such as diagrams or drawings, but these are generally superstructures based on numerical or otherwise abstract codifications.

that belongs to a particular historical period, i.e., the epoch of industrial society — and one might perhaps expect that it would gradually loose its hold over
the imagination of researchers as this form of production loses its dominant status in society.

In fact, the strict separation between planning and execution is scarcely typical anymore for production processes in the advanced sectors of almost any
modern economy. On the contrary, it is semiotic competence that is now increasingly in demand. Thus it ought to amuse sociologists of science that digitalism
now seems increasingly challenged in science by models of embodiment — i.e.,
models where the bodily anchoring of cognitive or biological functionality are
seen as essential to that functionality (Damasio 1994: van Gelder and Port 1995;
Deacon 1997; Etxeberria 1998; Rocha 1998; Lakoff and Johnson 1999).

The development of clever computers and robots has undoubtedly been facilitating this change of perspective (Hendriks-Jansen 1996; Clark 1997; Ziemke
and Sharkey 2001; Sharkey and Ziemke 2001a, 2001b). The understanding that
cognitive and biological processes cannot really be comprehended in isolation
from their character as somatic processes reflects a growing awareness (in the
computer age) of the fact that digital pre-specifications are essentially *dependent*
on the agency of autonomous structures and mechanisms acting in space and
time. That such relations are best grasped from within a semiotic frame of reference seems obvious to this author (Hoffmeyer 2002a), and the rejection of theoretical positions that ignore the decisive significance of our bodily existence for
cognitive function has been, of course, a central theme in the biosemiotic tradition (Hoffmeyer 1996b; Brier 2000; Kull 2000; Danesi 2001; Emmeche 2001;
Favareau 2001).

Code-Duality

In 1991, Claus Emmeche and I suggested that life at the most fundamental level
may be characterized by a dynamic trait that we called *code-duality* — i.e., a
recursive and unending exchange of messages between analog and digital coding surfaces (Hoffmeyer and Emmeche 1991; 2005).

As analog codifications, organisms recognize and interact with each other in
ecological space, whereas as digital codifications (genomes), they are passively
carried forward in time from generation to generation (in sexually reproducing
species, after recombination via meiosis and fertilization). Seen from this perspective, life must be understood as *semiotic survival* — survival via a fundamental code-duality.

Our idea of code-duality was much inspired by Gregory Bateson's grandiose
attempt in the book *Mind and Nature* to draw a connection between thought

and evolution as deeply related processes (Bateson 1979). We were also inspired by the American biophysicist Howard Pattee's suggestion that living systems must necessarily operate through the interactions between two complementary modes — a time-independent or symbolic (linguistic) mode and a time-dependent dynamic mode (Pattee 1972; 1997). I shall return to a discussion of Pattee's idea of an *epistemic cut* shortly, but preliminary to that, let us here consider the original presentation of the idea of code-duality.

The idea was introduced in the context of a discussion regarding what kind of status one should ascribe to life. In answering this question, we took as our point of departure the Batesonian understanding of life as systems possessing the ability to process information in the sense of "differences that make a difference." The following passage is a slightly modified version of the original presentation in Hoffmeyer and Emmeche (1991):

> *Who is the subject to whom the differences worked on by such a system should make a difference?* If one admits at all that living systems are information-processing entities, then the only possible answer to this question is this: The system itself is the subject. Therefore a living system must exist *for* itself — and in this sense it is more than an imaginary invention of ours. For a system to be living, it must create itself, i.e., it must contain the distinctions necessary for its own identification as *a system*. Self-reference is the fundament on which life evolves, it is its most basal requirement. (Tellingly, this fact does not pertain to nonliving systems; there is no reason for the hydrological cycle to know itself. Thus, rivers run downstream due to gravity, water evaporates due to the solar heat, nowhere does the system depend on self-recognition).Another way to express this whole matter is to say that differences are not intelligible in the absence of a purpose. If nothing matters, matter is everything.

> But what is the basis of this self-reference that is at the basis of life? We shall suggest here that the central feature of living systems allowing for self-reference — and thus the ability to select and respond to differences in their surroundings — is *code-duality,* i.e., the ability of a system to represent itself in two different codes, one digital and one analog.[5] (Symbolically, this code-duality may be represented through the relation between the chicken and the egg.)

> Thus, self-reference clearly depends on some kind of re-description. The system must somehow be able to construct a description of itself in order to perpetuate itself (Pattee 1972; 1977). This description furthermore must stay inactive in — or at least protected from — the life processes of the system, lest the description should change, and thereby ultimately die with the system. In other words, the function of this description is to assure the identity of the system through time; it is the mem-

---

5   Two decades ago, I suggested a less fully developed version of the idea of code-duality (Hoffmeyer 1987).

ory of the system. In all known living systems, this description appears in the digital code of DNA (or RNA) that is eventually distributed to the germ cells.

We suggest that it is by no means accidental that the code for memory in living systems is of a digital type. For what must be specified through the memorized description is not the specific material details of the presently living system, but only their structural relations in space and time.[6] If such abstract specifications could be expressed through an analog code, only very simple systems would be possible and those would probably not survive. For a parallel, if human communication and memory depended solely on analog codes (e.g., exclusively on the ability to mime) — if, in other words, we did not posses the digital code of language — our cultural memory would be as short as that of chimpanzees, and our social structure would be accordingly as simple.

And in a complementary fashion, in order for the system to work the memorized description in the digital code must be translated to the physical reality of an actual living system. For this translation (the developmental process) to take place, the fertilized egg cell (in sexually reproducing species) must be able to decipher the coded instructions of the DNA[7] as well as to follow its instructions in a given way. This need for the active participation of cellular structures in realizing the digital codification of the DNA — into its analogue protein and organism form — shows us that a sort of *tacit knowledge* is present in the egg cell (Polanyi 1958; Pattee 1977). And the existence of this tacit knowledge inherent within the cellular organization must be presupposed by, rather than materially built into, the DNA description. Thus, the digital re-description alone is far from a total description of the organism.

The realization in space and time of the structural relations specified in the digital code determines what kind of differences in its surroundings the system will actually select and respond to. In Uexküllian terms, these sepcifications determine not

6   For example, if the gene codes for a distinct enzyme, the rate of degradation of this enzyme will determine how long the enzyme will remain active in the cell, and thus also determine the concentration of metabolites that are available for catalytic reaction. Since the concentration of metabolites will often have a regulatory influence on other cellular processes, the gene indirectly also determines the temporal relations between these different processes (see Figure 5.5). However, the gene cannot, for good reason, determine precisely where in the cell a given molecule will be located at a given time. This problem may perhaps be illustrated by imagining a protein that is the size of a family car. The cell would, in relation, be the size of Copenhagen (although spherical rather than flat). And, due to the intricate internal structure of the cell, the freedom of movement of the proteins inside the cell would hardly be any bigger than the freedom of movement of a car in the street web of Copenhagen.

7   Please notice that there is no sequence of nucleotide triplets per se — just an endless string of bases whose "reading" *determines* what will be acted upon as actual triplets. The base sequence, CAGTCAAAGAAC, might for instance be read as composed by the triplets: CAG-TCA-AAG-AAC. But in another reading frame it might be read as C-AGT-CAA-AGA-AC. (See Chapter 5 for further details on the semiotics of the genome.)

just the anatomical and physiological buildup of the organism but also the kind of Umwelt the organism will get through this realization. A new phase in the perpetuation of the system is initiated: the phase of active life. One might say that in this phase — the analog phase — the *message* of the memory is expressed.

Eventually, the system will survive long enough to pass on its own copy of the digitalized memory (or part of it) to a new generation; this corresponds to a back-translation of the message to the digital form. But this latter process takes on its true significance only when seen at the level of the population. For it is the population (rather than the single organism) that passes on messages about conditions of life to the memory of the collective (the gene pool). The population could in this sense be considered a codification that *itself* expresses a message. This codification, however, is necessarily analog — since it has to interact with the physical surroundings, and thus must share with these surroundings the properties of physical extension and contiguity.

The chain of events that sets life apart from nonlife — i.e., the unending chain of responses to selected differences — thus needs at least two codes: one *code for action* (behavior) and one *code for memory*. The first of these codes necessarily must be analog, and the second very probably must be digital (Hoffmeyer and Emmeche 1991).

Looking back at this first presentation it must be noted that many readers, unfortunately, felt this concept of code-duality to be confusing, and presumably this confusion has two main sources. One has to do with the way that the term *code* is used, since the term has quite different connotations in different disciplines (e.g., jurisprudence, genetics, computing). The other source for confusion is more principled and concerns the conception of living creatures *as* codes. This idea may easily switch on a warning signal that I am arguing for a kind of universal pansemiosis. Is *everything* in this world just signs, then? Am I claiming that there is no reality beyond that of signs?

Let me immediately answer these two questions in the negative. The claim that the world contains nothing but signs is as reductionistic and unfruitful as the claim (in the epistemological sense) that all the phenomena of the world can be reduced to their quanta of matter and energy.

Admittedly, what Peirce called *brute force*, or the category of Secondness (see Chapter 3), is an irreducible aspect of our world that cannot in a meaningful way be thought away, as little as chance and Firstness can be neglected. And because of this, John Deely has introduced the concept of *physiosemiosis* as a designation for semiotic processes taking place in inanimate nature (Deely 1990; see also Christiansen 2002; Taborsky 2001a; 2001b).

For example, the meteoric storms of the past are still visible today through the traces they left as craters on our moon where craters may be seen as an enmattered memory about those distant times. Now, it takes an intelligent being, of course, to decode this connection between real-world states and events and to interpret the cavities on the surface of the moon as meteor craters. But the possibility of such an interpretive act was nevertheless latently present in the very mark of the indentation itself — a mark, to be sure, that nobody took an interest in before human beings started wondering about the puzzles of the universe.

Nature's "forming of habits" — in other words, its tendency to develop new regularities as the result of its own ongoing interactions — has been at work at all times. Semiosis — "the action proper to signs," in Deely's words (1990) — should not be seen as an either-or phenomenon that suddenly appeared on Earth when life began. A more sensible view is that the origin of life initiated new avenues for semiosis to follow, avenues that implied new and unheard of kinds of freedom. Living systems are anticipatory, in the sense that they systematically recognize and exploit (interpret) important regularities (causal relations) in their surroundings, and in doing so, living systems gain access to the world of genuine triadic sign processes.

One should therefore distinguish between the kinds of semiotic processes that occur in physical, biological, and psychological systems. Semiotic freedom is much more pronounced in the latter two than in the former. In dealing with purely physical systems, one can in almost all cases get away with disregarding the semiotic dimension, with no lack of explanatory sufficiency. But this quickly becomes absurd if human nature is one's concern. Biology falls somewhere between these two.

However, the word *pansemiosis* carries connotations to German *Naturphilosophie* and especially to Friedrich Schelling's conception of the world as *panpsychic*. And regardless of what Schelling exactly did mean by this — and, as this is uncertain, I shall not pursue the matter here — it does seem to me inappropriate to confuse natural processes with human psychology. But it may not follow that this quite legitimate rejection of the overwhelmingly anthropocentric notion of *panpsychism* should be conflated with, and thereby also infect the reception of the concept of *pansemiosis*.

For this latter concept, explicitly, does *not* extend humanness to penetrate the universe at large — but rather sees the human being as one among infinitely many instantiations of a universal semiosis. But since this confusion seems inescapable, the term *pansemiosis* tends to block understanding more than it advances it. The relevant answer to the accusation of pansemiosis, therefore, is to make clear that the project of biosemiotics neither subscribes to nor advances the claim that there is nothing in the world but signs.

Likewise, the concept of *code-duality* implies neither a reductive pansemiosis nor a naïve anthropomorphism. Having made this clear, let us now confront the question of what *is* meant by code in the expression *code-duality*.

In its everyday meaning, the word *code* refers to the customary use of distinct entities or actions for communicative ends, such as when we speak of a dress code or about different social milieus having different behavioral codes. In semiotics, the term has had two rather different designations. Under the influence of Information Theory, it was used in the '60s and '70s in the sense of "a context-free set of rules for the encoding, transmission, and decoding of information" (Thibault 1998, 125). As discussed in the section on the Central Dogma, the DNA code has been conceived as a code in this sense.

But in attempting to apply this concept, complications abounded even in these years. For even in the simple cases, where we limit ourselves to considering just the essential information transfer from a gene segment of DNA to a functional protein, it has proven to be impossible to speak of a genuinely context-free and unbreakable set of rules that in all cases will assure the correct transfer of information. For, as we shall see in Chapter 5, the cellular (or organismic) context exerts a number of causal consequences at several key steps in this process, often in unpredictable ways (via RNA editing, for example, or via ambiguities in the "correct" reading of those stop-codon triplets that determines the end point for the addition of amino acids to the growing protein chain). Therefore, when the shorthand term *genetic code* is now used — as it is, more often than not — to express the idea that genes are coding *for* certain specific phenotypic traits (such as missing eyes in salamanders or schizophrenia in human beings), we are of course much farther from a position where one could talk about a context-free code in the precise sense of Information Theory.

Yet the information-theory concept of *code* has been, and still is, of great use in engineering the transfer of data (as distinct from information) in telecommunication systems, and underlies much of the efforts to build "intelligent machines." Modern semiotics, however, has abolished the conception of a code as a "simple mechanism for pairing of concept and reference."[8] The focus in recent years has been on understanding the concept of code *as a vehicle for creation of meaningful activity*. Winfried Nöth (2000, 216) puts it thus: "*Als Kodes erforschen Semiotiker kulturelle Zeichensysteme der verschiedensten Art, von der Verkehrszeichen bis zur Mode, vom Morsekode über der Kode der Heraldik bis zu den Kodes des Theaters. Häufig ist mit Kode nichts anderes als ein Zeichensystem gemeint.*" („*Code* has been the designation under which semioticians have studied cultural phenomena of the most diverse kinds, from traffic signs to fashion, from the Morse code and heraldry to the theater. *Code*, in this context, often means simply 'sign system.'" — freely translated by Winfried Nöth.) From this point of view, *a code is a semiotic resource* that enables us to create and express certain types of meaning but not others. Seen as a sign system, body language,

---

8 One exception is the die-hard tradition based on Noam Chomsky's ideas of a *generative grammar* (Chomsky 1965), and in particular Jerry Fodor's *language of thought* (Fodor 1975) or Steven Pinker's *mentalese* (Pinker 1994).

for example, is very well-suited to express emotional content such as disgust, joy, or ennui. It may also, when more artfully used, be narrative — as we know it from mime. But, as I shall discuss below, it can only with much effort, if at all, express abstract and logical connections such as nonexistence or denial.

Digital Codes

The term *digital codes* thus covers a whole set of codes, that are alike in that they are all based on discrete sign tokens as well as an arbitrary (conventional, historical, or customary) relation between the sign token and the signified. This property endows digital codes with certain special advantages that are unique to the phenomena we call life and culture. Of these advantages, three in particular must be emphasized (Hoffmeyer 2001a). The first is that *messages expressed in digital codes do not have to observe the limitations of freedom imposed by natural laws.* Possible as well as impossible messages may be expressed in digital codes. Thus, in a novel, Meryl Streep might well have lunch with Socrates; a mega-size cod might ravage the streets of Oslo; or the wives of pilots might start giving birth to children having wings. Conversely, it is hard to see how any of these impossible events should be communicated via analog codings such as mime — or at least not without additional use of conventional signs of many kinds, including symbolic cultural gestures and sign language, both of which are fundamentally digital codes.

Too, digital codes are, of course, always based in material processes and can for that reason be destroyed: books may be burned, computers may be smashed, or the freedom of speech might be restricted. But this kind of destructive action does not change the reality of that content which has (at least once) been coded, only the material carrier of the code.

And exactly the same is the case for the digitally coded *message content* of genes. This, too, has the freedom to be impossible in the sense that the fertilized egg, by executing such coded instructions, could produce a nonviable individual. And in fact, this happens all the time (among other reasons, because of the genetic crossing-over processes, whereby the hereditary material is recombined in new — and not always viable — patterns). It is this property of digital codes that explains the surprising evolutionary *creativity* of living systems, the incredible combinatorial capacity and the consequent incessant testing of the eventual limits for possible combinations.

The second advantage of digital codes is their *time independence* and consensual *objectivity*. Digital codes are ideal codes for memory. Only because Plato wrote down the dialogues of Socrates do we know them today. Had the dialogues not been coded in written language, but instead in mimed episodes, they would probably not have survived intact for more than a few generations at most.

The key to digital codes' objectivity is that the codes depend on a shared convention. The Canadian communication theorist Anthony Wilden (1980, 173) has pointed to one critiacl difference between analog and digital codes (among many others) as being that "a digital code is 'outside' the sender and receiver and mediates their relationship; an analog code *is* the relationship which mediates them."

At first look, this definition might seem to fit poorly with the understanding of the genome as a digital code, because the genome is transferred through cellular divisions (mitosis or meiosis) and therefore spatio-temporal continuity is maintained between sender (the parent organisms) and receiver (the zygote, or fertilized egg). But the topological detail that chromosomes travel, so to speak, from generation to generation *inside* a cell shouldn't detract us from seeing that it is exactly the principle of separation between germ cell and body processes (see the discussion on Weismannism, above) that is the deeper reason why Lamarckian inheritance does not work as an evolutionary principle.[9] Thanks to its secluded existence in protected isolation from the metabolic jungle of the cell — and thus due to its very passivity — the DNA code is capable of conserving experiences (in the sense of nucleotide sequences) shaped by past survival outcomes under the then prevailing ecological conditions. Such structures are inherently signs of these past relations, and this is exactly why genes are not functional in themselves, but must be unfolded through the operation of an interpreting agency.

The third advantage of digital codes is that they can be used as *tools for abstraction* — and this is why they are necessary for making meta-messages, messages that deal with the way other messages should be understood. Gregory Bateson (1972, 177–93) pointed out that such meta-messages may well be communicated in the analog, and he used the example of young monkeys engaged in "play." In these circumstances, when the monkeys snapped at one another while creating a simulation of a combat situation, this snap would actually signify the following meta-message about itself: This is not a bite.

The negative message — the absence of a bite — cannot be directly communicated in the analog, so instead it is announced as a positive message: the presence of a bite-*signifying* snap. The snap is an indication of something that is *not* there. Bateson also commented that this is probably as far as an analogically coded communication event can go in the direction of the abstract category of "not." For real abstractions to take place, digital codes and their more arbitrary and detached conventions are needed.

9   Lamarckian inheritance — by conflating the analog and the digital — loses the fertility of their interplay, which, seen semiotically, is the key to evolution.

Thus, while we still do not know the full syntactic structure of the genetic code, we know that what are called the regulatory genes, for example, function as such meta-messages — and the regular occurrence of atavistic reappearances such as the three-toed horse indicates that actual deletion is not the only way to get rid of outmoded ontogenetic instructions. Active negation of what still remains may suffice. *Abstraction* in this sense thus furnishes a kind of creative plasticity — in the absence of which, the evolutionary process might perhaps not have become as rich as it actually is.

Analog Codes

"If you say to a girl 'I love you,' she is likely to pay more attention to the accompanying kinesics and paralinguistics than to the words themselves," writes Gregory Bateson (1972, 374). He continues, "We humans become very uncomfortable when somebody starts to interpret our postures and gestures by translating them into *words* about *relationship*. We much prefer that our messages on this subject remain analogic, unconscious, and involuntary." Analog codings such as body language go much further back in evolution than spoken language does and are also much more strongly anchored in human emotional constitution. Wilden (1980, 163) says,

> The analog is pregnant with *meaning*, whereas the digital domain of *signification* is, relatively speaking, somewhat barren. It is almost impossible to translate the rich semantics of the analog into any digital form for communication to another organism. [On the other hand,]
>
> what the analog gains in semantics it loses in syntactics, and what the digital gains in syntactics, it loses in semantics. Thus, it is because the analog does not posses the syntax necessary to say "No" or to say something involving "not" that one can *refuse* or *reject* in the analog, but one cannot *deny* or *negate*.

It must be admitted that the classification of coding strategies into digital and analog derives at least part of its attraction from the exaggerated weight that formerly was ascribed to such purely formal aspects of human communication as its grammar, its logical structure, its rationality, and so on. And more than perhaps anybody else of his time, Gregory Bateson managed to uncover the paralinguistic and paralogical dimensions of communication in humans as well as in animals.[10] He thereby acquired and advanced an acute understanding of the much overlooked importance of analog codings in natural systems.

10   More recently, the related fields of Interactional and Conversation Analysis have disclosed that an amazingly high proportion of everyday, moment-to-moment linguistic communication

As a semiotic category, however, *analog coding* is perhaps not quite satis-factory, and I shall use it here primarily as a counter-concept to *digital coding*. Specifically, I will use *analog coding* as a common designation for codings based on some kind of similarity *in the spatio-temporal continuity*, or on internal rela-tions such as part-to-whole, or cause-and-effect. *Digital coding*, in contrast, will be used to designate sign systems where the relations of sign to signified are due to a demarcation principle of purely *conventional* or *habitual* origin.

In Chapter 7, we will have a closer look at endosemiotic analog codings as indexical and iconic sign processes inside the organism. On the ecosemiotic level, we have already seen how Jakob von Uexküll occupied himself with what I am here calling analog codes. Uexküll sometimes referred to these as *contrapun-tal duets* and noted numerous examples in nature, such as the relation between flies and spiders or between birds and butterflies decorated with spots as icons for eyes. He wrote, "By opening their wings they chase away the small birds that pursue them: These birds automatically fly away at the sight of the eyes of other small predators that may suddenly appear" (Uexküll 1982 (1940), 59). Such spots are an example of analog coding in nature.

Code-duality therefore implies that the *singularis* of the digital code is placed on equal footing with the *pluralis* of analog codings that make up the biosemi-osis of life. This positioning of the single digital code as in a sense equal to the totality of analog codings is justified by the unique properties of digital codes that I have discussed above. Moreover, it is precisely the play between these two types of coding that makes evolution possible, as analog and digital coding are two equally necessary forms of referential activity. They appeared, I would argue, as twins in the individuation process that gave rise to life's internal logic (see Table 2.1). For as I have written earlier,

> Had it not been for digital coding there would have been no stable access to the temporal world — i.e., the unidirectional continuum of pasts and futures — and therefore there could have been no true agency or communication. On the other hand, had it not been for the analog codes there could have been no true interac-tion *with* the world, no other-reference, and no preferences. To claim that only the digital twin is semiotic, whereas the analog twin remains in the sphere of classi-cal dynamics, is to block the only possibility for transcending the *epistemic cut* of Howard Pattee. Code-duality and semiosis open up a dimension of our world and its evolution that is left underdetermined by thermodynamics. Organismic "context space" expands at an accelerating rate in proportion to the increase in the semiotic sophistication of species;[11] for, simply put, there are so many more different ways to

---

is grounded in such paralinguistic interaction. For a biosemiotic perspective on this work, see Favareau (2002; 2007).

11   That "context space expands" in our biosphere is in accordance with the analyses given by Stuart Kauffman (2000, 151), in the book *Investigations*, where he shows that the *adjacant*

be smart than there are different ways to be simple (and this may be the reason why the speciation rate among mammals is five times higher than the speciation rate among lower vertebrates) (Hoffmeyer 2001a, 128).

The concept of code-duality thus illuminates the semiotic core at the heart of the evolutionary process[12] and thereby also the evolving semiotic dynamic that leads life on Earth toward the development of life forms possessing still more sophisticated kinds of semiotic freedom (Hoffmeyer 1992).

Analog and digital codes are tricky concepts, however, because a code that in one context is analog may in another context be digital and vice versa. Computer games, for example, are of course operating on digital codings and yet the image presented on the screen is an analog representation. And the converse is often true for hieroglyphs, where the isolated signs may (at least in some cases) be taken as analog codings — whereas the same signs, when interpreted as part of a text, become symbols making up a digital code.[13] The latter case is comparable with that of a painting that, when seen in isolation, might perhaps represent an aspect of the inner state of the painter's mind, but which, when seen on the wall together with other paintings at an exhibition, becomes instead one of a series of discrete signs that, as an ensemble, make up a higher order message. In the terminology of Bateson, this represents a shift in logical type[14] in the sense that the painting as physical object per se and the painting as an exhibited artifact do not belong to the same logical type. One might perhaps even say that it is only in the context of the higher logical type that the picture is indeed a painting.

Paintings, at any rate, would be a very different kind of thing if exhibitions did not exist. The exhibition digitizes the work of the artist and at the same time conserves it as a painting. Through this operation, the painting becomes simultaneously freer and less engaging. It may now be recombined with other objects in collections, in art-history works, or even in interior design. But this increase in freedom is paid for by a loss of individual meaningfulness — which may be the reason why artists often do not like to sell their paintings.

---

possible — i.e., the set of states that could possibly be realized in the next step of the ongoing material reconfiguration of the biosphere — exhibits exponential growth: "Our biosphere and any biosphere expands the dimensionality of its adjacent possible, on average, as rapidly as it can."

12   Code-duality may also be seen as the semiotic core of cultural evolution (Hoffmeyer and Emmeche 1991; 2005 (1991)).

13   The analog-digital *gestalt shift* may in some cases be observed in individual hieroglyphs that may function both as ideograms and as phonograms. A hieroglyph resembling an eye may, depending on the context, signify either the notions *blind, awake,* or *weep* — but it may also signify simply the sound *ir,* because the name for an eye is *irt* (*Den store danske Encyclopædi* vol. 8, p. 438).

14   Bateson is here referring to Bertrand Russell's type theory (Russell and Whitehead 1910–13).

Examples of such analog-digital shifts are numerous in the biological world,[15] and they are presumably always involved in the kind of phenomena we call emergent. Analog codings, for example, may be digitized when brought to bear in the processes of *quorum sensing*. The word *quorum* is taken from legal language, and is used in connection with meetings (typically general assemblies) where there have to be a certain number of attendants for the meeting to be legally competent to transact business. When this is the case, a quorum is said to apply.

In biology, *quorum sensing* has become the designation for a kind of communicative activity in bacteria where the density of bacteria present is a causal factor. In short, quorum sensing is due to a process where each single bacterium excretes a certain chemical compound such that the concentration of this compound in the medium will reflect the number of bacteria per unit of volume. Quorum sensing occurs if the compound, after having reached a threshold concentration, binds to a regulatory protein in the cell and thereby initiates the transcription of specific genes. An interesting example of quorum sensing occurs in a species of squid, *Euprymna scolopes*, which hunts small fish by night on the coral reefs off the coast of Hawaii. Moonlight causes the squid to cast a shadow that makes it easy catch for predators. As a defense strategy, *E. scolopes* has evolved a sophisticated way of emitting light that effectively hides its own shadow. *Counter-illumination* is the name given to this kind of camouflage, and it is only made possible by the squid's symbiotic relationship with luminous bacteria called *Vibrio fischeri* that live in the mantel cavity of the squid. Living off of food provided by the digestive system of the squid, the bacteria emit light of the exact same intensity and color as the light reaching the squid from the moon, and this prevents predators from seeing the squid from below (McFall-Ngai and Ruby 1998).

In the morning, the squids bury themselves in the sand and excrete 90–95 percent of the bacteria, which brings their density well below the threshold level. Apparently the squid is in full control of bacterial growth rate by adjusting the supply of oxygen, and at sunset the population of bacteria reaches the threshold level once again.[16]

Interestingly enough, the same compound — N-acyl-homoserine lactone (with a variable acyl group) — seems to be used as a signal in many different manifestations of quorum sensing. In *V. fischeri*, high concentrations of this

15   Stjernfelt (1992) has observed that the linguistic concept of *categorial perception* perhaps might be extended so as to cover transformations from iconic to symbolic representations quite generally, and this idea concords with the significance I have ascribed here to the analog-digital shift. An alarmone, as will be discussed in connection with endosemiotics in Chapter 7, is yet another example of such a shift.

16   A thorough treatment of the biosemiotics behind this phenomenon has been given by Luis Bruni (2002; 2003).

compound[17] cause an induction of what is called the lux gene, and thereby a one-hundred- to one-thousand-fold increase in bioluminescence. The concentration of N-acyl-homoserin lactone, quite generally then, operates as an analog coded message about bacterial density — a message that can then be digitized to become a sort of an either-or switch by which, in this case, the squid switches on and off its "false" light.

The opposite situation, where natural digital codes are transformed to analog codes, is known for instance from the nervous system. Here, every single neuron makes up an either-or mechanism in the sense that, either the cell delivers a series of action potentials or it doesn't. In a typical nerve, however, there are large numbers of axons running in parallel, and the accumulated signal will come as a continuously varying signal reflecting the fraction of neurons that at each time emits an action potential. And, as noted by Favareau (2002), the threshold at which a neuron does or does not fire its digital (all or nothing) action potential, is mediated throughout by the analog presence of the ever-changing number and kinds of neurotransmitter molecules that are currently occupying the synaptic gap. Code-duality, then, seems to be as much a property of neuronal transmission as it is of genetic transmission.

The Epistemic Cut

As already mentioned, one of the sources that inspired Emmeche's and my work on code-duality was the American biophysicist Howard Pattee's theory that life is characterized by its operating simultaneously in two complementary modes, a *dynamic* mode and a *linguistic* mode. In this connection, Pattee refers explicitly to Niels Bohr's application of the principle of complementarity to the phenomenon of life (Pattee 1977). In later papers, Pattee talks about a semiotic mode rather than a linguistic mode, but in principle the idea is the same. In 1997, Pattee discussed his ideas by quoting John von Neumann (1955): "We must always divide the world into two parts, the one being the observed system, the other the observer . . . . That this boundary can be pushed arbitrarily deeply into the interior of the body of the actual observer is the content of the principle of the psycho-physical parallelism — but this does not change the fact that in each method of description the boundary must be put somewhere, if the method is not to proceed vacuously." Pattee then explicates his idea as follows:

> Von Neumann defines a physical system, *S*, whose detailed behavior must follow from the fundamental laws of physics, since these laws describe all possible

---

17   The acyl-group in this case is a 3-oxohexanoyl group.

behaviors. But if the particular behavior of S is to be calculated, we must measure the initial conditions of S by a measuring device, M. Therefore, the essential function of measurement is to generate a computable symbol, usually a number, corresponding to some aspect of the physical system.

Now, the measuring device also must certainly obey the laws of physics, even in the process of measurement, so it is possible to correctly describe the measuring device by the laws of physics. One must then think of the system and measuring device together as just a larger physical system $S = (S + M)$. But then to predict anything about S we must have a new measuring device to make new measurements of even more initial conditions. Obviously, this way of thinking gets us nowhere except an infinite regress.

The point is that the function of measurement cannot be achieved by a fundamental dynamical description of the measuring device, even though such a law-based description may be completely detailed and entirely correct. In other words, we can say correctly that a measuring device exists as nothing but a physical system, but to function *as* a measuring device, it requires an observer's simplified description that is not derivable from the physical description. The observer must in effect choose what aspects of the physical system to ignore and invent those aspects that must be heeded. This selection process is a decision of the observer or organism, and cannot be derived from the laws (Pattee 1997, http://www.ssie.binghamton.edu/pattee/semiotic.html)

A few moments reflection as biologists shows us that not only humans but all living creatures are fundamentally engaged in processes of measuring, and therefore Pattee's epistemic cut follows automatically from the above consideration: "We must define an *epistemic cut* separating the world from the organism or observer. In other words, wherever it is applied, the concept of semantic information requires the separation of the knower and the known. Semantic information, by definition, is about something" (ibid). Pattee is anxious to underline that his epistemic cut is not another version of Cartesian dualism, but only a *descriptive dualism*. In the perspective of this book it remains, however, a big question, as to whether or not Pattee is right in this — for central to the current investigation is the question: Through what processes exactly could a dynamic functional mode possibly *become* a semiotic functional mode? Pattee is not very specific when it comes to this question.

As Bohr — and later, von Neumann — had, Pattee has taken a far and courageous step forward in facing the *limit paradox* that necessarily arises when one subscribes to an ontology of natural law, i.e., that conception of the world that Pattee formulates quite unambiguously in the first few lines of the quote given above, when he claims that the laws of physics "describe all possible behaviors."

To transfer Bohr's complementarity principle of physics[18] to the biological sphere of life seems not, however, to be particularly helpful in solving the limit paradox. For if complementarity is thought of as being ontological, then we are right back in dualism, which Pattee explicitly condemns. But if complementarity is thought of as being epistemic, then we must consider it as an assertion to the effect that we cannot describe the semiotic dimension of the world in the same language that we use to describe its dynamic aspects. And this, supposedly, is due to the shortcomings of language or of thought as such: The semiotic aspect of life is, so to say, just a glimmer that we cannot think ourselves apart from because we are, in an existential sense, wrapped up in it.

This manner of thinking is, in fact, quite widespread and is related to the position, recommended by the philosopher Daniel Dennett, called "our taking of the intentional stance." Briefly stated, this view holds that we cannot understand the life of other humans (or of animals) without describing those lives as guided by, or woven into, intentionality. This does not mean that these creatures *possess* intentionality as a real property — rather, the thesis states only that we cannot understand these creatures unless we *pretend* that they do.

That scientists and philosophers willingly accept such an absurd conception of what human life is about in its deepest or most fundamental content — that, in other words, one accepts that the feelings, experiences, aspirations, sorrows and desires of human beings (to say nothing of the experiential world of other forms of animal life) is all just chimerical hot air — bears witness, in my mind, to how deeply anchored the ontology of natural law is in the self-understanding of the scientific mind. I hope the reader will excuse me for the comparison, but it really does remind me of the evermore complex (and increasingly less likely) sets of epicycles that Ptolemaic astronomers had to introduce into their explanations of the planetary orbits in order to uphold the belief in the geocentric system. Rather than seeking shelter in such powerless conceptions about what, for all of us without exception, is the deepest and most real content of our lives — i.e., the fact that such life is being *experienced* — the author shall suggest that it is instead the ingrained belief in the *exclusive* ontology of natural law that ought to be given up.

This does not mean that we disregard the reality of those laws, but that we refrain from reifying them as ultimate explanations and first principles. And it means that we, as Peirce recommended, should understand the natural laws that we see existing as phenomena that are themselves in need of explanation.

18   The complementarity principle of quantum theory refers to effects such as the wave-particle duality, in which different measurements made on a system reveal it to have either particle-like or wave-like properties. In Bohr's understanding, complementarity reflected the weaknesses of human language and not any deeper property of reality.

Much in the same way that Einstein recontextualized the "universal" laws of Newton by showing them to be the local products of more general principles, Peirce saw natural laws as the secondary products of a more general tendency in the universe to generate regularities (or *habits* as he often called them). Rather than seeing the universe as characterized by lawfulness, its primary state, according to Peirce, is indeterminacy and chance. And thus the formation of regularities, such as for instance natural laws, must be explained by other means — for natural law is thus a product of evolution, and not its source.

The tendency of nature to generate regularities — or, as it is more commonly described today, to self-organize — may be understood as the very first exposition of the principle that will develop through cosmic evolution to become semiosis, the ability of living systems on planet Earth (and possibly many other places in the universe) to form *interpretants* (or self-maintaining and self-perpetuating habits, if you like).

This change of basic viewpoint will allow us to reach a solution to Pattee's paradox of the epistemic cut, because now we can assume that not only the symbolic functional modes (related to DNA function), but also the dynamic functional modes (related to the functional cytoplasm) are both, in the end, semiotic functional modes. For the difference between the two modes is, at bottom, a difference in the kind of semiotic dynamics involved. Thus, the sign processes characteristic of the dynamic functional mode — i.e., the protein world, so to say — are indexical and iconic (i.e., analog-coded) rather than symbolic or digital as are the sign processes connected to DNA function. The analog-coded signs correspond to the jumble of topologically organized indexical and iconic sign processes in the cells that are responsible for the interpretation of the digital genetic instructions as well as for the execution of them. Or, as we wrote above, "To claim that only the digital twin is semiotic whereas the analog twin remains in the sphere of classical dynamics is to block the only possibility for transcending the epistemic cut."

Replicators and Interactors

Pattee's distinction between the dynamic and the symbolic (or semiotic) domains forms a strange and certainly unintended parallel to Richard Dawkins's distinction between vehicles and replicators, discussed above. As we saw, *gene vehicles* is, in Dawkins's terminology, a designation for organisms and thus for metabolically driven activities including movements and growth, or in short, the dynamic domain. Replicators, however, are the genes themselves (in whatever way such genes may be defined in Dawkins's context), and these, of course, correspond to the symbolic domain in Pattee's view. But there is an enormous difference between the two theories, for the whole point of Dawkins's construction is that replicators are the real *agents* at the scene, whereas the contribution from organisms (vehicles) are just that of passively assisting in the processes of copying and spreading as many replicators as possible. Yet for Pattee, for the symbolic domain to make any sense at all, the measuring processes (i.e., the processes whereby organisms are sensing and interpreting their environments) are presupposed; to imagine the relation between the symbolic and dynamic domains as a master-slave relationship makes no sense. But in Dawkins's way of thinking replicators become masters while the organisms/vehicles are obedient slaves whose untiring efforts only serve to deliver to the replicators the much sought-after prize of multiplication.

Metaphorically speaking, one might say that where Pattee thinks that the organisms of this world use words in order to get along and to produce more organisms, Dawkins thinks that the words use organisms to produce more words, and that the organisms are just tools for this self-promoting process. Seen from the perspective of both Pattee and biosemiotics, Dawkins commits a logical error in that he treats symbols as if they were things (replicators) — and, worse yet, not merely things but *things with purpose*, for the very term *replicator* implies a sort of agency, but an agency that one can hardly imagine to be a property of a DNA molecule (Deacon 2002, 122). The gene may be legitimately termed to be *a replicative unit*, but it can not, in the normal use of language, be a replicator.[19] The American philosopher David Hull (1980) attempted to clarify this discussion by substituting the term *interactor* for the term *vehicle*. In Hull's terms, *replicators* are units that make more units like themselves (more or less reliably) and thus conserve and transfer information across generations via the processes of natural selection. *Interactors*, on the contrary, are units that spontaneously interact with their surroundings in ways that may either promote or restrict the spread of replicators. Replicators thus form lines of descendants and have kinship relations with one another. Interactors have causal effects on their environments in the here and now, only during the course of their own lifetimes. In most versions of genetic Darwinism, organisms are *interactors*, whereas genes are *replicators* (Depew and Weber 1995).

For years now, this dualistic terminology has managed to penetrate and, in fact, to dominate debates on evolutionary theory. One may then choose to put special emphasis on the fact that replicators effect interactors by supplying them with information that may provide them with a competitive edge relative to other interactors (as in theories of *gene selectionism*). Or, one may instead put emphasis on the fact that interactors influence replicators by their decisive role in determining

19   Dawkins would perhaps reject outright the notion of a symbolic domain as something really existing. But then, how would he propose to solve the von Neumann-Pattee paradox?

which information winds up being, in fact, transmitted to the next generation (as in *classical selection* theory). The question resolves to whether replicators or interactors should be ascribed causal priority in life's evolution. Seen from the perspective of the present book (see Chapter 3), the answer is that we are dealing here with two different kinds of causality, where replicators exert formal causality, while interactors exert efficient causality. Final causality, *sensu* Peirce,[20] makes use of both tools and is, in this context, connected to life's code-dual structure.

In short, the replicator-interactor terminology relies on an inherent reification that serves to restrict discussions to a nearly rachitic[21] understanding of causality. And thus, by re-framing the evolutionary processes as the interdependent interactions between digital and analog codings (wherein the contents coded belong to the historical subject of the lineage, which is both the product of, and the subject for, selection), biosemiotics de-privileges the theme of the competition motive that has, almost to the point of obsession, dominated the understanding of evolution within contemporary Darwinism.[22]

As an alternative to the fetishization of Darwinian competition, the semiotic perspective makes us turn our investigation toward the processes whereby new significative patterns are generated and exchanged. Such processes seem to hold the key to the general evolutionary trend towards the appearance of creatures that increasingly depend for survival on their semiotic sophistication.

Organisms

Amusingly enough, it was Dawkins and his kindred souls who helped paving the way for a more biosemiotic understanding of life processes. For in the attempt to procure the necessary *lebensraum* for gene selectionism, they provided a number of sharp arguments to show that positing the organism as the basic unit of biology is an irreparably anthropomorphic construction.

20   ". . . that mode of bringing facts about according to which a general description of result is made to come about, quite irrespective of any compulsion for it to come about in this or that particular way, although the means may be adapted to the end. The general result may be brought about at one time in one way, and at another time in another way. Final causation does not determine in what particular way it is to be brought about, but only that the result shall have a certain general character" (*CP* 1:211).

21   Young readers may not know the disease *rachitis* — or rickets (which, by the way, in Denmark we call "English disease") — that is caused by a deficiency of vitamin D, and which, in the childhood of this author, was still sufficiently common for everybody to know the characteristically hollow-chested look of children having suffered from this disease.

22   Ironically, this is an obsession that plays right into the hands of precisely those radical social-constructivist theories (e.g., those claims that reality, or at least the scientific model of it, is *nothing but* a social construction set up to perpetuate unequal power relations) that many scientists so despise.

For example, in most biological models, as well as in everyday folk psychol-ogy, the prototype *organism* remains essentially a vertebrate, like ourselves. Vertebrates are always well-integrated, coherent organisms with well-defined forms. They consist of genetically uniform cells and have well-defined life cycles, starting with a single cell and ending in reproduction via the transmission of germ cells.

However, by far most organisms of this world are *not* vertebrates — and most of them do not obey the aforementioned criteria very well. Neither plants nor fungi have the kind of individual identity and autonomy that vertebrates have. Their life cycles do not necessarily start from (or as) single cells, and they are not as genetically homogenous as are the vertebrates. Many invertebrate ani-mals also deviate from our conception of how a typical organism lives. Thus, for instance, many insects undergo metamorphic changes that give them totally dif-ferent body shapes from that of their earlier life stages — and many corals swim freely around as individual larvae, but end up as colonies where individuality is completely extinct.

The idea that most seriously undermined the classical concept of the organ-ism was probably Dawkins's notion of the extended phenotype — for this idea implied dissolution of the supposed unity between the organism and the gen-otype. By the term *extended phenotype*, Dawkins referred to situations where genes are selected for because of their effect on a *different* organism than the one that is carrying the gene. The phenotypic *effect* of the gene, in such cases, is played out in a foreign organism.[23] The most dramatic examples of this mechanism are probably found in parasites. An example is a species of fungus, *Enthomophtora muscae*, which infects and kills ordinary house flies. In addition to killing the flies, however, the parasite also causes dead females to develop a set of special traits, 'such as distended abdomens, that acts as sexual attractants to the male flies . . . that are subsequently infected and killed by the fungal parasite' (Moeller 1993, cit. in Sterelny and Griffiths 1999, 72).

Similarly, Stephen Jay Gould has discussed the case of the parasitic barnacles of the crustacean *Rhizocephala*, which completely take over behavioral control in the crabs that they parasitize. These parasites suspend the crab's internal molt cycle (which might otherwise allow the crab to shed the parasite) and succes-sively transform the brood care behaviors of the crabs in such a way that they will start nursing the parasitic eggs instead of their own (Gould 1996, 15–16).

Morever, not only parasitism, but symbiosis, too, may be of a mutualis-tic kind, as we saw in the case of the bladderwort in Chapter 3. Such inter-active mutualism is often so tightly interwoven that it feels more natural to talk about one *joint organism* or a *superorganism* — as in the case of *lichens* (a

23   Vehkavaara (2003) has suggested the term *externalized purposes* for this survival strategy.

general term for a mutualistic ensemble of a fungi and algae). Another example is the fungus-growing-and-harvesting ant. In this case, the fungi produce sterile fruit that the ants use to feed their larvae. The subterranean fungus gardens of such ants may be twenty meters long, and the fungi are nursed with a meticulousness that would awake indignation in the unions of agricultural workers. This particular example of inter-species collaboration must be called an evolutionary success story, for it occurs in more than two hundred versions, involving different species of ants and fungi respectively. And in some cases the integration is so complete, that neither the ant nor the fungus could survive without it.

Similarly, the partnership between higher plants and fungi that is found in root modules, *mycorrhiza*, is also of vital importance for both partners. In this case, the fungi extend out, as mycelium, into the soil and thereby provide their plant host with improved access to nutrients and water. In turn, the fungi profit from access to the organic compounds that are produced by the plant. The mycelium can also interconnect different plants, even of different species. Alan Rayner (1997, 63) writes, "By providing communication channels between plants, mycorrhizal mycelia are thought to enable adult plants to 'nurse' seedlings through fungal 'umbilical cords,' to reduce competition and to enhance efficient usage and distribution of soil nutrients."

However, there is also a risk connected to this arrangement: "On the other hand, they can be piratized, as demonstrated by the yellow bird's nest plant, *Monotropa hypopitys,* which by tapping into mycorrhizal networks is able to divert resources from the trees that participate in these networks (ibid.).

Another of Rayner's examples is the rot that is found in hollow trees. Rot is intuitively perceived as a disease, but in trees it is often just one link in a normal recycling process — i.e., a process whereby resources are redistributed from parts of the tree that no longer take active part in the life processes to those other parts of the tree where active life processes still occur. The growth of trees typically proceeds in a thin layer just inside the bark, and as the tree ages, more and more dead wood is left behind in the middle of the tree. Fungi then degrade the dead wood in the middle, and thereby initiate a hollowing out of the tree (called *heart rot*), which then offers habitats for a wide range of different animals. By sending roots into the resulting compost, the tree actively recycles the material derived from its own dead wood.

Hardly anybody has a full overview of all the species that coexist in an old tree but, counting microorganisms, the number may well approach one thousand. Yet this should not surprises us, as the human organism, as is well-known, is itself the domicile for numerous species of microorganisms that do useful or even vital work for us — in the saliva, in the skin, in the intestinal tract, and elsewhere.

Accordingly, the biochemical and physiological interactions necessary to the maintenance of these many symbiotic relations is equally quite comprehensive. In the aforementioned case of quorum sensing, the newly hatched squid early on develops a ciliated microvillus field that will tend to potentiate the bacterial inoculation — and that is capable of preventing all bacteria other than *V. fischeri* from colonizing the light organ. Yet as soon as the light organ has been inoculated with *V. fischeri* a programmed massive cell death is initiated, whereby the ciliated surface coating is again eliminated.

This "program" is released thanks to a short-lived and irreversible signal emitted by the bacteria, writes McFall-Ngai and Ruby (1998). In brief, cells in the cavities in the light organ where the bacteria attach themselves respond by undergoing comprehensive cytological alterations, during which some of them swell to four times their former volume; at the same time, a marked increase in the number of *microvilli* sets in. At this point, the cells start excreting a substrate that is rich enough to assure a very short bacterial generation time (time lapse between divisions) of approximately thirteen minutes. In the course of ten to fifteen hours, the density of bacteria is stabilized at a level that is determined by the host organism's regulation of the oxygen concentration. When the bacteria have reached their maximal density, they stop producing the proteins intended for formation of flagella — which may indicate that they use their flagella for the sole purpose of getting access to the light organ, for after having been so admitted, the flagellas then become an unnecessary burden (though it is not known at this time where the signal that blocks the formation of flagella comes from).

This slightly pedantic description of the mechanisms behind mutuality in the squid-*V. fischeri* symbiosis is meant to illustrate how very subtle the integration between symbiotic organisms often is — and has to be. When organisms that are so fundamentally different that they belong to different species (or even to different kingdoms) are going to cooperate, they are forced to overcome a host of obstacles concerning the establishment of unambiguous reciprocal interactions at all levels, from chemistry to social behavior.

The Darwinian calculus of gross reproductive advantages and disadvantage seems strangely poor in this context, because it reduces almost to mystery the minutiae of practical solutions to all the challenges of incompatibility. The naked end result, the "increased fitness" that is supposed to explain the eventual fixation in the gene pool of mutualistic interaction patterns, tends to hide the complex reality of the moment-to-moment lived reality. Or to say it another way, by lumping the numerous subtle interaction processes into the one single conceptual viewpoint of selection, one has already, in advance, ascribed priority to a retrospective view of things. But since such an end point must always be unknown to the agents while the process is actually going on, this perspective implies a systematic distortion of our understanding of what really happens *when* new behavioral patterns appear.

From the biosemiotic perspective, it is rather obvious that the "control" mechanisms overseeing all the numerous reciprocal processes that necessarily have to be in place, in order to scaffold an initial mutualism between organisms from different species, need not be digitally and unambiguously coded at localized loci (i.e., on the chromosome — where, not inconsequentially, mutational events might spoil the cooperative interaction without being selected against). Yet outside of biosemiotics, the idea that a nondigitized but relatively stable semiotic scaffolding of these integration mechanisms might establish itself is a strongly underestimated possibility, if it is considered at all.

The Analog Coding of Epigenisis

Before expounding the biosemiotic view, however, we must first consider a critique of the classical Darwinian focus on the individual that does not, as the previously described criticisms do, deal with symbiotic relations (either parasitic or mutualistic) — but instead claims that ordinary classical *individuals* should be understood as *conglomerates of competing cell lines* — and that these cell lines, rather than individuals, constitute the units of natural selection.

Yale biologist Leo Buss proposed just this theory in his book entitled *The Evolution of Individuality*, wherein he delivers a frontal attack on the doctrine of Weismannism. In it, Buss (1987, 3) writes, "While Weismann's inheritance theories were ultimately proved fictional, their corollary — that the individual is the sole unit of biological organization — was nevertheless incorporated as a tacit assumption in the modern synthetic theory of evolution." But, Buss continues, as far back as when the fundaments of the neo-Darwinian synthesis were put down, data that contradicted Weismann's doctrine were already known by embryologists. The problem was that embryologists were strangely absent from the discussions leading up to the modern synthesis — and so it happened that Weismann's theory was canonized as a synthesis, in spite of its being in disagreement with known developmental patterns. (In fact, development exhibits significant variation from one taxonomic group to another, and in some cases Weismann's hypothesis does indeed hold — but in most cases it does not.)

Buss, after reviewing the data on all the main taxonomic groups, shows that the most common form for development is *somatic embryogenesis* — where there is no autonomous germ cell line. In somatic embryogenesis, one and the same cell line may participate in somatic functions (as stem cells) and yet, throughout ontogenesis, conserve the competence to form germ cells. This mode of development is possessed by all plant and fungus groups and, with a single exception, also by all protoctists (unicellular eukaryotic species). And even in the animal kingdom, we find somatic embryogenesis dispalyed in no fewer than nine taxonomic groupings (among them cnidarians, bryozoa, and even some annelida species (ibid., 21–22)). Thus, the particulars of *preformationistic embryogenesis*, where Weismann's doctrine does obtain, are in no way typical for life forms on our planet.

And once Weismann's theory is removed from center stage, *the vertebrate individual* loses its favored position in biology as well. With the loss of the individual as the natural center for the evolutionary process, a new space is opened for Buss's own alternative perspective: "The thesis developed here is that the *complex interdependent processes* which we refer to as *development* are reflections of ancient interactions between cell lineages in their quest for increased replication. Those variants which had a synergistic effect and those variants which acted to limit subsequent conflicts are seen today as patterns in metazoan cleavage, gastrulation mosaicism, and epigenesist" (ibid., 29).

Buss's book takes us on a fascinating (and essentially semiotic) journey into the webs of cheating, humbug, bluff, and out-maneuvering that — according to him — were the means that individual cell lines used in their reciprocal competition to become multiplied as much as possible. (Chapter 7 examines some of these processes in more detail.) On the embryological process, he writes,

> Following a variable period of maternal control on embryonic cell fate, the metazoan embryo becomes organized into one of several discrete bauplans via interactions between embryonic cell lineages. The principal epigenetic interactions defining cell fate — those of induction, competence, and cell death — are all interactions in which one cell lineage acts to limit the replication of another, while enhancing its own. The fact that embryos develop by epigenesis is *prima facia* evidence that these very "programs" represent interactions between variant cell lineages arising in the course of the ontogeny of ancestral forms (ibid., 30).

Buss's heavy emphasis on *the competition motif* as the preferred explanatory tool, is, as the reader should surely know by now, not shared by the author of this book. But in placing *the developmental process* in the center of his evolutionary thinking, Buss calls attention to an epigenetic perspective that underlines the autonomous significance of a long neglected domain — i.e., the domain that I have in this book called *the analog code*. And this domain (or coding) is, in fact, nothing more nor less than the semiotic loops coupling embryonic cell lines together into a unit that — in spite of all internal competition — presses a shared destiny upon them all.

We can conclude then, that Dawkins, Buss, and many of the others who have challenged the classical conception of the individual organism as the uncontested unit of selection, do indeed have convincing arguments. This, however, does not persuade us to accept that genes or cell lines should then be automatically installed in the role that the individual has surrendered.

On the contrary, it confirms us in seeing evolution in the light of code-duality, for if organisms are not natural units of selection, they are at least natural units of communication. Via the digitally coded messages in their genome, they are literally in a line of communication with both their ancestors and their eventual offspring. And via the multiple analog coded messages of their current bodies, they take part in the local semiosphere and interact with other creatures, whether conspecific or not. The organism must then be seen as a nodal point in a semiotic landscape — one that gradually changes through time (evolution) and thereby leaves marks (genetic changes) in the conservative DNA code.

It is to some extent a matter of taste if you would like to think of these remaining marks by calling them *selected*. Since in any case, the proper *efficient causality* at issue here lies in the web of analog coded communication, Dawkins's position would actually presuppose that formal causality should be seen as *primary to* efficient causality. (But this is a discussion that reminds me more than a little of the medieval arguments about the gender of angels.)

The Life Cycle as a Unit of Evolution

In 1985, the American psychologist Susan Oyama of John Jay College in New York, published an important book in which she recommended that we give up altogether the idea that the developmental process is directed by some special information that is transmitted by genes (Oyama 1985). The information that manifests throughout the life cycle of an individual, Oyama claimed, is rather information that is constructed by and with the developmental process itself. With a pertinent choice of words, she named her book *The Ontogeny of Information*.

Oyama's point — which is closely connected to the one taken in the present book — is that developmental processes depend on "inform-ation" from a range of different sources, and that genes are only one of these sources. The conception of a special genetic program that unfolds its predetermined logic through embryogenesis is mistaken because this program — if the program metaphor should be valid at all — is not self-reliant, but only works at all because it is played out in a context that is derived from elsewhere.

And, indeed, there is no reason to believe that an organism could be pre-programmed to solve the multiple challenges it will meet in its lifetime. In this respect, the organism confronts the same horizon of troubles that has finally convinced the would-be creators of "autonomous agents" to change their paradigm from one of rational preprogramming (the old artificial intelligence approach) to one of training and learning (Clark 1997).

From the very beginning, embryogenesis presupposes that tissues in the growing embryo have the capacity for selecting and responding appropriately to relevant stimuli (which is exactly what an autonomous robot to some extent may *learn* to do but cannot be *instructed* to do). The developmental process, in other words, is not an *instructional* process, but a process of *self-calibration*. As such it is, of course, supported by the ever-present availability of necesssary protein resources that the cell at each moment may derive, by activating relevant sections of the genetic library. But basically it remains a process where individuals create themselves in a self-organizing interaction with their environment (the metaphoric here is my own, not Oyama's).

Therefore, the ordinary expression — that a gene codes *for* this or that trait — tacitly takes for granted that the conditions under which such a trait can be expressed are the conditions that are normal. In cases where such conditions are not normal, however, the trait in question may perhaps not be developed, even though the gene is there. And as Paul Griffiths and Russell Gray (1994) have pointed out *normal conditions* is not an unambiguous concept. Genes in an acorn, for instance, are supposed to code *for* the development of an oak tree — but by far, *most* acorns do not sprout but are eaten or just rot away on the forest floor. So in this case, what are the "normal conditions" that this gene finds itself in? (See Ruth Milikan's discussion of functionality in Chapter 3).

This phrase — *under normal conditions* — actually hides a whole lot of interesting things, so that on the one hand we have the genes and on the other hand we have *all the rest*. But here is an interesting symmetry. For just as one might say that a gene codes for a trait *when conditions are normal*, one might equally justifiably say that each and all other developmental resources code for a trait *when genes are normal*.

Many bird young, for example, must learn the song of their own species by first listening to the song of their parents. To say that their song program probably is coded for in the genes if things are normal implies, with the invocation of *normality*, that the bird parents do, in fact, sing. But one might as well say that the song of the young is coded for in the song of the parents — presupposing, of course, that the necessary proteins (and thus, the genes) needed for constructing the relevant anatomical structures in the throat are intact.[24]

---

24   Molecular genetics has increasingly undermined the simple genotype-phenotype relationship that was so passionately believed in just a few years ago. For it has become increasingly apparent that genomic systems exhibit unexpectedly *integrative* aspects. The lactose-positive phenotype in *E. coli*, for example, presupposes not only that the *lac*-operon proteins are expressed, but also that the genes that code for adenylate cyclase and for the cAMP receptor protein are expressed. "In many cases," writes James A. Shapiro (1999, 25), "it is really impossible to assign a specific

This gene-environment symmetry argument is the central component in a complex of ideas that Oyama, Griffiths, Gray, and others have developed under the name of *Developmental Systems Theory* (DST) (Oyama et al. 2001). According to DST, species-specific traits are formed with the help of structured sets of developmental resources within self-organizing processes, and there is no need to appoint any centralized information source. Some of these developmental resources are genetic, while most others — from the cytoplasmic machinery of the fertilized egg to the social structurings that influence human psychological development — are nongenetic.

It may perhaps come as a surprise to some readers that such parity between information sources is suggested at all — for such is the extent to which the rhetoric of modern genetics has managed to make us identify heredity and genes as two sides of the same coin, such that our eyes have become closed to the obvious. For organisms do, of course, inherit a lot more things than just their genetic material. Above all, they inherit a suitable milieu or habitat (without which, such genetic material would remain inert).

Putting it plainly, mice are not born at the bottom of the sea, and fish are not born on land. Or, here is a more appropriate example: Cuckoo young are, as is well-known, hatched in nests belonging to birds of other species, and it has therefore been believed — following a suggestion by Konrad Lorenz — that the cuckoo, contrary to what is the case in other bird species, did not need to be imprinted on by conspecific adult birds in order to mate with cuckoo partners. It was supposed, in other words, that correct cuckoo behavior was genetically buffered. A recent study, however, has shown that, at least in the cuckoo species *Clamator glandarius*, the converse is true. Adult cuckoos in this species do actually look for nests with newly hatched young — not necessarily their own — and then sit in a nearby tree singing their songs. And this, apparently, is enough for the cuckoo young to take on the cuckoo song — or to be reimprinted with it — in those cases where they had already become imprinted with the songs of the foster birds (Soler and Soler 1999).

The above serves as a good example of the point that I want to argue, because with regards to it, nobody quarrels with the idea that genetic imprinting is involved. What is noteworthy (and far too often overlooked), however, is that the interdependence between genetic expression and learned behavior is so subtle. The latter is not merely an addition to the former, as is often implied. Rather, the individual organism only emerges out of a process of their dynamic interaction, and the full significance of such alternative types of *inheritance* only shows up when one learns to see the intimate interplay between genetic heredity and environmental heredity.

---

organismal phenotype to a particular locus, because its gene product(s) can participate in the execution of multiple cellular or developmental programs."

Similarly, Konrad Lorenz's concept of *innate behavior* was heavily criticized as far back as the 1950s by the American psychologist Daniel S. Lehrman, who pointed to many cases that contradicted Lorenz's concept of instinct. In this regard, we shall consider here the very illustrative case of prenatal *learning by doing* in ducklings, as studied by developmental biologist Gilbert Gottlieb and as discussed in Griffiths and Gray (1994, 279).

Ducklings normally acquire a preference for their mother's species-specific call signal from the very beginning of their lives. Gottlieb discovered, however, that ducklings would not develop this preference if they were prevented from themselves vocalizing while still in the egg. Apparently, duckling embryos have to be exposed to their own prenatal call sounds before they can develop a preference for the quite different sounds of their mother's call (Gottlieb 1981). It is important to underline here, as Lehrman did and Gottlieb repeated, that "these sorts of facts do not show that all traits are 'learned.' They show, rather, that reliable developmental outcomes occur because of reliable interactions between the developing organism and its environment" (Griffiths and Gray 1994, 280).

A consequence of this increased emphasis on the importance of the nongenetic resources in the expression of "genetic traits" is that Weismann's doctrine becomes directly misleading. For not only is the postulated separation between the germ cell line and the somatic cell lines, as we have seen, the exception rather than the rule in the biosphere — but even in those cases where it does apply, it cannot really be said to play the exclusively decisive role ascribed to it by many geneticists. Rather, it appears that the supposed dead-end organisms — i.e., the phenotypes — via their interaction with the environment and with their offspring, yield a rich causal input to the welfare of the next generation.

In order to include this richer conception of causality in our understanding of evolution, Griffiths and Gray (1994, 296) suggest that neither genes *nor* individuals should be seen as units of evolution, and that such a role should be ascribed rather, if at all, to the whole life cycle: "The individual, from a developmental systems perspective, is a process — the life cycle. It is a series of developmental events which forms an atomic unit of repetition in a lineage. Each life cycle is initiated by a period in which the functional structures characteristic of the lineage must be reconstructed from relatively simple resources. At this point there must be potential for variations in the developmental resources to restructure the life cycle in a way that is reflected in descendant cycles." Within the framework of such a process perspective, the whole replicator-interactor distinction becomes rather misleading, for it is based on the unnatural dichotomization of the developmental process into *genes* and *all the rest*. Dawkins (1982, 98), for instance, says, "When we are talking about *development*, it is appropriate to emphasize nongenetic as well as genetic factors. But when we are talking about *units of selection* a different emphasis is called for, an emphasis on the properties of *replicators*" (italics added).

But this idea only makes sense if one disregards the fact that many of those factors that, in the course of the life cycle, are responsible for the success of an individual, are replicas of factors that were responsible for the success of the life cycle of its parents. A DNA segment — to stay within the terminology of Dawkins — is not even itself a *replicator* in the strict sense of *something that is capable of replicating itself.* For the replication of DNA segments cannot take place without the cellular machinery of proteins and membranes that — considered as topologically ordered structures — are in themselves extragenetic hereditary factors. And a range of other developmental resources are likewise furthermore required for the life cycle to be realized in a viable way.

Griffiths and Gray (1994, 300) therefore conclude that "if we insist that a replicator have the intrinsic power to replicate itself, there will be only one replicator, the life cycle. But if we allow the status of 'replicator' to anything that is reliably replicated in development, there will be many replicators."

Developmental systems theory is largely consistent with a biosemiotic conception of ontogenesis and evolution, and probably the most radical consequence of DST is that the dichotomizing into organism and environment is severely challenged. In the traditional view, evolution is due to the exposure of individual variants to selective forces caused by an independently existing milieu. But according to DST, such variants (here, life cycles) are by necessity already deeply integrated into the environment, and thus the conception of differential success is no straightforward matter: "One variant does better than another, not because of a correspondence between it and some preexisting environmental feature, but because the life cycle that includes interaction with that feature has a greater capacity to replicate itself than the life cycle that lacks that interaction" (ibid.). Griffiths and Gray therefore suggest that "life cycles still have fitness values, but these are interpreted not as a measure of a correspondence between the organism and its environment, but as measures of the self-replication power of the system. Fitness is no longer a matter of 'fittedness' to an independent environment" (ibid., 301).

And this conception is not too distant from the concept of *semiotic fitness* that I suggested in 1997:

> Instead of genetic fitness, evolutionary biology should try to develop a concept of *semiotic fitness*. After all, fitness depends on a relationship; something can be "fit" only in a given context. Genes may be fit only under certain environmental conditions, or environments might perhaps be said to be fit in the sense that their self-sustaining dynamic capacity has been adapted to the actual genotype resources offered to them. But if genotypes and envirotypes (Odling-Smee and Patten 1994) reciprocally constitute the context within which fitness should be measured, it seems we

should rather talk about the fit in its *relational* entirety, that is as a semiotic capacity. The evolutionarily relevant fitness concept of "semiotic fitness" *should ideally measure the semiotic competence or success of natural systems in managing the genotype-envirotype translation processes.* The optimization of semiotic fitness results in the continuing growth and depth of interpretive patterns accessible to life (Hoffmeyer 1997a, 68).

Genes Are Indeed Special

Although Weismann's doctrine certainly is not waterproof in its substantial sense, it did at least illuminate an important aspect of evolution that is still in need of much clarification. This aspect concerns the general role of digital codes in evolution.

Because Weismann and several generations of biologists after him have reified the digital code — first as germ cells, then as chromosomes or genomes, and now, as DNA segments or replicators — it has been thought that this reified code could carry the weight of exerting efficient causality in the evolutionary process. And by framing this strange efficient causality as that of a *transport process* (where what was transported was called *information*), the mystification had finally been made complete. The missing substance in the concept of *genetic information transport* has been thoroughly analyzed by Sahotra Sarkar (1996; 1997) who points out that cracking the genetic code was the result of brilliant experimental work and was not, to any significant extent, helped by information-based reasoning.[25] "At the very most," Sarkar (1996, 199) says, the concept of a code carrying information "provides a succinct look-up table on the basis of which one can predict the sequence of the polypeptide chain that would be determined by a particular DNA chain provided at least five conditions, discovered since 1966, are fulfilled. Unfortunately, if prediction is the goal, these conditions are quite debilitating" (for further details, see Emmeche 1999b).

I claimed in Chapter 2 that the digital code appeared with the first living systems as a mechanism for the description of central constituents of the holistic arrangement in prebiotic systems. This description (in DNA code) implied a self-reference in the absence of which a living system could not become a *self*

---

25   I agree with Sarkar (1996) in his demonstration of the inconsistencies inherent in the *information concept* of molecular biology. From this, however, Sarkar draws the conclusion that we had better stick to strictly chemical-biological terminology. Biosemiotics draws the opposite conclusion and introduces an explicitly *semiotic* understanding, seeing *information* as the exchange of signs or sets of signs, i.e., coded messages. A thorough discussion of this understanding is given in Emmeche (1999b). Sharov (1992) also has recommended a semiotic understanding of biological information, and Jablonka (2002), as we saw in Chapter 3, uses a concept of information that is nearly indistinguishable from the Peircean sign concept.

Table 4.1 Comparison of gene selectionism to developmental systems theory and bio-semiotics (with inspiration from Claus Emmeche).

	Gene selectionism	Developmental systems theory	Biosemiotics
Anti-preformationism	No	Yes	Yes
Causal parity between genes and other resources for the ontogenetic process	No	Yes	Yes
Privileged causal role ascribed to genes in evolution	Yes	No	Yes: combinatorial freedom and temporal autonomy

(in the sense of a system possessing a stable coupling of self-reference to other-reference). The role of the digital code is connected, then, to the vertical (temporal) continuation of the self-specifying capacity of life. And in this chapter, we have seen that, towards this end, the digital code manifests three important characteristics: temporal stability (memory), rich potential for combinatorics (renewal), and capacity for abstraction (formation of meta-messages).

Digital codes, in our understanding, are therefore something special, and we must consider the possibility that there may yet hide a potential misunderstanding in the DST conception of complete parity between the genes and other resources of the developmental process. We certainly support the dethronement of genes from their relatively uncontested position as deterministic executants of control within the ontogenetic process. But on the other hand, their role as digitally coded sign systems provides them with a unique status that ought to be recognized. And by seeing life, as suggested here, as being based on the incessant semiotic interactions between code-dual systems — organisms in life cycles — it is possible to respect the special role that genes do in fact play, without thereby privileging them relative to all other factors that contribute to semiotic control within the life cycles of the biological world.

Decoupled from the dynamics of cellular life, the genome is at one and the same time conservative and promiscuous. From the one side it is protected against changes, and from the other side, it becomes recombined all the time under the formation of new genetic combinations. In the next chapter, we shall have a closer look at such genetic semiotics.

5
The Semiotics of Heredity

The Logic of Mortality

Because living systems are mortal, they cannot therefore persist in the same way as rocks or celestial bodies persist — e.g., by internal cohesion or chemical resistance enabling them to withstand the unending flow of physical change. Yet even rocks and celestial bodies will disappear eventually — and when they do, that is definitively the end for those systems as such. Living systems, on the other hand, have found a way to continually persist — not as the autonomous physical entities that are living at any one moment, of course — but as templates that can be used to recursively and adaptively re-create the system. These DNA templates function as *messages* that are read and reread over time, and are lodged within small capsules (such as eggs), each of which contains a fully capable set of interpretive tools that allows them to retreive the hidden message coordinating the ontogenetic process, resulting in the birth of a new creature of the same kind. The innermost principle of heredity is thus *semiotic survival* — survival via a digital description coupled with an analog (or time-space continuous) set of functionally interpretive tools.

A big advantage of *this* kind of persistence (relative to that of rocks or celestial bodies) is its flexibility and adaptive capacity. The flip side of the coin is that each single living entity must die; *survival* concerns the living system as *type* not as *token*. Yet even as a *type*, such systems may not continuously survive in the long run, since each one always runs the risk of becoming extinct or, suffering the form of death characteristic to flexibility — i.e., irreversible change.

The proviso that living systems must necessarily die, however, is fortunate when seen in the light of *organic renewal* and *adaptation*. And because organisms, such as the reader and myself, depend on the uninterrupted functioning of a subtle interplay of some fifty trillion cells, it is obvious from a biochemical point of view that such finely coordinated systems must necessarily be fragile.

According to evolutionary biologist Stanley Salthe (1993), there is also a more fundamental *infodynamic necessity* that dictates the death of living systems. *Infodynamics* is the designation for a set of ideas that integrate thermodynamics and information theory in the attempt to describe biological evolution in light of the known physical dynamics of the universe's development (Brooks and Wiley 1986–88; Goodwin 1989; Weber et al. 1989; Salthe 1993; Weber 1998b). The essence of infodynamic theory, as it has been developed by Dan Brooks and Ed Wiley, is that the irreversible expansion of the universe involves an ongoing generation of new systems — in *far-from-equilibrium* states — that consequently possess the potential for self-organization.

For Salthe, this core insight is closely linked to Peirce's notion that nature tends to reinforce habits (*CP* 7:15). The self-organizing tendency that infodynamics sees in nature implies a continuous establishment of new informational constraints at the same time that the space of possibilities tends to become steadily diminished due to the growing sum of such constraints.[1] In the beginning we have vagueness and flexibility, but gradually *habits* (decreasing the possibilities of what may happen next) grow denser as the system becomes more and more rigid and ultimately dies. Salthe sees the universe as emerging through an endless succession of developmental processes where the fundamental theme is the same: a movement from immaturity through maturity to senescence and death — which then becomes the substrate for new beginnings. This *phasic* scheme plays out on all levels of complexity, that is to say, not just for individuals, but also for ecosystems and species, families, and even higher taxonomic categories.

As far as the semiotics of heredity is concerned, however, Salthe's very generalized perspective points to a potential inconsistency — for the finely elaborated dynamics of *heredity* that one finds at the level of individuals does *not* seem to be paralleled in other types of dissipative systems. A digital code that can bring back, in a sense, no-longer-existing system states, is a very special phenomenon that is *not* found outside of the life sphere — and even there, in organic evolution, such digital codes have been "invented" only rarely. A dissipative system such as, say, a tornado may perhaps be said to exhibit some kind of self-reference, but here, as in other nonliving systems, the self-referential activity is not based on digital codifications (Hoffmeyer 1998a) — and thus tornados always die out without leaving offspring capable of exhibiting individualized kinship relations.

Seen as a dynamic mechanism, code-duality is the essential link that connects individual and population. Reproduction implies that in each generation, the interplay of analog-coded phenotypes in the population's ecological space leads to the formation of cohorts of digitally coded genotypes (or kimfloks,

1   Milan Kundera paints a nice picture of this relationship in *The Unbearable Lightness of Being*, when he compares life to a symphony. In the beginning the symphony is wide open, so that many thematic strands still can be taken up and woven in, however, as the composition proceeds through stages, the composer becomes more and more bound to choices already made. By the end of the symphony's last movement, freedom has become fairly limited.

a somewhat specialized term that I will explain below). Each of these then becomes recoded or "back-translated" into phenotypes (phenotypic codes) via the ontogenetic process. The mechanism behind these perpetual codings and recodings is the subject matter of the present chapter. Here I shall claim that these processes of coding and recoding do indeed reflect a general *info-dynamic* tendency in our world — but also that the "invention" of the digital code instituted a whole new creative dynamic which is unique for life and cannot be described in the generalizing terminology of currently formulated info-dynamic theory.

Rather, events in code-dual systems are individualized to an extent that makes analysis based on quantitative measures (such as averages or concentrations) inadequate. And because of the incredible combinatorial power of digital coding systems, organic evolution emerges as an open, semiotic, and (in a deep sense) unpredictable system. This openness cannot be captured in the infodynamic description that instead operates on the idea of *inherent attractors* in our universe as the ultimate directional powers behind the evolutionary process.

To understand evolution, I shall claim, requires a biosemiotic conceptual frame.

Acetabularium acetabulum

The unicellular algae *Acetabularia acetabulum* is an absolutely singular creature. Biologist Brian Goodwin (1995) spends almost a whole chapter discussing this animal in his book on the principles of morphogenesis in biological systems, *How the Leopard Changed Its Spots*, and as can be seen in Figure 5.1 this unicellular organism has a life cycle that brings it through a sequence of distinctive forms. The full-grown alga is three-to-five centimeters long and its bottom consists of a root-like structure (called the *rhizoid*) by which it attaches itself to rocks in shallow, subtropical waters. This rhizoid is the zone of the cell where the nucleus, and thus the genome, is localized. Through an outgrowth from this area, a stalk is formed, and at the other end of this stalk, there is a graceful cap with a diameter of about one-and-a-half centimeters. An analysis of this gigantic cell throws an interesting light upon the relation between genes and cellular structure.

If an experimenter makes a cut through in the middle of this stalk, the lower part (with the genome) will regenerate a new cap identical to the old one — and it will continue to do so, even if the experiment is repeated many times. The upper part, on the contrary, will only live for a few weeks longer and will then die. This result corresponds well to conventional wisdom, since the information about morphogenesis is thought to reside exclusively in the digital code of the genome. If, however, the algae is cut through in *three* parts — rhizoid, stalk, and cap — the cap and the rhizoid will again behave as already described, but now the stalk exhibits a capacity to regenerate a new cap. Still, under these conditions, the stalk cannot repeat this performance and will also die after some time.

The conclusion to be drawn from this experiment is clear enough. The morphogenetic process is *not* the direct product of the genes alone, since, in the stalk, it continues to go on without the genes. The stalk, on the other hand, is dependent on gene *products* (proteins) that can only be produced when a nucleus is present (i.e., only in the rhizoid). When the stalk has used up its store of protein resources for the morphogenetic process, this process comes to a standstill. The *digital code*, thus, is necessary for the continuation of life, because it is needed for the formation of new proteins — but the *analog codes* in the cytoplasm of the stalk are responsible for the concrete execution of morphogenesis.

As is depicted in Figure 5.1, the life cycle of *Acetabularia* begins with the fusion of two *isogametes* thus named because they are all of the same size and structure (*iso*=alike, *gamete*=germ), contrary to the relatively tiny spermatozoa's and huge egg cells found in species with sexual differentiation. The product of this fusion is called a *zygote*, and when this zygote has been formed, it will slowly sink to the bottom where, thanks to a gluey substance excreted right

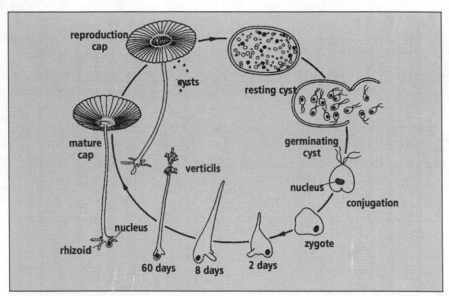

Figure 5.1 The life cycle of *Acetabularia acetabulum* showing the main stages of development from zygote formation through stalk growth to the formation of the mature form, followed by cyst formation and release of isogametes as the cycle starts again (from Goodwin 1994, 75, with permission).

after fertilization, it will attach itself to whatever rock it happens to land upon. From the bottom of the otherwise symmetrical zygote is now formed a little outgrowth destined to become the *rhizoid zone*, and from the opposite end the *stalk* will appear. The difference in light between the two ends of the zygote is the releasing factor for this differentiation process. Goodwin explains, "There is something about the internal dynamic organization of an egg that makes spherical symmetry unstable, so that any perturbation, an internal fluctuation of ions, say, or an external stimulus, will get it started. What light and the other stimuli do is influence where the axis will appear, but it is not the cause of axis formation itself. It is just a trigger that initiates something that is poised and ready to go, like a sprinter at the start of a race" (ibid., 41).

The self-organizing semiotics of the cytoplasmic structure is thus in control of the morphogenetic process right from the earliest beginnings of new life. The protein specifying resources of the genome are needed for this process to succeed, but considered, as it so often is, as a self-sufficient *blueprint for ontogenesis*, something very essential is missing. Digitally coded descriptions are, as already stated, only partial: they support the journey of the zygote through its life cycle in much the same way a musical score supports the work of a symphony orchestra. In both cases, the digital codifications must be there, but they are not themselves in charge of events (and, in fact, can themselves *do* nothing at all, as they are both inanimate). Certainly, they do not specify their own *interpretation* in the real world of spatio-temporal continuity. This is where living, analog codifications must take over. (A noteworthy difference to the orchestra metaphor, it should be pointed out, is that — unlike the performance of a symphony orchestra — the ontogenetic process doesn't need a centralized conductor. The bodily semiotics of cellular interaction is in this regard considerably more self-sufficient than the semiotics playing out among musicians in an orchestra.)

The Semiotics of Reproduction

It is this tight interdependence of the digital and the analog that the geno-centric perspective systematically overlooks. Reproduction does lead to changes in the gene pools of populations — but these changes do not, as such, tell us much about what goes on at the locus where organisms actually live and die. Here, we find that the concrete result of reproduction is, as we saw, the formation of new cohorts of genomes, each of which is embedded in the nucleus of an actual zygote. It was just these cohorts that Claus Emmeche and I tried to make visible by giving them a designation of their own, *kimflok* — from the Danish, *kim* = germ, *flok* = crowd (Hoffmeyer and Emmeche 1991). The result of reproduction, as seen in the light of evolution, is simply the formation of ever new unique kimfloks — for the term *kimflok* both includes the concept of life cycle at the population level, and constitutes the basic temporal unit of a lineage (the generation).[2] And it is at *this* level that the process that is retro-actively called *natural selection* takes place — for it is not *nature* as an abstract entity that selects.

Darwin was careful to underscore the fact that that natural selection was not similar to artificial selection as known from breeding stations, for natural selection is a selection without a selector. Nature has no overseeing breeder. The very expression,

> *natural selection*, thus contains a semantic contradiction that may well be the main source for the continued confusion that Darwinism has occasioned — inside as well as outside of biology. On one hand, science vehemently maintains a conception of nature that makes it nearly unthinkable that anything like value could possibly be a part of it. And then, on the other hand, scientists easily talk about this odd concept, *natural selection*. But how could one (i.e., the presumed subject of the term, *nature*) select without somehow basing this selection on criteria of one sort or other, and thus on some underlying system of *value*? Is it possible that the extreme metaphoric strength of the designation *natural selection* is due to its successful concealment of this blatant semantic contradiction — one which in our unconscious imagination (and in our emotional fundament) endows the concept with causal and volitional capabilities that, as a matter of fact, we perfectly well know that it does not possess?

Since most of us in biosemiotics do not share the ordinary scientific reluctance to deal with the eventual occurrence of value as a genuine relation in the workings of the natural world, we don't have any troubles in placing the value where it belongs, which is at the level of the code-dual system, the lineage. It is the lineage — seen as a historical and transgenerational subject[3] — that acts as the selective agent via its overall reproductive patterns. By virtue of the genetic texts carried forward by its individual members (i.e., the kimflok) the lineage is maintaining — and continuously updating — a selective memory of its past that in most cases will be a suitable tool for dealing with the future.

To say this differently, the lineage's present is evaluated relative to an internal standard based on past successes — and the logical structure of such a process is very much like the logical structure of the human act of selection. When

---

2   The word *lineage* here is used in the conventional sense of a succession of ancestors and descendents, parents and offspring.

3   *Populations* are, contrary to *lineages*, coterminous with the concrete individuals living here and now. The *lineage* is a *historical subject* and has a kind of *collective agency* as such, since its destiny as a temporal integrative structure is continuously influenced by the collectivity of interactions of its single units with their environments — much in the same way that multicellular organisms are integrative structures interacting with their surroundings via the activity of individual cells.

a teacher, for example, selects the boys for the class football team, he naturally bases his selection upon a set of values that reflects his expectations of how to optimize the chances of winning matches. The present (the set of pupils at hand) is evaluated relative to the past, as reflected in the teacher's previous experiences coaching football games. So if we accept the metaphor about selection *in* nature, the lineage (i.e., the spatio-temporal continuum of changing kimfloks) must be the entity occupying the logical position of the selector in the process.

The activity of the kimflok in its dealings with a variety of conditions will be reflected in the overall outcome of reproduction and, thereby in the geno-type pattern of the next kimflok. The lineage thus, by way of its kimfloks, inces-santly scrutinizes the ecosemiotic niche conditions and through this activity automatically brings about the formation of new interpretants in the form of a concrete reproductive pattern as manifested by the next generation's kimflok. Seen in this way, it turns out that natural selection is at heart a semiotic process, whereby single organisms — both enabled and constrained as they are in their activity by their belonging to the overarching project of the lineage — act as semiotically competent agents in the survival game of the code-dual system.

Seen in this light, it follows that *speciation* (the processes whereby new species come into existence) does not necessarily have to be a gradual process whereby new hereditary material (e.g., mutations) slowly accumulate in the gene pool of a population. It might as well, or perhaps even more likely, occur via the estab-lishment of new intra-specific *interpretive* patterns — as I will be discussing in Chapter 6.

In our discussion so far, however, we have not focused on cases of asexual reproduction such as budding, root suckers, or offshoots. But even in spe-cies with widespread vegetative reproduction, there is most often some point in the life cycle where sexual reproduction does take place. Apparently, sexual reproduction results in a considerably more extended reshuffling of the genetic material inside the population. And from a semiotic point of view, sexual repro-duction is particularly interesting because a variety of subtle semiotic interac-tion patterns have evolved to make sure that relevant mating partners (or their sex cells) are recognized and attracted while less relevant mating partners are ignored or rejected.

This is the case, for example, in two species of tree-living crickets with closely overlapping habitats in the eastern United States: *Oecanthus quadripuncta-tus* and *Oecanthus nigricornis*. In general, male crickets attract females by the sounds of the "songs" that they create while rubbing their specialized forewings against each other, producing series of fast pulses that endure for several min-utes at a time. Yet the members of these two species reproduce at the same time of the year. It is therefore important that females do not confuse the songs of their own males with the songs of males from the other species. The key to this problem turned out to be pulse rate; the songs of *O. nigricornis* had a much higher pulse rate than the songs of *O. quadripunctatis*. The pulse rate is augmented in both species when temperatures are rising, but this doesn't bother the females — rather, they merely shift their preferences accordingly, and now are safely attracted to even faster pulse rates (relative to one another) than before (McFarland 1987, 339–60).

It is alpha and omega for the efficiency of sexual reproduction that the two sexes find each other and mate at the right time, and the literature is actually bulging with descriptions of intricate semiotic mechanisms to assure this. Let me here just mention the surprising case of cooing in ring doves of the species *Streptopelia risoria* (Cheng 1992) that we have elsewhere described as follows:

> Before a female ring dove lays her eggs, she and her mate go through a series of courtship displays. As courtship proceeds, hormonal changes in the female trigger the growth of follicles in her ovaries, each of which eventually bursts to release an egg. Recent experiments have shown that if a female dove is operationally hindered in making the so-called "nest coo," she will not be able to ovulate, despite enthusiastic courting by males. In control experiments, tape recordings of nest coos were played to females with no males present. Now follicles began to grow. The conclusion is simple: Female doves that coo during courtship are not (only) cooing at the male — they are cooing at their own ovaries, to trigger the release of eggs.

> One can only guess, of course, how such a strange mechanism has actually evolved. But at least it seems to indicate, that the set of habits involved in courtship and thus the cooing was in fact more safely correlated with the timing of mating, than any simple endogenous release system would have been. The obvious although speculative explanation is that the cooing behavior is a measure of the state of the relation between two birds, which is superior to a simple measure of the state of the female organism itself (Hoffmeyer 1998c).

The ring dove example is only one of probably millions, but it highlights how the subtlety of synchronizations and chemical communication that must be reciprocally tuned between organisms involved in social behavior is stunning — yet scientists at present are only scratching the surface of this new area of biosemiotic study.

Epigenetic Heredity

It is a strangely overlooked fact that genes are not the only important hereditary contribution from one generation to the next. Environmental conditions and the social heritability of learned behaviors often prove to be just as vital hereditary factors, insofar as the very viability of the offspring will often depend critically on both the parental choice of breeding place and parental care in general. But there is also another kind of heredity that is directly connected to reproduction and yet is not caused by genes. This kind of heredity is called *epigenetic heredity.*

One of the first to track this kind of heredity was the American biologist Tracey Morton Sonneborn (1905–81), who for most of his life worked on the little unicellular protozoa *Paramecium* (Figure 5.2). In *Paramecium,* reproduction is mostly vegetative: cells simply divide into two daughter cells. At intervals, however, *Paramecium* will enter a process called *conjugation,* where two *Paramecium* cells create a cytoplasmic bridge between them. They then use this bridge for an exchange of haploid (half chromosome) *micronuclei* — a process that formally corresponds to sexual reproduction.

Sonneborn noticed that some *Paramecium* strains, which he dubbed *killers,* produced a toxic compound that might kill other *paramecium* strains, and, through breeding experiments, he showed that the competence to produce this toxic compound was dependent on genes in the cell's nucleus. But at the same time, his experiments showed that a cytoplasmic element called *kappa* was involved. This cytoplasmic element had the effect of making the killer strains resistant to their own toxic compound. When *kappa* was lost from the cells, those *Paramecium* strains would then become sensitive to the toxic compound. Many years later, it was learned that *kappa* was, in reality, a symbiotic bacterium capable of sustaining life inside the cytoplasm of *Paramecium* (Preer 1996).

This phenomenon is called *endosymbiosis* and occurs in many species. In fact, all modern cells, called *eukaryotic cells,* are endosymbionts — since it has been shown that important cellular organelles like mitochondria and chloroplasts (and possibly several others) are all but certainly descendents from once free-living bacteria that have now lost the ability to survive and reproduce outside of a eukaryotic cell (Margulis 1970).

The inheritance of cytoplasmic *endosymbionts* may at first seem to be a somewhat unusual kind of cytoplasmic *inheritance,* yet it may possibly be quite widespread. Sonneborn, certainly, showed genuine cases of cytoplasmic inheritance in *Paramecium,* and his discovery is worth considering in detail here.

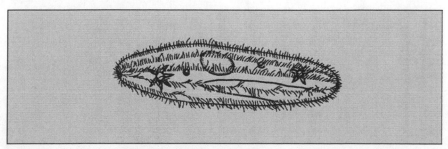

Figure 5.2 *Paramecium.* The surface is covered with rows of cilia (from Goodwin 1994, 7, with permission).

As can be seen in Figure 5.2, the surface of *Paramecium* is covered by *cilia* — hair-like protrusions that can move around like small whips. There are hundreds of these cilia sitting in long rows from the front to the back of the organism, and thanks to the coordinated movement of all these cilia, the *Paramecium* can move freely around in the water. Eventually, Sonneborn discovered that one of his *Parameciums* had an unusual stripe that, upon closer scrutiny, turned out to be caused by an inversion of one of the rows of cilia. Sonneborn then showed that this new trait (dubbed *melon stripe*) was heritable.

The really interesting discovery, however, came when he tried to see what would happen if he operationally turned around one of the rows of cilia in a normal *Paramecium*. Surprisingly, he found that in this case the new trait was inherited by daughter cells, even though here the genes were untouched by the intervention. What happens in this case is that when a *Paramecium* is extending its surface structure in preparation for the division, it uses the *already existing structure* as a template. New molecular units are assembled using the model of the old surface, so that the inversed row of cilia will automatically be copied to the daughter cells. So here again we find, as we did in the above-mentioned case of *Acetabularium,* that the concrete 3-D morphogenetic processes are not exclusively predetermined by the genome, but are controlled equally (if not more so) by extragenetic factors in the cytoplasm or membrane structures.

This should not surprise us, really, as the cells in a multicellular organism are often very different from one another and only a very limited fraction of the genes is actually transcribed by any one of them. Early on in embryogenesis, almost all cells loose their *totipotentiality* — i.e., their potential to become virtually any kind of cell at all. A *determination* (more properly, a series of determinations) takes place whereby the cells, bit by bit, become more and more specialized with each divison, so that after a certain point in time, a given cell can no longer become, say, a liver cell, but only a nerve cell. Sophisticated mechanisms exist to make sure that once such a determination event has occurred, it will be conserved in that cell line.[4] Thus, the Italian embryologist Marcel Barbieri has observed that in addition to the genetic DNA memory, there also exists an epigenetic cell memory (Barbieri 2001, 109).

The mechanisms that bring about this stepwise epigenetic cell memory are not well-known, but they depend, among other things, on enzymatic modifications of the *chromatin* — i.e., the complex of proteins and DNA whereby the DNA double helix is fixed into a more or less compact 3-D superstructure. DNA sequences that end up inside this compact superstructure are quite inaccessible to the enzyme complexes involved in transcription. Thus, according

---

4   A *cell line* consists of any given cell and *all* the daughter cells that ever derive from it.

to the current view, genes that are not supposed to be transcribed in a given cell line may perhaps be "put away" — coiled into those compact regions of the DNA (see Figure 5.3).

The classic example of such *chromatin marking* is what is called *parental imprinting,* whereby one of the two X chromosomes is permanently inactivated in this way. Specifically, mammalian females are born with two copies of the X chromosome, one inherited from the father and the other from the mother, but one of them — either the father's or the mother's — is always inactivated right from the beginning by chromatin marking. This inactivation is then passed on to the daughter cells so that in the end, the X chromosomes of all the cells in the body will be of either maternal or paternal origin.

In addition to chromatin marking, there also occurs a widespread methylation of the DNA — an enzymatic process whereby a methyl group (CH_3) is attached to specific "letters" (bases) in the DNA. This methylation seems to be an essential mechanism for the tissue-specific inactivation of distinct genes throughout embryogenesis — and thus for the determination process whereby different cell lines in the growing embryo end up as different types of tissue. How cells manage to re-methylate newly formed chromosomes at the exact right positions after each cell division is not known. Yet, so they do.

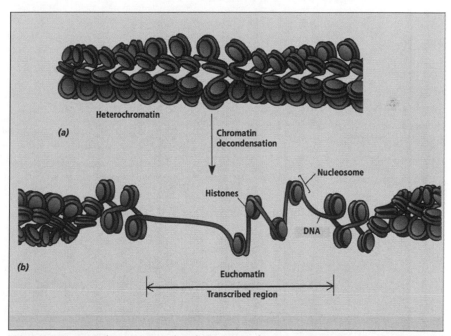

Figure 5.3 Heterochromatin is the designation for inactive regions (nontranscribed regions) of the DNA that are coiled up into a compact structure (modified from Solomon, Berg and Martin 1999, 301).

It was long believed that epigenetic inheritance systems only applied to somatic cells, and that the meiotic division leading to the formation of gametes (haploid sex cells) would wipe the slate clean (i.e., remove methylations and chromatin markers). But increasingly this fine picture is starting to crack. At least in plants, chromatin markings may sometimes survive meiotic purification to influence the expression pattern of the genes in the next generation — a phenomenom called *paramutation* (Hollick, Dorweiler, and Chandler 1997; Sternberg 2000).

From a theoretical point of view, it might be noted that what was not too long ago characterized as the *self-replicating* and deterministic *entity* of the genome has increasingly become understood to be a changeable and flexible point in an ongoing *process*. Yet, after all, what should have hindered nature in profiting from the obvious possibility for evolutionary innovation that springs from chromatin reconfigurations' vast potential for inheritance?

Chromatin reconfigurations exhibit an interesting case of *analog coding* that supersedes or builds upon a prior digital codification. The analog coding of the DNA superstructure doesn't *destroy* the underlying digital coded specifications. Rather, it (in effect) *plays* with them, tentatively organizing them into new patterns of decipherment. And only time and investigation will show how far such mechanisms have been realized in living systems.

The Genome

The genome, i.e., the *totality of genetic hereditary material* in a cell of a given organism, is an impressive piece of chemistry. In eukaryote cells, the genome is divided into a number of distinct units (chromosomes) each of which is constituted by a linear DNA molecule (a double helix) looped around spherical complexes of protein units (with a diameter of eleven nanometers) called *nucleosomes*. Certain domains of the helix (the *heterochromatin*) are packed into an even denser *hypercoil* structure with a diameter of thirty nanometers (for comparison, the diameter of the DNA helix itself is only two nanometers). Yet, in spite of the compact folding, it remains fairly astonishing that approximately three meters of DNA can be packed into a cellular nucleus with an average radius of no more than one one-hundredth of a millimeter.

Only a minor part — some 5 percent — of the genome is occupied by actual genes in the form of DNA sequences that will eventually be transcribed into proteins, and it is far from clear what the rest of the DNA is good for. Yet as we have already seen, there are some regulatory potentialities connected to the macrostructure of chromatin which might contribute to the need for an abundance of DNA. Perhaps there might also be unknown tasks of DNA connected to cellular division. Furthermore, some nongenetic areas of the DNA are taken up by *pseudogenes* that have interesting potentials for the evolutionary process, as we shall see later in this chapter. In any event, research on DNA sequences and structure proceeds extremely rapidly in these years, and answers to such questions may soon be forthcoming.

What has been learned recently is germane to our discussion at hand, for biochemists have been surprised to find that there is very little difference among taxonomic groups concerning the total amount of DNA per cell. Apart from bacteria (which are prokaryotic cells), and yeast and fungi (that clearly have less DNA), there is no systematic increase in the amount of DNA per cell to be found amongst the other main groups of living systems. A toad, for example, has more DNA than a human. And even though the average content of DNA in insects is lower than it is in mammals, it is easy to find insects that have much more of "the glorified compound" than we have ourselves. One could perhaps think that this apparent disproportion might be due to the low percentage of protein-coding DNA, but as we shall see, the numbers don't get much better if we shift to a comparison of the content of the *true genes*[5] across the groups.

Considered as a digital code, DNA consists, as is now well known, of sequences of four nucleotide bases, conveniently abbreviated to A, T, G, and C. Thanks to the *base pairing mechanism*, whereby A and T, and G and C, respectively, are paired with each other through weak forces (hydrogen bonds), one may — at least in principle — explain how new strings of DNA or RNA can be formed as replicas of the already existing strings. (It should be noticed, however, that DNA molecules, when seen from the standpoint of biochemistry, consist of electrically loaded three-dimensional atomic *shapes*, rather than just of *sequences*.)

When considered at this level, the difficulties connected to explaining how a gene may be transcribed to RNA and later translated to protein are conspicuous. To take just the first of these two processes, it involves, first of all, the disentangling of the DNA string from the chromatin complex it is bound to in order to allow the enzyme RNA polymerase II (which catalyzes the formation of the primary RNA transcript as a copy of the DNA) to get past the nucleosomes. The DNA helix must then be cut through at the exact place where the gene is situated and the two strings must be physically unwound from each other to keep pace with the movement of the enzyme down along the DNA strings.

The accomplishment of these tasks presupposes a precisely coordinated spatio-temporal interaction between the active surface domains of many proteins and transcription factors — as well as the precise spatio-temporal interactions

---

5 As we saw in Chapter 4, there is no simple definition of a gene — thus, by *true genes* in this context I simply refer to the biochemical genes as traditionally defined: those sequences of DNA nucleotides coding for a protein.

between transcription factors and the specific area of the surface of the DNA helix. Sometimes this implies a cooperative interaction between elements of the DNA molecule that may be separated from each other by up to a thousand base pairs (see Figure 5.4).

The formation of the initiation complex itself (Figure 5.4) is semiotically controlled in ways that are yet only barely understood. First and foremost, a range of extracellular signals may eventually be bound to receptors at the cell surface, and thus release an intracellularly mediated activation or inhibition of the transcription factors. In some cases, hormones can pass directly through the cell membrane and take part in the process via their bindings to transcription factors — while proteins released through transcriptive processes at other localities at the DNA molecule may inhibit or enhance the formation of the initiation complex.

The reason why I have been dragging the reader with me through the minutiae of biochemical complexity involved in the regulation of transcription has

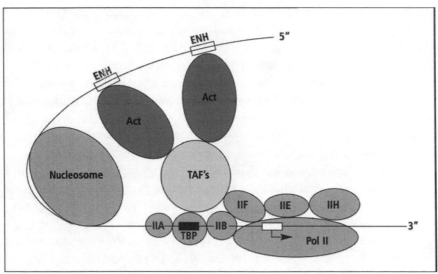

Figure 5.4 Initiation complex formation for a structural gene in a eukaryote cell. The DNA thread is seen to loop around a nucleosome (an aggregate of proteins) so that in one direction (towards the 3'-end) it is bound to a transcription complex consisting of the transcription enzyme polymerase II (or, POL II), and a range of different transcription factors. Among these should be mentioned the TBP (TATA-binding protein), that bind to the sequence of the DNA string called the TATA box, which defines the locus for initiation some twenty to thirty base pairs away (here indicated as a small white rectangle encircled by POL II). Thanks to the looping around the nucleosome, other sequences of the DNA string, called enhancers (ENH) — separated from the TATA box by perhaps several thousands of base pairs — are brought close to the initiation complex, which they may possibly influence through the mediation of regulatory proteins (TAFs and Act) (Modified from Mathews, van Holde and Ahern 1999, 1094).

been to make it clear that digital codes necessarily are extremely complex — and that their *modus operandi* is dependent on difficult and energy-consuming systems. Again, were it not for the fetishization of the gene as a kind of self-replicating controller (a thoroughly incorrect conception on every count, as I argued in Hoffmeyer 1996b), none of this should be surprising. Indeed, we can recognize the true dynamics of the process from our own experience with such codes. It usually takes a child approximately two years to learn the digital code of spoken language and — as observed by the neurobiologist Terrence Deacon (see Chapter 8) — this *always* takes place in a strongly supportive linguistic milieu including other active interpreters — in the absence of which, one might well doubt that a child could learn to master the art of linguistic reference at all.

Similarly, when the concrete spatio-temporal complexities involved in digital-analog recoding processes are considered in detail, it becomes much harder to accept the everyday jargon of *genes for* this or that trait in human psychology (e.g., promiscuity, alcoholism, and xenophobia). For as soon as one tries to invent concrete biochemical models to explain how a fraction of a molecule (here, DNA) could possibly be *the* cause of such disparate and long-distance effects, the conceptual difficulties begin piling up exponentially. Recently, the biochemist and philospher Lenny Moss has suggested that we distinguish between two different concepts of the gene. Moss calls these concepts *gene-P* and *gene-D* — where *gene-P* is the ordinary phenotype-related gene concept, such as the gene for blue eyes (or for a distinct defect of the drosophila wing), while the *gene-D* concept refers to the gene perceived as a transcriptional unit, inside of which is contained a molecular template serving as a resource base for the production of *gene products* (RNA molecules or proteins). This is the gene that draws our interest from an ontogenetic point of view, as D genes are determined by their molecular sequence of nucleotide bases, but undetermined with respect to their phenotypic effects.

Now, it could happen that occasionally a P gene and a D gene may be one and the same — but such a case would turn out to be a lot rarer than many people would expect (especially given the widespread tendency to mistakenly think of D genes *as* P genes). For example, the gene for *cystic fibrosis* may appear at first glance to be a gene-P — but this gene-P would in most cases, as we now know, be related to a gene-D that describes the sequence of amino acids of one of a family of proteins that are used for trans-membrane ion channels (Moss 2001). While gene-Ds are often molecularly well defined, gene-Ps are most often postulated from very indirect evidence (e.g., purely statistical evidence) and the confusion of the two gene concepts into just one concept of the gene tends to conceal the vague status of those gene-Ps (e.g., "the gay gene") that so often appeals to the public imagination — and that thereby spreads the general misunderstanding that genetics has a much firmer grip on human psychological traits than in reality it has.

For almost a century now, genetics has been focused almost exclusively on the *transmission* aspect — i.e., the question of how the genes become transmitted and their frequencies changed down through the generations — at the expense of the *functional* aspect — i.e., the question of how genes actually bring about those traits of the organism that they are claimed to cause. It is only now — now that we are finally beginning to get a clearer picture of our human genetic constitution through a number of big mapping projects like the Humane Genome Project — that we are gradually becoming aware that the gene does not fill its not-too-long-ago supposed role as the ultimate (bordering on *magic*) explanation for biology.

Caenorhabditis elegans

In a way, the simplistic picture of genes as unambiguous determinants for phenotypic traits was already beginning to crack as work progressed in mapping the relation between genes and behavior in the little one-millimeter-long nematode *Caenorhabditis elegans* (usually called *C. elegans*, since almost nobody remembers this long name). In the mid-sixties, Sydney Brenner — one of the leading scientists behind the detection of the genetic code — selected *C. elegans* as a model organism for the genetic studies of behavior, on the basis that its short life cycle and small size (less than one thousand cells in total) made it an excellent candidate for genetic analysis. Brenner's ambition was to make a complete mapping of the structure of the nervous system in this worm, and then to relate this map both to the organism's genes, as well as to its behavior. Brenner (1974, 72) optimistically put it this way: "In principle, it should be possible to dissect the genetic specification of a nervous system in much the same way as was done for biosynthetic pathways in bacteria or for bacteriophage assembly" (cited in Schaffner 1998) — and this quote may well serve as a prime example of the reductionistic credo that dominated biology in the second half of the last century. That the *dynamics* of the nervous system should, at least in principle, be "inscribed" in the genetic machinery in the same way that metabolic processes are in bacteria, was the basic premise for the project. But Brenner saw that if one was going to prove this, the fruit fly — with its hundred thousands of nerve cells — would already be much too complicated. In *C. elegans,* on the other hand, there is, as we know now, exactly 302 nerve cells.

Today, forty years later, it can safely be said that *C. elegans* is the most fully scientifically described multicellular organism on Earth. More than two hundred research teams — or approximately one scientist per cell — have helped in studying this little worm. Every single cell has been named and described and even all of the approximately five thousand synaptic connections between nerve cells are known. (These synapses are uniform from animal to animal, but not totally identical with one another.) Moreover, a complete map of the organism's entire hereditary material has been constructed, partitioning its 19,900 genes in the precise sequence that they appear upon its mere six chromosomes (Schaffner 1998).

The project has been closely followed by the philosopher of science Kenneth Schaffner, and here we shall stick to the thorough review of the project that Schaffner (1998) gives in *Philosophy of Science.* An important aspect of this work concerns the response of the worm to various chemical compounds — that is, taste and smell and the behavior connected to these senses. As intensely studied by Cornelia Bargmann and her team, *C. elegans* is able to distinguish between at least seven classes of compounds which either attract it or repel it. Only two pairs of sensory neurons are engaged in this task, and only approximately twenty genes were shown to be necessary for the normal functioning of these neurons. Still, however, it has not been possible to establish any simple *correspondence* between gene and behavior in this case. In 1993, Bergmann's group reported the following:

> One way to identify genes that act in the nervous system is by isolating mutants with defective behavior. However, the intrinsic complexity of the nervous system can make the analysis of behavioral mutants difficult. For example, since behaviors are generated by groups of neurons that act in concert, a single genetic effect can affect multiple neurons, a single neuron can affect multiple behaviors, and multiple neurons can affect the same behavior. In practice these complexities mean that understanding the effects of a behavioral mutation depends on understanding the neurons that generate and regulate behavior (Avery, Bargmann, and Horovitz 1993, 495, cited in Schaffner 1998).

Based on the available evidence, Schaffner distilled eight general principles, or rules, said to account for all the possible relations between genes and behavior in "this extraordinarily well-worked-out simple organism" (Schaffner 1998, 220), and he sums up these rules in the table given here as Table 5.1.

Under Schaffner's conception, these eight rules — *rules* that, in fact, only generalize the already known principles of genetic pleiotropy,[6] genetic interaction, neuronal multifunctionality, plasticity, and influences from the environment — ought to count as standard assumptions for the continued study of the relations between genes and behavior in more complex organisms as well.

Yet most interesting to note in a biosemiotic context is the fact that, even in this extremely simple organism, one cannot completely explain behavior without also considering the aspect of *learning.* Thus, for instance, *C. elegans* exhibits a remarkable plasticity in its behavior when it is starved, or if the density of the populations grows too great. "Water-soluble chemicals that are strong attractants to naive animals are ignored by crowded, starved animals," writes Sengupta

---

6   *Pleiotropy* is the designation for the phenomenon whereby one single gene has an effect of several different phenotypic traits.

Table 5.1 Rules for the relation between genes and behavior in C. elegans (modified from Schaffner 1998, 301).

1.	many genes ! one neuron
2.	many neurons ! one type of behavior
3.	one gene ! many neurons (pleiotropy)
4.	one neuron ! many behaviors
5.	stochastic development ! different neural connections *
6.	different environments/histories ! different behaviors *
7.	one gene ! another gene ... ! behavior (combinatorial effects)
8.	environment ! gene expression behavior

The ! can be read as "affect(s), cause(s), or lead(s) to"
* in prima facie genetically identical (mature) organisms.

et al. (1993, 243), adding that "these changes induced by crowding and starvation persist for hours after the worms are separated and fed" (cited in Schaffner 1998). It has furthermore been shown that if *C. elegans* is exposed to fatigue or diminished food intake during its development, it may enter a *dauer state* — meaning that the larval phase is extended, so that the worm may survive for four to eight times longer than it would be able to in the normal (post-larval) state. And pheromones emitted by the worm itself are implicated in the process.

One interesting aspect of *C. elegans*'s biology that Schaffner does not himself address is that the worm is a very "brainy" creature: nearly one third of its cells are nerve cells! This is significant, because it takes a lot of energy to run a nerve cell. Humans at rest, for instance, spend 18 percent of their metabolic energy just to maintain the nervous system — even though it occupies a mere 2 percent of their total body weight. The burden of maintaining a brain is relatively more demanding in small animals like birds and monkeys, since the brains of those animals make up nearly 10 percent of their body weight. Brain size, in fact, most commonly correlates with the social and behavioral complexity that the animal has to cope with, and only reflects body size to a limited extent. It simply takes a certain amount of neurons to coordinate a given activity such as breathing, running, or singing — and this burden can often be so high that the brain size of an animal species will be limited by the amount of food energy the species typically is able to procure for its progeny (since early brain growth makes such particularly high demands on the access to metabolic energy). The evolution of suckling in mammals some two hundred million years ago, therefore, was very likely a decisive prerequisite for the later development in these animals of much bigger brains.

It is therefore no surprise that the number of neurons in *C. elegans* is very accurately controlled. The loss of a single nerve cell in a nematode may make it unable to leave offspring because it consequently cannot lay eggs. But an increase in the number of nerve cells is also potentially lethal, because the increased drain of energy leaves the worm without sufficient strength to leave offspring. Accordingly, *C. elegans* does what nearly all other animals do. It produces far too many nerve cells relative to what it needs and it then gradually kills those cells that show themselves superfluous through the process of programmed cell death or *apoptosis* (which I will discuss below). Neurons that do not yet exhibit sufficient synaptic activity after a certain period of time has elapsed are treated as superfluous, and only those neurons that have proven to be functional are kept alive. The integration of this communicative logic into the general growth dynamics of the larva apparently is so efficient that the number of surviving neurons invariably ends up as exactly 302. It is worth noting that this *apparent* determinism does not depend on rigid genetic predetermination, but results from an organizational principle rooted in the semiotic interplay of neurons in the growing brain, in their ever-changing contextual situation. We shall call this organizational principle *semiotic emergence* and examine it in some detail in Chapter 7.

For now, it will suffice to consider what Schaffner calls "the principal take-home lesson" in his summation of the analysis of the *C. elegans* project — the realization that "genes act in complex interactive concert and *through* nervous systems, systems that are significantly influenced by development, and exhibit short- and long-term learning that modifies behavior. The environment plays critical roles in development and also in which genes are expressed and when. Characterizing simple 'genes for' behaviors is, accordingly, a drastic oversimplification of the connection between genes and behavior, *even when we have the (virtually) complete molecular story*" (Schaffner 1998).

In short, it seems that the reductionist credo that Brenner expressed in 1974 — "In principle, it should be possible to dissect the genetic specification of a nervous system in much the same way as was done for biosynthetic pathways in bacteria or for bacteriophage assembly," a credo that apparently has nourished the extraordinarily generous flow of funding to such gigantic research programs for the last forty years — was, it turns out, not only a gross oversimplification, but was false. The dynamics of the nervous system cannot reasonably be said to lie inscribed in the genes in the same way metabolic processes in bacteria are inscribed in *their* genes. And to get to the bottom of what really may be going on here, such simplistically causal assumptions and methodologies are going to have to be replaced with much more sophisticated ones, as I shall argue at length below.

The Human Genome

The assumption that a map of the human genome could be equivalent to the book of life or even to the holy grail of developmental biology is a message that has been industriously circulated by both grant-seeking scientists and story-seeking journalists. Again and again, we have been told that our genes are the hidden keys to our psychic and physical well-being or lack thereof. And still — long after most scientists have abandoned their initial overestimations for the project — newspapers regularly announce in big headlines that now scientists are close to identifying this or that gene for major health threats such as schizophrenia, bipolar disorder, alcoholism, or even cancer and vascular disease (in addition to even more outrageous claims linking genes to such nonbiological phenomena as racism, musical talent, and likability).

And the message has gone through. For now a large majority of people in the Western countries unquestioningly believe in this myth. Apparently, the general public did not see the significance of a February 11, 2001 news release issued by the scientists behind the Humane Genome Project — the project that for over a decade had been engaged in mapping the humane genome, gene by gene. The announcement made that day revealed that not only did the sum total of human hereditary material contain thirty thousand genes (or less than one-third of the expected number) — but even more striking, that it was found that most human genes are not unique to our species at all, but are also identically present in other species. Indeed, it now turns out that only a few hundred human genes will *not* also be present in closely related forms in other mammals, such as the mouse. Craig Winther, president of Celera, one of the leading forces involved in the Humane Genome Project, admitted, "This tells me that genes can't possibly explain all of what makes us what we are" (Strohman 2001). With as few as ahundred unique genes at our disposal, one does not need a deep study of the catalog of human shortcomings to conclude that there cannot possibly be one diseased gene for every sin.[7]

The hegemonic status of "DNA-ism" probably owes a lot to some rare but nasty diseases, such as muscular dystrophy, that do in fact depend on a deficiency in only one gene. The disease gene in this case specifies a deficient protein, *dystrophin*, which muscle cells cannot deal with in a normal manner, and so the carriers of the *dystrophin* protein-producing gene become seriously ill. But monogenetic diseases such as muscular dystrophy are atypical. Indeed, they make up less than 2 percent of our total human disease load. To the extent that genes are "responsible" for major disease patterns, many more than one gene in isolation are always involved — *and* these gene complexes form part of a lifelong *dynamic interplay* with the body and its myriad of ongoing, nongenetic processes (Strohman 2001).

Moreover, it has been shown that even a clearcut monogenetic disease like PKU (phenylketonuria) exhibits a considerable range of variation as it is manifested at the phenotypic level. For instance, not all patients suffering untreated

7   Perhaps, rather, there is one for each human *virtue*, as Claus Emmeche has playfully commented.

PKU will demonstrate any disturbed cognitive development. This is so because the concentration of *phenylalanine* in the brain is dependent on a diversity of factors that influence the efficiency of the blood-brain barrier — and so not all patients' brains reach toxic levels of the nondegradable amino acid (Scriver and Waters 1999).

Given all this begs the question of how a few hundred genes can determine whether an egg will become a baby or a mouse? The explanation must be that genes do not — as previously supposed — *correspond* to distinct functions in the organism. Rather — and both embryological and neurobiological research bears this out overwhelmingly — genes function as signposts in a dynamic interplay with each other and with the network of proteins and membranes in the growing embryo. It is not the genes per se, but their interplay and *interpretation* in the cell that counts.

And again, current research in molecular genetics increasingly shows us that this is so. Noncoding repetitive sequences of DNA are everywhere involved in the coordinative processes that *decide* (in a genuine sense — more on that forthcoming) which genes in a cell will or will not be transcribed at a given time. James Shapiro (1999, 26) has suggested that these repetitive sequences are the parameters that determine the "system architechture" characteristic for different species and he claims that "resetting the genome system architecture through reorganization of the repetitive DNA content is a fundamental aspect of evolutionary change ... [a] resetting process that can occur without major changes in the protein coding sequences."

To illustrate how an understanding of the workings of genes as *signposts* (rather than as *causes* or *blueprints*) facilitates an understanding of why it does not necessarily take many unique genes to assure that a given fertilized egg develops into a human being rather than into a mouse, let us consider the growth of an embryonic tissue layer in the virtual organism *Scitoi mesoib*. We shall specifically observe the dynamic growth of a neural tissue layer called *sirap* that normally is regulated by the concentration of a signal molecule, *sirapin,* which gradually builds up due to its excretion from an underlying embryonic cell layer (Figure 5.5, top and middle). Normally, when *sirapin* reaches a certain concentration inside the cells of the *sirap* layer, further cell divisions — and thus further growth — will come to a stop.

Now, suppose that a mutation hits the gene *sir* which specifies the enzyme (*sirapin synthetase*) that catalyzes the formation of sirapin in the underlying tissue layer. Such a mutation might eventually cause a slight decrease in the catalytic activity of the enzyme, and the mutated embryo would therefore produce sirapin at a slower rate. Consequently, the sirapin concentration in the sirap layer will stay below the threshold value for an extended period and, accordingly, the relative size of the sirap neural tissue will be increased (Figure 5.5, bottom).

Early phase

The formation of a new *sirap*-layer is indicated in dark. Cells in the newly formed layer grow and divide unless growth is inhibited by the signal molecule *sirapin,* which is produced in the underlying layer (light cells)

Late phase – Normal

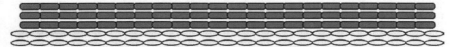

Sirapin production is catalyzed by the enzyme Sirapin Synthetase (ST) and the gene specifying ST is called sir. *Sirapin* will slowly diffuse into the upper *sirap*-layer and block any further growth

Late phase – Mutation

A mutation in *sir* has caused slightly decreased Sirapin Synthetase activity leading to a decreased production rate for sirapin. As a consequence, sirapin concentration in the newly formed tissue only reaches the inhibitatory level at a later time allowing for extended growth of this tissue.

Figure 5.5 Fetal development of sirap tissue in the virtual organism *Scitoi mesoib* (see text).

This simple yet realistic model easily explains how small mutational modifications might eventually provoke severe changes in morphology or instinctual behavior in an individual or a species. Large-scale morphological change may therefore take place without any new genes being involved at all — and in cases where such changes affect the eventual formation of brain architecture, it follows that *new* behavioral patterns may appear simply through the modification of *already existing* genetic material.

Very probably, the addition of new genetic functionality had been a major theme in early evolution — say, for the first three billion years. Yet when those lineages that would become, respectively, *H. sapiens* and *C. elegans* separated some hundred million years ago, most of the fundamental genes that modern species carry were already present. Protein evolution since then has mostly been variations upon the same ten thousand to twenty thousand basic themes, where small adjustments in the properties of proteins have changed the interplay of the settings for embryogenetic semiotic control.

The key here is the idea that the ontogeny of morphological and behavioral traits is secured through a sophisticated mechanism of *semiotic scaffolding* — and thus is only in a very indirect and mediated way, based essentially on genetic scaffolding. For to modify genes is to change the way that different *processes* are "tuned to one another" — and the product that results from such a process bears no simple relation to the mutation itself. From a semiotic point of view, what happens in the *sir*-mutant is a weakening of the *signal* involved in fitting the growth of one cell layer to another. Thus, in the case of the mutant *Scitoi mesoib*, the cells in the sirap tissue layer do not recognize the signal in time, and this delay in the interpretive response has the effect of allowing the growth of this to continue for an extended period of time.

Genes, then, do not specify traits in the adult organism, and in a way they do not even specify proteins. They do, of course, mechanistically determine the amino acid backbones of given proteins (although it should be remembered that a plurality of processes — such as RNA editing, protein folding, and protein targeting — all contextually interfere in the simple causality of this otherwise seemingly deterministic process). However, since proteins are not just molecules, but are always also semiotic tools, *what the genes really do specify is the efficiency of semiotic modulators.* They serve to adjust the screws, as it were, in the biosemiotic machinery of the organism. Therefore, in order to adjust the predominant metaphors by which we think about the relations between genetic regulation and the semiotic reality of life, let me suggest a model of the genetic system as *a computerized inventory-control system* — i.e., a system for ascertaining that the appropriate stock of molecular resources are always present when the cell or organism needs them for its never-ending semiotic interactions. In this inventory-control system, individual genes appear like the ever-present (though largely unseen) application options that only become available for use through the activation of different menu bars in our computer software applications. In these "menued" gene ensembles, there are application options for chromatin structure, application options for the proteins involved in the cell-division apparatus, application options for catabolic and anabolic enzymes, application options for membrane related proteins and other structural components, etc.

Decisions on which application menus should be activated, as well as which options made available through the activation of these "menued" gene ensembles should actually be realized at any given time, will depend on the intricate semiotic *interplay of the total cellular system* as it may be biased in any given moment by the relative concentrations of important modulatory proteins. This context-dependent event is most properly referred to as a *decision* in this sense, for that term highlights the fact there is no one element deterministically causing one thing to happen over another, but that the selection is a *function* of the system state and its needs at any given moment.

In this heuristic, every activated application option functions as a release mechanism whereby the production of a new resource is initiated and stored in the proper department of the cell. Mechanistically, this initiation corresponds to the RNA polymerase reaction whereby a particular gene is transcribed into RNA. As long as the given menu option is activated, the corresponding mRNA production will continue and protein synthesis will go on. But the moment it is interrupted, mRNA degradation will naturally ensue and this will rapidly bring any further synthesis of that particular protein to an end.

Spam

In addition to the operational genes that most people equate with the very concept *genes*, the genome contains huge stores of hidden, never sought-after menu items — what scientists call *pseudogenes* — that potentially may be brought into circulation, and thus may come to play a role in evolutionary changes. Pseudogenes are areas of the genome that are *nonfunctional* in the sense that they are not available to the transcription process — but which nevertheless exhibit remarkable similarity to known *functional* genes in their base sequences.

Thus, before the mid-1970s, DNA was considered a very static molecule. But it has become very clear since then that there exist quite a number of mechanisms whereby sequences on the DNA molecule may be duplicated and even change positions on the chromosome. Such mechanisms have led to the formation of *gene families* — i.e., families of narrowly related functional genes, as well as *pseudogenes*.

Pseudogenes are of evolutionary interest because, unlike functional genes, they are not expressed and, therefore, are not subject to natural selection. Since they do not in any way contribute to the success or failure of their carrier organisms, they are free to undergo mutation without incurring the selectional consequences that normal genes do. (For normal genes mutate just as frequently, but those mutants that express functional deficiencies are accordingly sorted out.) Pseudogenes therefore represent a tacit *resource base* of latent proteins with unexplored properties — from which natural selection may eventually pick up new functional genes.

The *hemoglobin gene family* nicely illustrates this dynamic. Since oxygen diffuses through tissue slowly, animals beyond a certain size must find a way to solve the challenge of getting enough oxygen into their tissues. Insects have solved the problem by developing small air channels (trachea) but this solution only works because insects' body mass is so small. (Conversely, this may be the reason why insects never become big.)[8]

---

8   It is interesting that in the Coal Age, more than three hundred million years ago, dragon flies with a wingspan of more than sixty centimeters flew around. The explanation might be that the atmospheric oxygen pressure (concentration) was higher then than it is now.

In vertebrates, however, the solution to this same problem is *hemoglobin* — a highly specialized protein molecule with a capacity to carry forward a small iron-containing molecule (a *porphyrin*) that will bind oxygen reversibly — meaning that it will take up oxygen wherever there is plenty of it (for instance, in the lungs) and give it off where it is lacking (for instance, in tissues). The delicate biochemical task that the hemoglobin molecule performs is to keep the highly reactive oxygen molecule safe inside the 3-D protein structure while it is transported around the body. In molecular models of hemoglobin, the iron porphyrin (called *hem*) can be seen to consist in a nearly flat structure enclosed within a hydrophobic depression of the protein molecule.

The oxygen molecule extends perpendicular to this plane — but is cut off from entering into unwanted interactions thanks to a well-placed unit of the amino acid *histidine* that safeguards the configuration. A corresponding *histidine* unit fixes the *hem* group on the other side of the plane. These two histidine units occupy position numbers 93 and 64, respectively, in the sequence of 150 amino acids that make up the hemoglobin chain (as counted from the N-terminal, i.e., the end of the polypeptide chain that carries a free amino group — NH_2). And it is exactly these two amino acids that have remained unchanged by natural selection throughout the five-hundred-million-year-long history of hemoglobin evolution (whereas changes have occurred on almost all of the other positions).

I have not digressed into the discussion of these biochemical details because I expect the reader to nourish a sudden longing for an understanding of hemoglobin biochemistry. Rather, I have done so because these details hold an important key to unraveling the intricate interplay between digital and analog codes, as disclosed by mapping the evolutionary history of this and other gene families. Human hemoglobin is in reality a complex (a *tetramer*) of four subunits each of which is a protein chain of just under 150 amino acids. All four subunits are derived form the same ancestral protein chain but pairwise they are now slightly different and are called *alpha* and *beta chains* respectively.

Significantly, the early embryo doesn't form any of these two subunits, but instead forms *ksi* and *epsilon chains* that deviate somewhat from the alpha and beta chains of the adult hemoglobin. Towards the end of embryonic development, a fifth chain is made called the *gamma chain* — this is because the oxygen supply for the embryo comes through the placenta, and it therefore needs a special hemoglobin tuned to this temporary — but life-sustaining for that time period — condition. And for a short period around birth, the newborn baby even produces a sixth chain of proteins called the *delta chain*.

Such variety in the kinds of hemoglobin chain formation found through the embryogenesis of one individual is a relatively recent phenomenon, in terms of evolution — as is disclosed by comparing hemoglobin chains from a number of different species, as in Figure 5.6. This figure shows that the original oxy-gen-carrying protein was a monomer not very different from the protein, *myo-globin* — a protein that nowadays is used for storing oxygen in muscle cells. Approximately five hundred million years ago, the gene involved in the protein synthesis of *myoglobin* underwent a duplication, where one of the gene cop-ies became the ancestor of the *myoglobin* genes found in all higher organisms today, while the other gene copy, in time, developed (probably as a *pseudogene*) to become a proper oxygen transporter — and *this* gene thus became the pro-genitor of all the world's hemoglobin genes today. Thus, all later species in this lineage have separate genes for myoglobin and hemoglobin.

In retrospect, it is easy to see this evolutionary development as a reflection of a new "size strategy" in evolution. In the small animals of early evolutionary time (e.g., the protozoans and flatworms), there was not much need for an oxy-gen transporter, since the distance to the surface of the animal would always be insignificant. But as animals increased in size, and as a result began to feed on smaller animals for their nourishment, oxygen transportation became a major challenge. Four hundred million years ago, a second gene duplication occurred through which the ancestral forms of *alpha* and *beta chains* were established. Since then, all organisms in the lineage have both *alpha* and *beta chains* in their hemoglobin. In natural history, this corresponds with the divergence into sharks and bony fishes, and the evolutionary line of the latter that led to the reptiles and eventually to the mammals.

Similarly, the gene duplications behind the divergence of the *specialized fetal hemoglobin chains* occurred some two hundred million years ago, which roughly corresponds with the appearance of placental mammals — where the particular oxygen-binding properties of these chains would be needed for the special tasks of ensuring an adequate supply of oxygen to the embryo. And thus the number-ing of the chromosomes in the upper part of Figure 5.6 refers to the positions of the different hemoglobin genes on the human chromosome as they are today. Like the musical motifs of a classical symphony that are interwoven in a variety of ways to form the experience of a unified whole, so we see how evolution lit-tle by little has managed to weave the hemoglobin themes into one another to form a functionally optimal unity. But contrary to the product that is the sym-phony, here there is no composer behind this evolution, and its dynamics must be explained by means of natural processes. Again, we shall search for the expla-nation in the dynamics of code-duality, the digital mutability and combinato-rics of *pseudogenes* coupled to the incessant process of trial and error exhibited by the kimflok in its interaction with its niche.

The code-dual dynamics that take place in this evolutionary process are not all that different from the code-dual dynamics that play out in cultural evolu-tion — and the modern-day phenomenon of e-mail spam throws an interesting

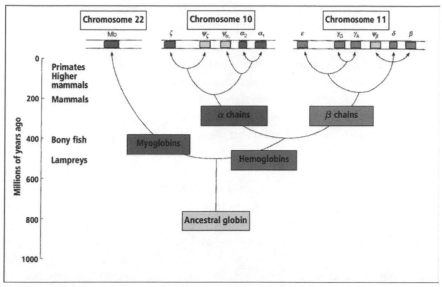

Figure 5.6 An evolutionary tree for human hemoglobin. Above is shown the arrangement of human hemoglobin genes over three different chromosomes. Sequences that are still transcribed are shown in dark gray. The very light areas are no longer transcribed in humans, but are now pseudogenes. Below is shown a time axis (in millions of years) indicating approximately when different hemoglobin genes appeared (modified from Mathews, van Holde, and Ahern 1999, 234).

light upon the dynamic operations of code-duality. All over the world, this word *spam* has come to signify the widespread interruption of privacy caused by the invasion of one's electronic mailbox by unwanted and unauthorized ads. Yet few persons outside of the English-speaking world know that this new signification of the word *spam* originated with Monty Python's Flying Circus. In that one of their comedy routines, a group of Vikings in a diner sang a chorus of "SPAM, SPAM, SPAM, lovely SPAM, wonderful SPAM" in an increasing crescendo, drowning out other conversation, much like unsolicited e-mail drowns out other e-mail in one's in-box. Likewise, the song's endlessly repetitive lyrics suggest an endless repetition of worthless text, similar to what is contained within the e-mail variety of spam.

Now, technically the term *spam* is a *primultima* (or telescope word) formed by the contraction of *spiced ham* and was invented as a marketing name for a line of canned meat. But although this term was consciously invented, it is safe to say that neither the original inventors of the brand name, nor the Monty Python troupe, had any idea of the particular worldwide fame awaiting the word as a *referent* for something which would not even exist until long after the time that these two groups were using this one term to serve totally different ends in two totally different contexts.

The metaphoric transformation of *spam* from the Monty Python setting to the Internet vocabulary was a creative act, but the reason that it stuck was that it happened to hit an unfilled locus in linguistic space — i.e., a nonverbalized but increasingly general experience in modern society. And through this metaphorical transformation onto the experiential space of the Internet, the term has now become a semiotic actor in its own right, generating a whole range of new spam-related habits. Because of it, we can now engage in making rules for internet services to eliminate spam or discuss and administer punishments for what might be called *spam perps*. And we may eventually expect further conceptualizations to develop on top of the original concept (e.g., *tele-spam* or *bio-spam*). In this way, new terms are *scaffolding devices* for cultural development.

The point is that the coining of the term *spam* was a creative response to the needs of a new cultural situation. But as such, it was not the result of some single thinker's conscious deliberations and action. Instead, it grew out of already existing linguistic resources, more or less spontaneously, by a sort of tacit consensus. And because of this, I want to argue that one can see the appearance of the term *spam* as a prototype case for the origin of new digitally coded signs not only in culture, but also in nature — most specifically, in evolution. The digitalization of the Monty Python sketch in one simple phonetic sequence served to *scaffold* a complex social experience by making it an easy general resource for communication. And this kind of informative and novelty-generating scaffolding, I suggest, is exactly what genes in general are good for.

New genes may often be formed through very much the same kind of scaffolding conversions that we have seen to be instrumental in furthering the inclusion of new words in a language. Thus, in the case of the new term *spam*, the decisive point was the conjunction of pointed meaning (submitted by Monty Python) and social need (created by the widespread experience). Likewise in the biological realm, we can suppose that gene duplication that was accompanied down through generations by what geneticists call the "hitchhiking" unexpressed copy as nonessential or "masked" genetic material (and thus prone to all kinds of nonlethal mutations) would assure the availability of a rich resource base for potential future genes.

The decisive cause of the birth of a new functional gene would be a lucky conjunction of two events: 1) an already existing nonfunctional gene might acquire a new meaning through integration into a functional (transcribed) part of the genome, and 2) the gene product would hit an unfilled gap in the semiotic needs of the cell or the embryo. In this way, a new gene becomes a scaffolding mechanism, supporting a new kind of interaction imbuing some kind of semiotic advantage upon its bearer. It would thus be proper to refer to this phenomenon as *semiogenic scaffolding*. For by entering the realm of digitality, the gene's new semiotic functionality becomes available not only to the cells of the organism carrying it, but also to future generations — and, if we allow for horizontal gene transfer, even to unrelated organisms. Digitality in the life sphere assures the sharing (objectifying) of functions (and, in the human case, ideas), and thereby also their conservation through time.

But this very function is itself dependent on the relative *inertness* of the genetic material and its very indirect and highly sophisticated way of interfering with the worldliness of cellular life. Genes, like human words, do not directly influence the world around them (which is why we do not believe in the power of spells), but have whatever causal efficacy that they have *only* when — and for so long as — some living system *interprets* them. And just as inert, intrinsically meaningless words serve to *support* human activity and communication, so do inert, intrinsically meaningless genes *support* cellular activity and communication. In the next section, I will explain in detail how this can be so — and why genes and words are the two most marvelous scaffolding tools ever created by nature.

Organic Codes

Translations of sequentially coded messages used to support dynamic activity in space and time — e.g., from the passive world of genes to the active world of proteins — are necessarily complex, multilayered processes. And so far we have only touched upon the initial complexities of this process: the *transcription* step from the string of DNA nucleotides to the complementary string of RNA nucleotides. Yet the *real* difficulty begins when one has to assure the correct coupling between molecular surfaces belonging to combinatorial sets of very *different kinds* of monomers — i.e., *nucleotides* and *amino acids*, respectively. For there is no chemical or molecular isomorphy between genes and proteins — and thus the problem of gene to protein *translation* has been solved only by the development of specific molecules, transferRNAs (tRNAs) that can build a bridge between them.

Embryologist Marcel Barbieri has suggested the term *organic codes* for these molecular "bridges" and in 1985 introduced his idea of *semantic biology*: "The function of an organic code . . . is to give specificity to a liaison between two organic worlds, and this necessarily requires molecular structures — the *adaptors* — that perform two independent recognition processes," writes Barbieri (2001, 99).[9]

9 Barbieri (2003, 94) notes the following three general characteristics of codes: "1) They are rules of correspondence between two independent worlds, 2) they give meanings to informational structures, and 3) they are collective rules which do not depend on the individual features of their support structures." An organic code is thus a more restrictive concept than the concept

Prototypes of organic codes in this sense are the small tRNA molecules that are often (conveniently) depicted as a kind of cloverleaf structure. One of its loops carries an *anticodon*, i.e., a triplet that is complementary (re base pairing) to a distinct signifying triplet, or *codon,* on the mRNA molecule — while at the opposite end of the tRNA, there is a binding site for the particular amino acid that has become associated with the codon in this case. For example, to codon UUU on the mRNA, there will be a distinct tRNA with the anticodon sequence AAA — which, on its opposite end, has a binding site for the amino acid *phenylalanine*. The base-pairing mechanism (AAA UUU) thus assures that this specific tRNA will be attached to UUU triplets (and nothing else) on the mRNA.

The role of tRNA molecules here is to act as *adaptors*. These molecules cannot themselves bring about the translation of mRNA to protein, any more than an electric adaptor by itself could connect a European plug into an American socket. Rather, somebody must see to it that the plug and the socket come into contact with the adaptor and with each other in the right way.

Likewise, to carry out the task of connecting mRNA to tRNA and its amino acids, cells create and keep at their disposal a specialized and very complicated aggregate of RNA and proteins called a *ribosome*, that (in synchrony with a host of interdependent enzymes) assures the correct ordering of temporal and spatial relations throughout the translation process (Figure 5.7). This *ribosome* travels down along the mRNA string at an amazing speed, recognizing each codon and arranging the specific tRNA molecule with the corresponding anticodon in such a way that the amino acid on the other end comes into a position to form a peptide bond with the amino acid from the preceding tRNA placement. The nascent and still growing protein chain is thereby gradually left free behind the ribosome as it travels its way along the string.

Considered as a *semiotic* process, this translation process whereby the mRNA sequence is converted to a sequence of amino acids (i.e., to a protein chain), is a strongly *ritualized* process that does not seem to leave much interpretive freedom to the ribosome, and therefore to the cell. But when studied in more detail, complexities appear that may change this initial and far too simple image.

For example, recent experiments have shown that the *termination* of the translation process is not the well-defined process it was once thought to be. Under certain conditions, the ribosome apparently may bypass the *stop codon*

---

of *analog codes* used in this book, and one might perhaps consider Barbieri's organic codes as an intermediary form between the digital and the analog. In the Peircean sign taxonomy, such organic codes belong to the category of *symbols*. The research horizon of biosemiotics, however, reaches far beyond just symbolic codings and includes all and every kind of semiosis going on in nature. Still, Barbieri's analysis does throw light on many of the same phenomena that we are concerning ourselves with here.

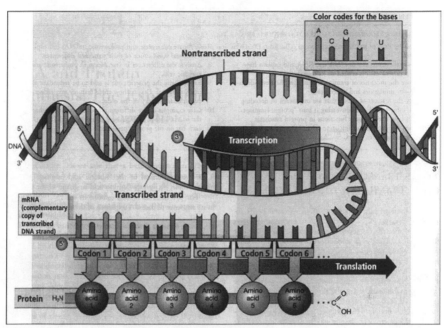

Figure 5.7 The translation process; see text (modified from Solomon, Berg, and Martin 1999, 268).

which would normally signify an end to any further prolongation of the newly formed amino acid string — and as a result, an unexpected tail of amino acids can become joined to the protein chain (Gallant and Lindsley 1998). Possession of this "tail," in turn, causes the protein to be very short-lived since the cell's "internal quality control system" (the *proteasomes*) will now more easily pick it out for scrutiny and degrade it. Moreover, depending on which protein we are talking about, many other effects may result, including the formation of a malignant cell line.

Another interesting locus for possible contextual interaction is at the initiation of the translation process. Topologically, the translation process usually takes place right outside of the nucleus, which is connected to the rough endoplasmatic reticulum. The immediate RNA transcript that is formed when the enzyme RNA polymerase II transcribes a segment of the DNA must therefore be transported to the cytoplasm through pores in the nuclear membrane. But because the genes of eukaryotic organisms contain a number of *introns* — sequences that do not code for amino acids "meant to" enter the protein — the RNA transcript must first undergo a complicated *editing* process whereby these introns are cut out, and the remaining parts of the RNA transcript (the *exons*) are successively spliced together in the correct order.

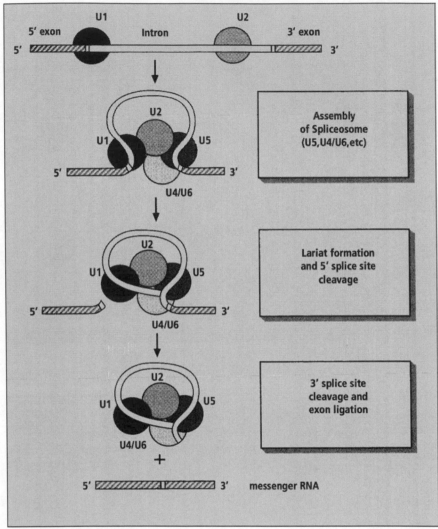

Figure 5.8 RNA splicing. U1, U2, U4, U5, and U6 designate different small particles that are themselves complexes of proteins with snRNAs (from Barbieri 2003, 103, with permission).

This process is taken care of by a *spliceosome*, an immensely complicated molecular machine that, again, is assembled by the cell itself and that in size resembles the ribosome. It also, like the ribosome, does its job through the use of a smaller *adaptor molecule* called small nuclear RNA (snRNA or *snurps*). Here, too, this process is analogous to tRNA translation, as snRNA likewise contains two independent *recognition* sites — one for the initiation point of an intron and one for the end point of the intron. "The two recognition steps are *independent* because

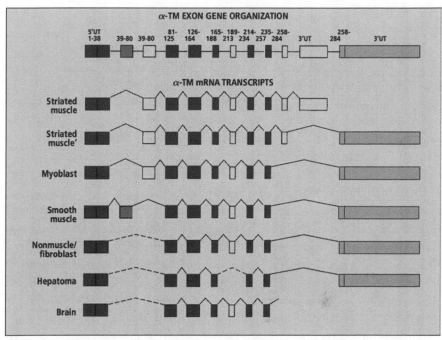

Figure 5.9 Alternative splicing. Splicing of the alpha-tropomyosin gene (in rats) illustrates how many exons may be spliced together in different ways to form different mRNA transcripts in different tissues (modified from Mathews, van Holde, and Ahern 1999, 1100).

the first step could be associated with different types of the second one, as demonstrated by the cases of *alternative splicing,*" writes Barbieri (2003, 104).

In *alternative splicing*, exons are spliced together in an order other than the one in which they originally may have appeared on the chromosome. Alternatively, one or more exons may be cut out in some cells, but not in others. A dramatic example of this is found in the gene coding for the protein *alpha-trypomyosin* — a protein that plays a regulatory role for the *actin-myosin* complex responsible for contractive processes in organisms, e.g., in muscles. As seen in Figure 5.9, the twelve exons contained in this gene may be combined in at least seven different ways in different tissues. How cells in striated muscles "know" that they are supposed to splice alpha-tropomyosin together in one way, while cardiac muscle cells "know" to splice the molecule in another way, is unknown — but presumably cell-specific proteins interact with the spliceosome. And the number of *known* eukaryotic genes exhibiting alternative splicing tendencies is presently well over a hundred and still growing.

In some cases, the process of *splicing* is supplemented with a real *editing* of the mRNA, whereby certain nucleotides are added to or removed from the string—or are *exchanged* with other nucleotides. Such an editing process is known, for example, from research into human *apolipoprotein B*—a protein that is excreted to the blood from the liver or, to a lesser extent, from special glands in the intestinal tract. *Lipoproteins* is the name for a group of proteins that assist in the transport of lipids in the blood stream. And while the gene *apoB*, which codes for apolipoprotein B, is edited to become a mRNA of 14,100 base pairs (resulting in a protein consisting of 4,536 amino acids) in the liver, the same gene in the glands of the intestinal tract is edited to become a mRNA consisting of 7,000 base pairs, resulting in a protein of only 2,152 amino acids.

The reason behind this, it has now been shown, is that the shortened length of the *intestinal* apolipoprotein B is the result of an enzymatic intervention that leads to the formation of a stop codon (i.e., a sequence that marks the end of the translation process) in the mRNA. This enzymatic intervention is a deamination of a Cytosine unit at position 6.666, which is thereby converted to a Uracil unit, resulting in the formation of a stop codon (Chen et al. 1987). The two versions of the apolipoprotein B have the same capacity as far as transport of lipids is concerned, but the short intestinal protein is removed from the blood much faster than the liver version, and because of this, the intestinal version does not have the additional role of activating another enzyme involved in lipid biosyntheses (LCAT = Lecithin:Cholesterol Acyltransferase). RNA editing thus is used to modulate the properties of gene products in order to tune them to the needs of each particular tissue

The Cellular City

While DNA spends most of its time sitting passively and rather tightly packed inside the nucleus, the vast majority of vital life processes actually take place in the many various structures of the cell that, somewhat misleadingly, are lumped together under the term *the cell's cytoplasm*. This term is inherited from a time when our technical instruments had not yet made it possible to see the complex inner architecture of cells in terms of organelles, membranes, and filaments that are present in this so-called plasma. And yet, it is precisely the subtle, never-ending interplay between the nucleus and these internal membranous structures that makes the cell a holistic organizational unit in itself. In the absence of these differentiated structures and their functions, the genomic specifications would not make any sense. One way to illustrate this complexity, as I have previously suggested, would be to compare the cell with a big city:

> If we think of the cell as a city like Manhattan, the nucleus of the cell would perhaps cover the area of Central Park while the membrane around the cell would be situated about as far from the center of the cell as the Bronx is from Fifth Avenue.

In such a city, a protein molecule would be the size of a family car. And it would have almost less freedom of movement within the cell than a car has in Manhattan. Membranes and stop signs steer every single protein round and about in well-ordered patterns (Hoffmeyer 1996b, 73).

The semiotics of this intracellular traffic is one of the most fascinating topics in modern biochemistry. We have, in the preceding pages, followed the route of the genetic specifications from the closed chromatin structure of the DNA to the ribosomal protein-synthesizing machinery bordered by the nuclear membrane. And precisely here is where two questions arise that, upon further scrutiny, show themselves to be closely connected: (1) How does the protein string fold up to form the highly specific three-dimensional structure that it does? (2) How do the proteins know how to get to the right spot in the cell — or, to put it in the figure we used above, how do the cars find their way around Manhattan? The answer to both questions appears to concern a class of protein complexes that have been given the name of *chaperones* — a suitable designation, since these protein complexes have, as their task, to make sure that newly formed (or, more generally, *unprotected*) proteins do not "get into trouble."

At first glance, one should think that the folding of a protein into the right conformation would be a nearly impossible process. The biochemist Cyrus Levinthal once calculated that the enzyme *ribonuclease*, consisting of 124 amino acid units, will have approximately 10^{50} possible conformations — and thus if one imagined that the molecule could "test" 10^{13} possibilities per second, it would still take 10^{30} years to work its way through a considerable part of this vast possibility space. And yet, it has been shown experimentally that ribonuclease *in vitro* folds up correctly in approximately one minute. The solution to Levinthal's paradox (as to many other paradoxes of this kind) seems to be that the folding process occurs through a series of intermediary steps that are *not given beforehand* — but which are all energetically "downhill" processes — thus leading to the correct conformation, as if the folding process itself had been led down through an imaginary and irregular funnel.

Newly formed peptide chains do nevertheless find themselves in a somewhat vulnerable condition in vivo, since their hydrophobic areas have an inherent propensity to aggregate with the hydrophobic areas of other peptide chains. This is where chaperone protein complexes enter the picture, as they protect newly formed chains against such unwanted interference. For example, if the virgin protein is intended to serve inside one of the cellular organelles (e.g., the mitochondria), its N-terminal end will carry a specific signalling sequence that signifies to the relevant chaperones that it needs help. The chaperones then "protect" the unfolded chain from aggregation with the nearby peptide chains, and deliver it at specialized receptor loci on the surface of the organelle where passage through the outer and inner membrane will be allowed. Such passage

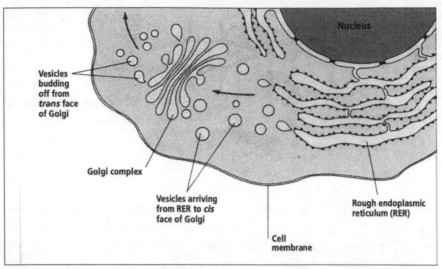

Figure 5.10 Protein targeting. Proteins intended to work in the cell membrane or lyso-zomes are processed in several steps that take them through the lumen of the endo-plasmatic reticulum and the membrane sacks of the Golgi complex (modified from Mathews, van Holde and Ahern 1999, 1107).

occurs through interactive *gate* mecahnisms in the membrane, with the help of specific transport proteins that discriminate between proteins destined for the cell lumen, membranes, or matrix. (In some cases, chaperones are also involved inside these organelles.)

Proteins intended for use in the cell membrane, or in the lysozomes — or that are destined for excretion into the extracellular space — undergo a rather more complex series of transfers. Here also, amino acid sequences at the N-terminal end elicit the relevant cue for other interlocutors in the process — which in this case is routed through the rough endoplasmatic reticulum (RER) and the Golgi complex, as depicted in Figure 5.10.

Let us take the opportunity here to briefly comment on these peculiar com-partments of the eukaryotic cell, the RER and the Golgi apparatus. While bac-teria send degradative enzymes out into their surroundings so that they literally swim in a soup of nutrients "like maggots in a piece of cheese" (to use Christian de Duve's pithy expression), eukaryotic cells instead ingest morsels of their surroundings and degrade the eventual nutrients they might contain through *endocytosis.*

In endocytosis, a confined cavity is gradually formed through the invagi-nation of the membrane. Then, through a combined fission and fusion of the cell membrane, the cavity is eventually released to the cell's interior as a vesicle (Figure 5. 11). Christian de Duve has suggested that endocytosis was an impor-tant ingredient in the evolution of the eukaryotic cell. For in the eukaryotic cell,

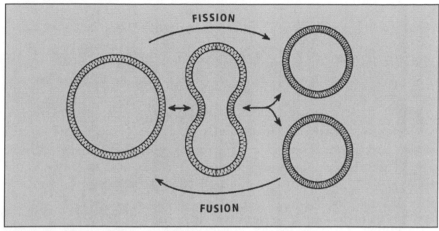

Figure 5.11 Vesicle formation through membrane invagination.

the ratio between surface and volume is a mere one-twentieth of the ratio that it is in bacteria. The eukaryotic cell therefore can only maintain sufficient flows of nutrients and waste compounds to and from the surroundings because of its extended internal area of membranes — these enclose sacks of internal space that topologically are *part* of the extracellular space with which it stays connected, through the constant flow of vesicles derived from the outer membrane.

These "sacks" are the structures that in the eukaryotic cell form the rough endoplasmatic reticulum (RER), the Golgi complex, and the endosomes and lysozomes. Thus, while the plasma membranes of bacteria are simultaneously engaged in virtually all of the cell's internal and external activities, eukaryotic cells have, thanks to these internal organelles, evolved a division of work so that the metabolic processes (including protein synthesis) are relegated to the *internal* membranes — whereas the *outer* membrane has been set free to take care of communicative tasks in interaction with the external surround.

> Most characteristic of the eukaryotic plasma membrane is that it is almost exclusively adapted to *communication* and *exchange* with the outside world. It is fitted with a variety of transporters, pumps, channels, and gates; and, especially in animals, it veritably bristles with receptors of various kinds hooked onto sophisticated transducing machinery. All in all, the eukaryotic plasma membrane appears as a boundary organelle of enormous complexity, in keeping with the almost irreconcilable requirements of cells obliged to be continually or intermittently open to a variety of substances and influxes and closed to many others. Such a development would probably have been impossible with a plasma membrane cluttered by biosynthetic and translocation systems, let alone respiratory chains and photo systems (de Duve 1991, 67).

The synthesis of proteins that travel through the RER begins, as usual, in the cytoplasm, but in this case the first sequence to be synthesized is a special

N-terminal squence destined to be recognized by a special *signal recognition particle* (SRP) that is itself a complex of both protein and RNA. SRP binding immediately stops the continued prolongation of the protein chain and thereby makes sure that the protein will not be formed at the wrong place — i.e., outside of the RER. And at the same time, SRP operates as a *chaperone* for the nascent peptide chain.

SRP assures the recognition of a definite protein at the RER that will be capable of binding the ribosome in such a way that the signal sequence at the N-terminal end is eased into the RER. Having brought about these benefits, the SRP is then released, allowing the translation process to be resumed. And at that point, the protein being synthesized is actually *pulled* through the membrane where it is finally set free and the signal sequence is eliminated.

While the newly formed protein is still inside the RER, a very important process takes place: this is the process of *glycosylation* — i.e., the formation of a covalent bond[10] between the protein and a carbohydrate (sugar). Carbohydrates are a richly varied group of compounds in which various sugar units are chained to each other. In *monosaccharides*, the sugar unit appears alone; in *disaccharides* they are chained two and two; in *oligosaccharides* there are only a few units in the chain; and in *polysaccharides* there are many. Carbohydrate chains are not necessarily linear, as are polypeptide chains making up proteins, but are often branching. When carbohydrates are bound to proteins, the resulting molecules are called *glycoproteins* — and these *glycoproteins* constitute more than half of all proteins in eukaryotic organisms.

The glycosylation of proteins implies a significant increase in their structural diversity and makes them suitable for use in communicative tasks. The intracellular and extracellular *recognition tasks* (whose complexity and importance to the organism is only now becoming a major focus of modern biochemistry) are often entrusted to glycoproteins. Likewise, it is the carbohydrate endings of glycoproteins "displayed" on cell surfaces that direct the interactions of cells both with other cells and with their surroundings — and that, therefore, are responsible for the precise migrations of cells during embryogenesis, for the control of cellular growth, for *endocytosis*, and for the often fatal *oncogene transformation* whereby a normal cell is converted to a cancer cell.

Glycosylation, as mentioned, starts little by little in the RER, but the major part of this task is carried out and completed in the multilayered membrane enclosures of the Golgi apparatus — where each enclosure is, in fact, a minature

---

10   A *covalent bond* is the name for an ordinary chemical bond in which two or more atoms share electrons — this in contrast to the much weaker hydrogen bonds characteristic of base pairing in DNA.

glycosylation factory of its own. The transport of proteins from the RER to the Golgi complex occurs via a process where a protein-filled vesicle buds off from the RER and begins traveling towards the closest unit of the Golgi complex (called the *cis-end*). After having been processed in the *cis*, Golgi-sack proteins are sent further on to the *next* Golgi sack, again by means of a vesicle, where they are further processed, and so on through the row of intermediary Golgi units, until the proteins finally arrive in the *trans-end*. From there, they are either delivered into the cytoplasm as functional organelles (e.g., lysozomes, peroxisomes, or glyoxysomes); to the cell membrane; or perhaps to be excreted into extracellular space.

Significantly, each and every one of these *transport* processes is accompanied by a *labeling* process so that each vesicle, on the one hand, contains a load of proteins to be used at an identical destination — and, on the other hand, will be both *recognized* and *accepted* at precisely this address. For that purpose, specific molecular protein signs are attached to the membrane of the vesicle, in functional anticipation of a complementary protein awaiting a corresponding segment of the membrane at the destination. This latter protein, in complement with the first protein, assures that the fusion occurs at the right position. These complementary pairs of proteins have been given the telling designation *SNARE proteins*.[11]

Signal Transduction

We have now discussed the *apparatus* at the disposal of the eukaryotic cell for the interpretation of its own genomic messages in some detail. But we have only sporadically dealt with the question of how the *cellular context* — the matrix of tasks a cell must confront at a given time in a particular tissue (or if it is a free cell, in a particular environmentally conditioned situation) — might itself exert true causal efficacy over this interpretive process.

For example, we have seen how the *ontogenetic determination* of cell lines for specific roles in the tissues may be caused by specific methylations of DNA, as well as by chromatin markings. We have also seen how a whole set of transcription factors have to work together just for the formation of an *initiation complex* that could bind to the DNA at the location where transcription shall begin — or, in other words, for the cell to get an answer to the question: Where on the chromosome is the message stored that describes the protein that is *needed right now*?

---

11   For those who like to know such things, these are called *soluble N-ethylmaleimide sensitive fusion protein attachment receptors*. Note, too, the two irreducibly biosemiotic terms *sensitive* and *receptor* embedded in this mainstream biochemical nomenclature.

We have seen that different tissues are capable of using the same gene as a recipe for quite different mRNA molecules; we have seen how this can be done by the complex editing and splicing of exons together in different ways; and in order to explain this latter process, we referred to the presence of cell-specific proteins.[12] And, finally, we have followed the widely distributed network of semiotically marked routes of the protein-synthesizing apparatus crisscrossing in and out through cellular organelles. Thus, we must conclude that there is no lack of "screws to adjust" if signals from the cellular surroundings should be allowed to interfere with the interpretation of genomic recipes.[13]

Collectively, the processes responsible for the canalization of messages from the surroundings to the interior of the cell are called processes of *signal transduction*. Luis Bruni has thoroughly analyzed the biosemiotics of signal transduction — and his work is one of the most fascinating examples of how a semiotic understanding of biochemical processes may open our eyes to the holistic-dynamic coherence of cellular life that remains rather hidden in the traditional descriptions (Bruni 2003, 2007).

Signal transduction is a remarkable process because it does not, as one might perhaps naively suspect, imply that *signals* from the outside somehow find their way into the cell, and then successively *instruct* the cellular machinery to initiate distinct activities.[14] Instead, what happens is much more interesting and semiotically complex, and typically consists in the following three steps: 1) the *primary molecular signal* is received at the surface of the cell because it is recognized by a specific glycoprotein, a *receptor*; 2) the receptor *responds* to the signal by *changing its conformation* which then activates a *mediator protein* in or at the cell membrane; and 3) the mediator again activates a *secondary signal* on the inside of the membrane, initiating the cascade of processes which will then take place

12   Recently it has been shown that microRNAs (miRNAs) have regulatory influence on how cells handle mRNA. Hundreds of microRNA genes have been found in different species. "With miRNA roles identified in developmental timing, cell death, cell proliferation, haematopoiesis, and patterning of the nervous system, evidence is mounting that animal miRNAs are more numerous, and their regulatory impact more pervasive, than was previously suspected," writes Victor Ambros (2004) in a review article for *Nature*. So, of course, the story of life, we keep finding out, goes deeper and deeper and deeper.

13   For, of course, there are also a multitude of other "screws" that I have refrained from describing to the reader. For example, *termination of transcription* factors were only briefly mentioned. Yet one very important factor in the regulation of how the cells make use of the protein resources available to them through transcription and translation is the *turnover times* of proteins — i.e., the question of how short- or long-lived they are. And as we can see from the footnote above, the list of mediating factors being discovered between the gene and its expression is growing every day.

14   Oddly enough, it should be noted, it is precisely this kind of description that is so rife with hidden anthropomorphism — assigning communicative and directive signal status to *things in themselves* — that is posited (without objection) in mainstream materialist science, and rejected in biosemiotics!

within the cell. Taken together, what occurs is a kind of translation of a *signal* caught at the cell surface to a *message* formed in the intracellular *molecular sign system* of the cell. This process may be seen as a case of what Barbieri has called an *organic code* — here, it is a signal-transduction code (Barbieri 2003, 110).

An even more surprising aspect of this process is that the secondary signal is normally one out of just five molecular agents — namely, cyclic AMP, cyclic GMP, inositol triphosphate, diacyl glycerol, and calcium ions. At the biochemical level, it therefore looks as if an abundance of riches on the *sending* side (where more than a hundred different primary signals are known, such as hormones, growth factors, and neurotransmitters) is met by a thought-provoking poverty at the receiving side. Yet, a bit of reflection shows that this circumstance is actually the strength of the system, because the key to the critical dynamics that are involved is a *semiotic* one, and not a merely chemical one.

When the body is stressed and orders the adrenal medulla to secrete epinephrine, a multitude of different things will occur simultaneously in different tissues. In the lungs, a relaxation of muscles will take place; in the liver, free sugar units will be mobilized from the carbohydrate stores of glycogen; fat tissue cells will start degrading their lipid stores; and cells in the intestinal canal will react by dampening peristaltic activity. One and the same molecule (epinephrine) thus releases a range of different responses and different activities in different kinds of cells, exactly as when the conductor of a chorus gives just one hand cue and gets sopranos, altos, tenors, and basses respectively to respond with four different intonations. And when one and the same sign can signify a range of different things to different cells, even though these cells are genetically perfectly homogenous, it is because all cells in the adult organism are descendents of embryonic cells that at some stage in development became determined for just one distinct cell destiny — a process that also comprised a fixation of the frame for the *semiotic receptivity*, or *interpretance*, of the cell line.

For illustration, let us now pursue the path taken by the signal-transduction process released by epinephrine. Whether we talk about muscle cells, liver cells, fat cells, or cells of the intestinal tract, the epinephrine molecule is always recognized by the same kind of *beta 2-adrenergic receptors* that, via a mediator called a *G-protein*, activates the formation of the secondary signal — which in this case, is always cAMP. In eukaryote tissues, cAMP has the quite general effect of activating a *protein kinase* — i.e., an enzyme that activates other enzymes by attaching phosphate groups at distinct amino acid locations. And this is the point in the process where the cellular "memory" — fixed, as it were, through the embryonic processes of cell determination — will make a difference.

For cells in different tissues often have very different *enzyme profiles*. Thus, whereas the protein kinase in liver cells causes activation of the enzyme *phosphorylase b kinase* — the activity of which causes an activation of the enzyme *phosphorylase*, an enzyme that catalyzes a cascading degradation of *glyco-gen* — the same kinase in fat tissue cells (*adipocytes*) will activate the enzyme *triglycerol lipase*, thereby initiating degradation of their lipid stores. The take-away lesson here is this: Instead of equipping cells with sophisticated sensitiv-ities towards multitudes of signals, evolution has chosen to provide them with an ontogenetically canalized interpretive diversity — based on a diversity of tis-sue subcultures, each with its own characteristic receptivity.

But exceptions to this rule do occur. *Steroid hormones* (hormones from the adrenal cortex and sexual hormones), for example, are solvable in lipid and therefore capable of penetrating the lipid bilayer of the cell membrane on their own. Inside the cell, the steroid hormone binds to a specific mobile receptor that helps it to penetrate the nuclear membrane, where the steroid molecule attaches itself to certain *hormone-responsive elements* on the DNA — and this allows the hormone to influence the transcription rate of nearby genes. In this case, we must conclude that early ontogenetic determination has created a tissue-specific sensitivity. The difference in the modes of operation between steroid hormones and other hormones reflects the physiological fact that the effects of steroid hormones are supposed to last for a long time.

Thus, while epinephrine takes care of bodily reactions to suddenly appearing situations that demand extraordinary efforts — situations that most often dis-appear as suddenly as they appear — a steroid hormone such as estrogen exhib-its a *cyclical* pattern over a period of several weeks. Correspondingly, the role of the adrenal cortex steroid hormones (*glucocorticoids*) is to assure long-term adjustments of the metabolism. The point is that the effects of epinephrine play out through an activation (and in some cases, an inhibition) of already available hormones — whereas the steroid hormones have to re-activate dormant path-ways or activate new transport systems.

We must now deal with a complication that we have so far kept out of the discussion, namely the concentration effect. Receptors of all kinds exist in vast numbers at the cell surfaces — on the order of 10^4–10^6 receptors per cell — and the effect of external signals caught by these receptors depend on the degree of saturation, i.e., the percentage of present receptors that actually bind the sig-nal molecule. Cytosemiotic signs are almost never single molecules, then, but the relative concentration of these molecules (Kilstrup 1998). But this does not imply that the cellular response is numerically proportional to the con-centration of the molecular sign. As we shall see below, biochemical response depends on a balance between synthetic and degrading processes, and enzy-matic activity normally has an S-formed dependence on concentration, where the steep midsection of the curve corresponds to a very small interval on the concentration axis. Concentrations below this tiny interval will elicit insignif-icant amounts of enzymatic activity — while concentrations above it cannot produce any further increase in the activity. The result is a kind of enzymatic digitizing of the response into only two possible options: response or no response.[15] It is often said that epinephrine starts a cascade of degradative processes, whereby glycogen is split into its individual sugar units — but the implied meaning always is that epinephrine is present in a sufficiently high concentration for the saturation degree of beta 2-adrenergic receptors to reach above threshold level.

Cellular constituents, signal molecules included, exist in a constant flux of degradative and synthetic processes, and in order to increase the concentration of a signal molecule (e.g., cAMP), the rate of its synthesis must exceed the rate of its breakdown. Thus, if the saturation degree of the receptors is below the critical value, the formation of new cAMP molecules cannot compensate for the constant cellular degradation — and protein kinase activity will then not be activated. The effect, rather, will be comparable to pouring water into a wash basin with no plug in it. In order to keep up even a minimum water level, the faucet must be turned up enough so that its input into the wash basin compensates for and exceeds the constant outflux down the drain..

Of special interest in this connection are the mediators that transmit the messages between the primary and the secondary signal molecule. Often these mediators consist of a G-protein (thus termed because it binds to guanosine diphosphate, or GDP). When the beta 2-adrenergic receptor is activated by epinephrine, the G-protein is stimulated to exchange GDP for GTP (guanosine triphosphate) and this elicits the conformational change in the G-protein that activates its neighboring enzyme, Adenylate cyclase, to produce the secondary signal molecule, cAMP. And here comes the point: The G-protein itself possesses a catalytic activity that results in a slow conversion of its newly acquired GTP to GDP, and it thereby inactivates itself. The result is that each receptor that is hit by an epinephrine molecule causes a short-lived peak production of cAMP. Consequently, the number of receptors hit by an epinephrine molecule per time unit is decisive for the release of an effect; either there will be a response or there won't. And this number directly reflects the concentration of hormone in the surrounding blood.

A dramatic illustration of the physiological significance of the G-protein is seen in patients suffering from cholera, a disease caused by the bacterium Vibrio cholerae. This bacterium excretes a compound that inhibits the GTP-splitting activity of the G-protein. This has the consequence that the G-protein of the intestinal cells cannot get rid of its GTP and, as a result, continues to stimulate cAMP synthesis. Among the many regulatory tasks of cAMP is to control the

15   The characteristic, but lonely, exception to this rule is the digital signs of the DNA molecule that are actively kept away from the metabolic processes of the cell.

intestinal excretion of water and sodium, and the effect is the virulent diarrhea so well-known in cholera.

Here too, and as we've seen in so many different signal-transduction processes, different kinds of G-proteins are available for one and the same receptor — and the same G-protein may service different kinds of receptors. The cellular response to a signal will therefore depend not only on the concentration of the signal molecule, but also on which other internal signal-transduction cascades may be occurring simultaneously at that particular time. This occasions an interesting increase in the interpretive variability and semiotic freedom of the cell (freedom understood here as a loosening of the rigid causal bonding of one signal to one response).

For now, however, we must leave behind the cellular city at this point, even though we have only skimmed the surface in describing some of the many aspects of its complex, sleepless traffic. Other interesting cytosemiotic themes that I have not touched on here include apoptosis (induced cellular suicide), the production of antibodies, and the cytosemiotic breakdown that leads to the formation of cancer cells. I shall however, return to treat these themes in the chapter on endosemiotics (Chapter 7).

Before leaving the fascinating level of the cell, however, it must be emphasized that even though the cell is usually thought of as a mere subunit in a multicellular organism, it is much more accurate to acknowledge that each individual cell, functions as an integrated unitary autonomous and autopoetic system[16] based on finely tuned sets of coordinated semiotic control loops.[17] The interpretants generated through the cytosemiotic processes must therefore be understood as holistic cellular configurations in time and space.

The Semiotics of Ontogeny

At this point, we have seen how genomic instructions are channeled into the chain of semiotic relations that, taken together, incorporate the cell into a destiny determined by its topological position in the early embryo. To complete our analysis of the semiotics of heredity, however, we must now attempt to follow the genomic instructions further on to the production of the kimflok — from which the instructions themselves were originally derived one generation earlier. To do this is essentially to analyze the semiotic relations governing the *developmental process of the individual.*

As a first step along this path, we must consider the process whereby the individual is formed — the process of *fertilization* or *conception*. The

---

16   I use the term *autopoetic* here in the sense given it by Maturana and Varela (1980).
17   Exceptions to this rule are the degenerated red blood corpuscles or cancer cells.

modern idea of the individual as the focal point of the life process is quite new. Traditionally, rather, the *family* (what biologists call the *genus*) was seen as the emotional or organizational center of life. As mentioned earlier, biosemiotics will posit the family — understood as the *lineage of related individuals* — as evolution's *individual*.

Single individuals are carriers of a code-dual dynamic that is played out in the semiotic field made up of both individual and population; this field cannot be reduced to comprise only one of these two units. This consideration is not meant to challenge the legitimacy of the individual as the central *unit of life*, nor shall I in any way deny my own personal acquaintance with the emotional roots of this idea. The consideration may, however, be justified to the extent that current debates more and more often invoke belief in "the inviolability of the individual" on biological conceptions of humanness. Such an invocation, for example, often occurs in the context of ethical debates on modern reproductive techniques, and especially as an argument against free access to selective abortion (which will be further discussed in Chapter 9). Seen from the biology of biosemiotics, however, a human life does not necessarily start at conception. For *conception* is the term designating the passage from a haploid to a diploid existence. For each individual *diploid* human being, this passage is of course decisive — because for him or her it initiates an irreversible journey that necessarily ends by death and annihilation. The same however, is not necessarily the case for *haploid* human life — i.e., life as a germ cell, whether sperm or egg. In fact, in our sex-cell-producing tissue, every one of us carries cells that — through an unbroken chain of cell divisions — descend from the earliest forms of life on Earth.

These germ cells *do change* down through generations (or else there could be no such thing as evolution), but considered *as cells* they constitute a rather infinite and unbroken chain of divisions. No sex cell in existence today came from anywhere other than this unbroken chain. For one of the first things to happen in human development is, as we have seen, that the cell lines destined to become sex cells (*gametes*) are segregated from the rest of the cells and kept clear of those determining processes that impact so decisively upon all the other cell lines. And if we have children, one or a few of these particular cells will live on for yet another generation. Thus, our haploid sex cells are, presumably, the closest we will ever come to eternal life.

The first task of the egg in relation to conception is to make sure that sperm from foreign species cannot penetrate the egg wall. Especially in species where conception takes place in an unprotected milieu — as, for example, in sea urchins — this poses a problem. The sea urchin egg is enveloped in a protective membrane of *vitteline* that is surrounded by a gel of the same kind of glycoprotein that we discussed in the previous section, and this membrane effectively prevents the penetration of sperm cells. However, since species-specific fertilization should not be so prevented, evolution has left a few cracks open in this fortification of the egg.

Sea urchin sperm cells are activated by certain compounds in the membranes of the egg, and these compounds also attract the sperm cells chemotactically. Cheered up by this promising reception, sperm cells now break down their own *acrosomal* membrane, releasing a bucketful of degradative enzymes that help the sperm cell travel through the protective glycoprotein gel and penetrate the vitteline membrane. Making the process even more fine-grained (and spectacular), the plasma membrane of the egg is equipped with multitudes of *microvilli* (small extensions from the membrane), and when the sperm cell approaches, some of these microvilli lengthen and then veritably *envelop* the sperm head and *fuse with it*, at the same time drawing the sperm head inside the egg, creating a *fertilization cone*. Within three seconds, the cell membrane is now depolarized, which prevents additional sperm from fusing with the plasma membrane — and thus prevents *polyspermy*. This fast-acting blockage is followed by a somewhat slower excretion of enzymes that, by different biochemical paths, establish an efficient and lasting blocking against further fertilizations (which might lead to chromosome numbers that the cellular machinery couldn't handle).

Only the cell nucleus of the sperm cell enters the egg, whereas its midsection — containing the mitochondria that furnish the energy for sperm movement — and its flagella are left outside. At conception, the two haploid nuclei fuse under the formation of a diploid individual, as waves of calcium ions are simultaneously released from stores in the endoplasmatic reticulum of the egg cell. These waves serve as an *activation signal* for the fertilized egg that it can now initiate the developmental process. Most importantly for our purposes, it should be noted that the interaction between the sperm and egg here is thus in no way a simple *fusion* process. Rather, the almost billion-years-old semiotic game of accomplishing conception here seems (if the reader will allow me the image) to possess that mixture of eagerness and thrill that will, much later in evolution, come to characterize the mating rituals in animal species — not the least of which, our own.

From Egg to Individual

In Chapter 2, we looked at Hans Driesch's famous experiments from 1892, in which he separated the first cells from a sea urchin embryo, and observed that each of them developed to become a *pluteus* — a normal if somewhat small, larva.[18] The results of this experiment posed a fundamental problem for biolo-

18   Since that time, Driesch's experiments have been reproduced in many laboratories the world over with positive results. But they are not valid for all kinds of organisms. In most invertebrates,

gists — for how can life attain its variegated multicellularity if each single cell in the embryo can *by itself* develop to become a whole organism? The answer to this question is simple but decisive: *Cells communicate.* Colloquially put, each single cell will, from its interactions with its neighbors, learn to give up its ambition of becoming a whole individual and, instead, be content to be a specialist in the task of serving as part of just one small area of tissue.

The phenomenon whereby a cell that can yet become anything becomes, in fact, this or that, is called *induction* — and the phenomenon plays out in numerous clever ways throughout the branching chain of events that constitutes the developmental process. *Positive induction* is at work, for example, when the early notochord (the embryonic vertebral column) in amphibians *induces* the overlying ectoderm layer to develop as nerve cell rather than as skin cell. Conversely, *negative induction* is at work when the inductive course of events is inhibited due to a too large number of neighboring cells (Gilbert 1991b).

I have already mentioned Leo Buss's fascinating theory about the evolution of multicellularity as a compromise between a survival struggle among cell lines and a survival struggle between individuals. Here it is in his own words:

> The thesis developed here is that the complex interdependent processes which we refer to as *development* are reflections of ancient interactions between cell lineages in their quest for increased replication. Those variants which had a synergistic effect and those variants which acted to limit subsequent conflicts are seen today as patterns in metazoan *cleavage, gastrulation, mosaicism,* and *epigenesis.* The conservatism of early ontogeny is held here to reflect the *interplay between selection at the level of the individual and selection at the level of the cell lineage, under the strong influence of ancestral constraint.* . . . The principal epigenetic interactions defining cell fate — those of *induction, competence,* and *cell death* — are all interactions in which one cell lineage acts to limit the replication of another, while enhancing its own (Buss 1987, 29–30; italics added).

In order to fully comprehend the logic of Buss's theory, one must be acquainted with a peculiar circumstance about metazoan cell biology — that is, the fact that the cells of these organisms cannot *both* move *and* divide. Eukaryote cells usually move by the use of either flagella or cilia[19]. These are whip-like extensions filled with bundles of long tube-shaped structures called *microtubules.* These microtubules are fastened at their base to a structure called the *microtubule organizing center* that operates like a motor to drive the flagella or cilia around. The micro-

---

for example, early embryo cells are already allotted their future destination, thanks to asymmetries in the original egg. Due solely to their positions in different zones of this egg, therefore, cells immediately receive their basic determination. Development in more complex animals, as we have seen, does not work so straightforwardly.

19   Cilia are shorter than flagella and execute coordinated movements a little like rowing the cell forward. Flagella function rather as propellers, with wavelike movements.

tubule organizing center is also sometimes called *the basal body*, or the *centriole*, or even the *kinetosome* — the plurality of terms reflecting the fact that this body may be located at quite different positions in different cells, and that researchers have not always been aware that these different findings referred to one and the same organelle!

Most relevant to our concerns is the fact that microtubules also function as hardware in the *internal cellular transport system* in a variety of ways, and most importantly as *threads* that extend from the microtubule organizing center to attach to the chromosomes to draw the two homologous chromosomes, each to their respective daughter cell, during the critical processes of cell division. Thus, both the formation of flagella *and* the mitotic division require the presence of a microtubule organizing center. Yet, as shown by Lynn Margulis, this structure does *not* have the ability to execute both tasks at the same time — at least not in those unicellular organisms from which multicellularity originally descended (Margulis 1981). (Indeed, this may be why ciliated cells from metazoans may well regain the capacity to divide, but to do so they must first get rid of their cilia.)

Many contemporary species of simple metazoans deliver their gametes directly into the surrounding sea water, where fertilization takes place — at which point the egg traverses the usual divisions that lead to formation of a *morula*, a solid ball of undifferentiated cells. At this stage, cells acquire cilia and organize themselves as *blastula* (Figure 5.12). And here is where the trouble begins; having developed cilia, the cell cannot divide anymore. (In some taxonomic groups, a solution to this problem was never found, writes Buss.) Instead, those organisms evolved ways to skip the blastula stage and complete their development without ever moving by cilia. More commonly, however, most metazoans *did* find a solution, the most popular of which consisted in a *functional partitioning* — whereby a minor fraction of the cells *don't* develop cilia and therefore maintain the ability for *division* instead:

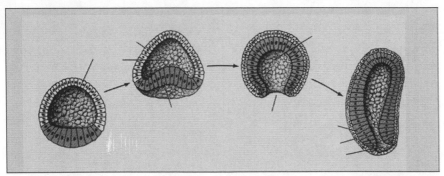

Figure 5.12 Development of blastula in amphibia. The figure shows the three layers, ectoderm, mesoderm, and endoderm. (Modified from Luria, Gould, and Singer 1981, 320).

These cells, however, can proliferate in only one direction. Since the primitive form is a sphere traveling in water, the proliferating cells cannot grow into the water column without producing a mass of cells which would certainly increase drag on the embryo, interfering with further movement. Neither can the proliferating cells grow over the surface of the sphere, for to do so would obliterate the ciliated surface necessary for locomotion. Only one alternative remains. *The only remaining place for these cells to move is into the sphere itself.* The movement and subsequent proliferation of cells from the blastular surface into the center of the blastular sphere is *gastrulation*. Animal gastrulation is the metazoan solution to the requirement of *simultaneous* development *and* movement (Buss 1987).[20]

In turn, the gastrula undergoes a *symmetry break* that divides it into three distinct layers with different destinies: *ectoderm, mesoderm*, and *endoderm*. Cell lines derived from the ectoderm become skin or nerve cells (including the cells of the sensory organs); cell lines from the mesoderm end up in bones and muscles (including the cardiac muscle, the heart) or in the vascular and reproductive systems; and cell lines from the endoderm come to constitute the structures of the intestinal tract.

Seen from Buss's neo-Darwinian vantage point, the premium in the ancestral competition between cell lines was to become a sex cell, with the implied chance of continued life for yet another generation. This favorable position, however, is only open for a small minority of embryo cells — the rest must suffer the wrong of differentiation and the loss of all further expectations for eternal life. This then leads Buss to pose the following question: "How could an organism evolve such that some cells in that organism *abandon their own capacity for replication*?" (ibid., 53).

Even if one does not embrace Buss's strong adherence to the competitive schema of evolutionary mechanics, this question seems basically to be a fruitful one. Individual cells in our body have no idea that we exist — but they nevertheless collectively reproduce us from second to second throughout the entire seventy-five-plus years of an average human life. Indeed, we would not be here now to ponder the question if this were not the case — but this doesn't make the enduring "unselfishness" any less strange.[21] For we are so accustomed to thinking of ourselves as unitary entities, that it may be useful sometimes to be

20   Most students of biology have wondered at this peculiar process whereby the blastula flattens at one pole and then invaginates itself under the formation of a gastrula (similar to what happens when you push a tennis ball to form a cavity in it — Figure 5.12). But according to Buss, this simply was the most effective way to solve the basic dilemma of the metazoan cells in our remote water-living ancestors.

21   *Selfish* in the biological sense means merely "that done for the continuation of one's own survival" — thus describing basically every single act that an individual cell normally does (e.g., grow, metabolize, reproduce, self-repair). How any of this could become "sacrificed" and "subsumed" under the workings of some *other* priority is the profound question at issue here.

reminded that each one of our fifty trillion cells possesses a significant degree of autonomy — a fact that is sadly confirmed by the eventual occurrence of diseases such as cancer. Cancer is the out-of-control population explosion that precisely appears when ordinary somatic cells stop letting themselves be persuaded to stay in their fixed and — seen Darwinistically — unpromising position in the tissue.

From our ancestors among the protists, each one of our body cells has inherited the entire machinery needed for *unicellular* life: the genetic code, transcription and translation enzymes, highly organized but flexible cytoskeletons, etc. And, as Buss writes, "The cells of a developing metazoan embryo follow the same rules governing any self-replicating system: They divide until some external force limits their further replication" (ibid., 77–78). This "external force" is *biosemiosis* and consists in the informative (and sometimes disinformative) messages emitted from other cell lines — messages that are capable of influencing growth processes in neighboring cell lines. Buss writes, "The metazoan innovation is the evolution of *epigenetic controls* on the growth of developing cell lines, which provide restraints on their inherent propensity for self-replication in a precise cascade such that cell lineages organize themselves into a functional *bauplan*" (ibid., 78, emphasis added)[22].

The establishment of these semiotic intercellular control mechanisms was not an easy process, as evidenced by considering how *rare* the appearance of the multicellular life form has been in evolution. The first eukaryotic cells may be dated back approximately nine hundred million years (Bitter Springs formation) and some two hundred million years after this, we find metazoans that are supposed to have belonged to extinct, but known, phyla. The property of *multicellularity* has arisen in seventeen out of the twenty-three monophyletic protist groups,[23] but only in three of these seventeen multicellular taxa (specifically, only in plants, fungi, and animals), has cellular differentiation developed in more than a handful of species (ibid., 70). The real difficulty for evolution, then, was thus not to develop viable multicellular forms of cellular life. The problem was to find a reproducible dialogical process *inside* the growing clone of cells descending from the same zygote — a semiosis that could make sure that some cell lines undertook (or perhaps more properly, could be fooled into undertaking) somatic duties on behalf of the common weal, even though they themselves would thereby die as individual existents.

In a certain sense, it is the differentiated multicellular organisms that introduce *death* as a common fate of living things. A *protist*, say an amoeba, does *not* die as a natural outcome of its life cycle — for it won't stop existing just

22   Such *epigenetic controls* form a subset of what, more broadly, may be called *semiotic controls*.
23   A *monophyletic* (Greek, "of one race") group consists of a common ancestor and all its descendants.

because it divides; rather, it now continues its existence *doubly* as two units (and then four, and so on). And even if individual amoebae regularly end up as food for other organisms, the surviving divisions of those very same amoebae will almost certainly have spread over the area. With the metazoans, however, this kind of survival is no longer possible. The differentiation process creates a new kind of "self" — one that is bound to a particular cellular architecture that *cannot* persist in the same simple way that free cells persist. Buss (1987, 78) writes, "Metazoan development today is manifestly a process of the sequential origin of, and interaction between, cell lineages arising in the clonal progression from the zygote. The fact that metazoans develop via a complex of epigenetic interactions between cell lineages is prima facie evidence that the principal modes of metazoan development arose as variants in the course of ontogeny."

As an illustrative case, Buss refers to the ontogeny of the eyeless mutant of the *axolotl* salamander — which, as the name implies, produces an individual without eyes. Here we see that, under normal conditions, the amphibian eye is produced by chemical interactions between the newly formed *optic vesicle* and the embryonic *ectoderm layer*. A chemical inducer produced by the optic vesicle is used for the scaffolding of this interaction. What happens in the eyeless mutant of the axolotl is that this step is disturbed because the ectoderm of the mutant does not respond properly to the *inducer* — so that no eye is formed and the mutant emerges blind. However, this is not the only problem that this poor creature has to cope with — for it is also born lacking the capacity for leaving offspring.

This incapacity results because the eyeless mutant develops a *secondary* deficiency in the region of the brain called *hypothalamus*, which will only properly develop through induction via signals that are sent to it from the developing eye. In the *eyeless* individual, no eye exists to direct the development of the hypothalamus, and thus the hypothalamus does not develop to the point that it can produce *gonadotropin* hormones — and in the absence of these hormones, the individual becomes sterile. I have commented upon this state of affairs elsewhere as follows:

> The deficiencies of the eyeless mutant clearly illustrate the tinkering ways in which ontogeny has become scaffolded by evolution. There is presumably no reason why the development of the hypothalamus should depend on the presence of a functional eye, other than that the eventual formation of the eye takes place in a location that happens to be anatomically close to that region of the brain where the hypothalamus is normally developed in this lineage. Making the development of the hypothalamus dependent upon the prior formation of an eye effectively assures that the hypothalamus will become constructed at exactly the right moment in embryogenesis. And this is precisely the situation that is not the case, of course, in the eyeless mutant. But rare mutants are statistically of little concern in evolution. Rather,

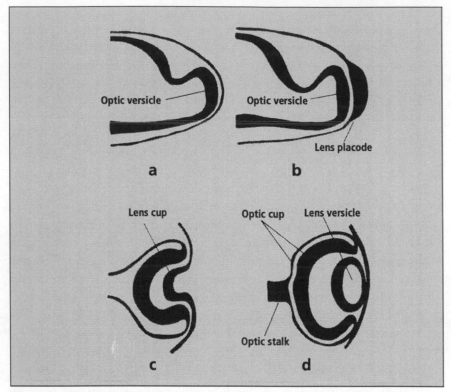

Figure 5.13 The normal sequence of steps in the development of the amphibian eye (modified from Coulombre 1965).

the axolotl eye just happened to be in the neighborhood of the nascent hypothal-amus region in *normal* individuals — and it is most likely for no other reason than this that evolution managed to exapt the eye for a secondary role as an ontogenetic switch for the initiation of proper development of a hypothalamus. As Buss says, "Ontogeny must reenact the interactions which gave rise to it" (Hoffmeyer 2007, 157; Buss 1987, 97).

In the terminology of this book, one might say that ontogeny is safeguarded by the myriads of semiotic scaffolds that depend on one another in long chains of successive and deeply interdependent steps.

The Individualized Brain

"Evolution is a process of tinkering rather than of engineering," as François Jacob (1982) famously proclaimed. And items constructed in a tinkering way typically bear witness to the detours taken on the path of their construction. A part that was formerly used for one purpose might somewhere along the path have been recruited for some other purpose without thereby loosing the structural property that made it functional for the original purpose. Likewise, in evolution, markers of the ways that organisms were originally constructed are often maintained in the present form of the organisms as testimony to the unique historical route that evolution took in creating them. And one might also formulate this realization the opposite way around, as Walter Garstang (1922) did when he turned Ernst Haeckel's famous dogma on its head by claiming, "Ontogeny doesn't recapitulate phylogeny, it *creates* it" (cit. from Gilbert 1991b, 346).

Induction plays a constructive role in evolution by facilitating evolutionary changes that affect sets of finely tuned relationships between different functionalities. Exactly because the formation of organs is coordinated via the inductive communication between parts (i.e., through *biosemiosis*), changes in one place in the process may often cause compensatory changes in other places. Raissa Berg has introduced the expression *correlation Pleiades* for this phenomenon (Berg 1960).[24] One fascinating example of a correlation Pleiades is the variety of snout forms in dogs — from the narrow jaw of the German Shepherd Dog to the flat face of the Chow Chow or Bulldog. These variations are genetically fixed and each of them represents a new harmonious arrangement of many bones fitted to each other as well as to the attached muscles. And, in spite of the bizarre forms that dog breeders have excelled in bringing into being, all these snout forms, even the Bulldog's, are capable of carrying out normal and critically important dog activities, such as moving the jaws, shaking the head, or barking.[25]

Thus we see that embryogenesis exhibits an inexhaustible fund of sophisticated semiotically controlled adaptations. Here, however, we shall consider just one: the growth of nerves in the embryo from their source in the spinal cord to the peripheral muscles and organs they eventually connect with. Nerve cells grow by extending their axons from an area in the tip called a growth cone that literally *feels* its way forward along the cell surfaces it passes by. This occurs with the help of small protrusions from the growth cone called *filopodia* that extend or contract themselves depending on the kind of signals that they receive from the surfaces of other cells or from the extracellular matrix (see Figure 5.14).

Developing nerve cells are not given any kind of geographic genetical information beforehand that could tell them *where* to go to and what to connect up with in the body. Nevertheless, all the muscles and organs of the body end up being correctly connected to their corresponding nerve cells. Considering the

24   The Pleiades is a fascinating group of stars that can hardly be discriminated with the naked eye — as though they belong together!

25   An exception to this rule, though, concerns dogs with a very short face (such as the bulldog) where, apparently, it has not been possible to compensate for the alterations in skin area through corresponding alterations in bone structure — which now leaves the dog with the well-known look of extra skin hanging off of its face.

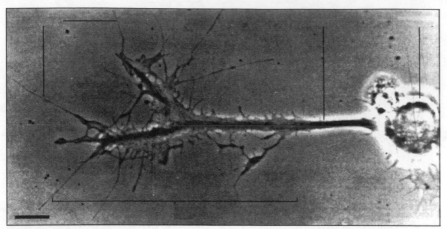

Figure 5.14 Filopedia (modified from Wolpert, Beddington, Jessell, et al. 2002, 354).

vast distances that nerve cells must sometimes overcome to reach the peripheral muscles, it is clear, however, that all this precisely targeted growth cannot be based just on local signposts. Thus, the initial phase of axon growth is safely guided via a trace of chemical attractors — and this explains why nerve cells initially are assembled in those thick bundles, that we call nerves.

As the growth cone approaches the peripheral tissues, however, genuine exploratory *search behavior* sets in, as nerve cells individually attempt to find a destination that needs ennervation. The embryo here uses a principle that is found throughout development: it creates a number of nerve cells that far exceeds the number eventually needed — and this causes a kind of Malthusian/Darwinian competition between nerve cells to find a useful connection. Correspondingly, organs or muscles that need ennervation excrete an aptly named *nerve growth factor* that attracts the growth cones via the filopodia. Not all nerve endings are fortunate enough to find a locus for ennervation and those that do not are quickly eliminated. In fact, they kill themselves. For each single nerve cell carries within it an internal "suicide program" to release a series of self-destructive enzymes dependent on conditions (intracellular or extracellular inducers). Thus, if the growth cone leading the nerve cell does not manage to hit a substantial concentration of nerve growth factor (and thus receive the suicide preempting signal), it will initiate its self-destruction program and die. The result of this amazing mechanism, called *apoptosis*, is that the nerves that do reach their targets for ennervation will survive, whereas the others will disappear.

The brain, and the entire nervous system, is therefore to a large extent shaped via an interactive, *online* process — a process governed by use and disuse, reflecting the local contextual situation of the embryo (and in human beings, the first few years of life). The strong overproduction of nerve cells that takes place in the embryonic brain also results in a corresponding overproduction of possible synaptic connections. And again the principle at work that will trim this chaos into order is *neuronal selection* — where nerve cells will only avoid programmed suicide (apoptosis) if synaptic connections are feeding them a sufficient amount of sign activity.

This was confirmed by experiments where kittens' right eyes were covered so as to admit no light during a critical period immediately after birth. Although the shades were later removed and the kittens again received reflected light through their right eyes, they would never become able to *see* with this eye — and, in fact, had become functionally blind in the right eye. What explains this result is the phenomenon of *synaptic competition*. Since the visual cortex only received impulses from the left eye, *all* the synapses that connected to the optical nerve from the right eye died away and were eliminated. And although the right eye later actually began to send signals again, those signals could not get through to the visual cortex because all the previously vacant cortical synapses were now occupied by nerve threads emerging from the left eye (Gilbert 1991a, 644).

And, of course, it turns out that the genetic *indeterminacy* of the vertebrate brain is actually its greatest strength. This is especially obvious in human beings, where the brain continues to grow and actually adds 250,000 new nerve cells per minute in the initial period after birth. The American neurobiologist Terrence Deacon writes,

> Cells in different areas of the brain are not their own masters, and have not been given their connection orders beforehand. They have some crude directional information about the general class of structures that make appropriate targets, but apparently little information about exactly where they should end up in a target structure or group of target structures. In a very literal sense, then, each developing brain region adapts to the body it finds itself in. . . . There need be no "preestablished harmony" of brain mutations to match body mutations, because the developing brain can develop a corresponding organization "on line," during development (Deacon 1997, 205).

In short, the developing human nervous system has the ability to take stock of the situation — i.e., to let its maturation process depend on the presence (and absence) of distinctive marks in the growing embryo. The biological logic behind the formation of nervous systems in nature is to delegate increasing amounts of power away from the level of the *species* and to the level of the *individual*. Genetic adaptations take generations to construct, whereas the brain can effectively wire itself up in the lifetime of the single individual. In the large scheme of things, the appearance of species with brains corresponds to a concretization or individualizing of the semiotic *subjecthood* that in more primitive species unconditionally resides at the level of the species. With the creation of more and more sophisticated brains, the *individual* increasingly takes power over its own actions and becomes an autonomous semiotic agent in its own right.

The Semiotics of Heredity

I began this chapter with a discussion of mortality as the logical precondition for heredity. But we have seen that mortality is perhaps a property unique to multicellularity, whereas unicellular life forms — lacking *individuality* in the sense that I have been discussing above — can only properly "die" by being exterminated at the species level.

In the same way, heredity is too, after all, mostly an invention that only multicellular organisms can really make use of. In unicellular organisms, the role of the genome as an efficient inventory-control system is quite obvious — but in multicellular organisms, these digitally coded specifications have the additional task of mediating an *ontogenetic* process, and this endows the organism with a *real heritability* in the sense of a temporal reconstruction of a complex spatial structure by the help of a linear set of digitally coded instructions.

We have seen that the fusion of the spermatozoon with the egg is followed by a wave of released calcium ions departing from the point of penetration, whereby the egg becomes activated to initiate this lengthy, complex process of genetically informed reconstruction. We have also seen in detail how the intracellular structures and membranes that the egg inherits from its mother organism — the endoplasmatic reticulum, the mitochondria, the Golgi system, and, not the least, the plasma membrane — are responsible for the long series of cytosemiotic processes that alone make this reconstruction possible.

The fact that the species-specific origin of the genome that we know is decisive for which organism the ontogenetic reconstruction will produce (e.g., that a snail genome will lead to the appearance of a snail-type living system and a wolf genome will lead to the appearance of a wolf-type living system) has implanted the conception in the minds of biologists that the genome *controls* the ontogenetic process, rather than filling the more modest role of being one tool among others for this process.

I find this former conception unfortunate, because it blinds us to the semiotic competences of the cellular system. And in my work in biosemiotics, I am trying to advance an understanding according to which the genome (both in unicellular and multicellular species) is, in effect, *an inventory-control system*, with the distinction that in multicellulars, access to the inventory-control system goes through a *user interface* — and that this interface is the product of hundreds of millions of years of evolutionary history in a narrow interplay between user (organism) and system (genome). New routines have incessantly been built upon older routines in such a way that the *menu items* that each cell can access is finely adjusted to the *menu choices* that this particular cell made in an earlier stage of its ontogeny.

In sum, I do not claim any Lamarckian interaction in the sense that the user organism can occasion the introduction of *purposive improvements* directly into the inventory-control system. But I am not afraid to adopt a modified version of the far more interactive Baldwinian scheme — as I will in the next chapter.

6
The Semiotic Niche

Zoosemiotics

"Images on the retina are not eatable or dangerous. What the eye of a higher animal provides is a tool by which, aided by a memory, the animal can learn the symbolic significance of events" (Sebeok 1979, 266). This observation, one that Thomas A. Sebeok ascribes to the ethologist John Z. Young, hits the nerve of the zoosemiotic conception that Sebeok launched as early as 1963 (see Sebeok 1963; 1972; Sebeok, Ed. 1977). Sebeok goes on to say that "Cephalopod brains may not be able to elaborate complex programs — i.e., strings of signs, or what Young calls *mnemons* — such as guide our future feelings, thoughts, and actions, but they can symbolize at least simple operations crucial for their survival, such as appropriate increase or decrease in distance between them and environmental stimulus sources" (Sebeok 1979, 43).

That beauty is in the eye of the beholder is a truth that cannot be repeated too often. Actually, as we now know, it is the brain rather than the eye that does the seeing for us. Don Favareau (2002, 10–11) has stated it this way:

> Significantly, recent research in the neurobiology of vision, especially the ground-breaking work of Semir Zeki (1993, 1999) demonstrate conclusively that sensory percepts such as *visual images* are not so much "received" from incoming photon impulses as they are semiotically and co-constructively "built" across heterogeneous and massively intercommunicating brain areas. Thus we find that sensory signification per se is intimately bound up with motoric processes of bodily and environmental interaction in an ongoing process of semiosis that cuts across the sub-systemic distinctions of brain, body and world.

What beauty exactly a cephalopod is seeing is of course unknowable, but whatever it may be, it hardly bears much resemblance to anything we would see in the same situation. However, the cephalopod certainly sees something and this something very probably is precisely the thing it needs to see. Conceivably, one might object that the cephalopod doesn't really *know* what it sees, and that it just reacts to what appears in its field of vision, not having the faintest idea of what it is about to do. But since our knowledge of what it means to *know*, neurobiologically speaking, is rather limited, such an objection likewise carries limited weight. Conscious knowledge is, for all we know, the privilege of a few big-brained animals, but most knowledge is probably unconscious, like the many routines one performs without paying attention to them.

The phenomenon of *blindsight*, for example, offers some surprising insights into the hidden reserves of knowledge that we all apparently carry around in our minded bodies. Blindsight may be observed in patients that have damaged their primary visual center so that they have lost access to a part of their visual field. If they are asked whether or not they can see an object placed in the blind area, their answer is, of course, no. And yet, if such patients are asked to guess where an object that they report they cannot see is placed, they may often point very accurately to its position. The explanation for this phenomenon is thought to be that visual impulses are divided into several parallel pathways on their way from retina to the brain, and some of these do not lead to the visual cortex but end up elsewhere in the brain. Here they obviously cannot produce conscious visual experiences, but the codified information is nevertheless still accessible to the analytic machinery of the brain. So, the patients see without seeing. Their vision is not accompanied by an experience of seeing — nevertheless they do, to some extent, *know* what their eyes tell them.[1]

Furthermore, many kinds of knowledge are purely embodied, such as, for instance, the immune system's *knowledge* of past infections, or the proprioceptive calibrations in the motoric system that explain why you are still able to successfully bike or swim, even though you haven't practiced any of these skills for years.

If you remove the brain from an earthworm (and this "brain" is nothing to write home about, as many readers may remember from this evil deed of childhood), the worm may still move forward as if nothing had happened. But when the worm comes to an obstacle, it is no longer able to pass beyond it: it continues again and again to push into the obstacle. It seems that it is no longer able to let the obstacle become knowledge in the sense that it uses the obstacle as a cause for changing its course. Compared to worms, the cephalopods have impressively well-developed brains. The octopus may have as many as 168 million nerve cells, half of them in the visual cortex, and is capable of at least limited associative forms of learning. Judged on this background, it may perhaps be permissible to say that the octopus does indeed *see*.

---

1   One might be tempted to call such people *clairvoyant*. And while literally this is of course rubbish, it may hit quite well into the heart of the superstition. For blindsight illustrates what clever people have always known — i.e., that undreamt of resources may be found outside of consciousness's little enlightened room. Or, as Pascal usually gets the honor of saying, *the heart has its reasons that reason doesn't know*.

Umwelt Theory

Early in the twentieth century, the Estonian-born German biologist Jakob von Uexküll saw, long before anybody else, that a biology that would be true to its subject matter would have to direct its searchlight explicitly on the perceptual worlds of organisms, their *Umwelts* as he called them.[2] The Umwelt, as Uexküll used the term, is the subjective or phenomenal world of the animal. The way Uexküll saw it, animals spend their lives locked up, so to speak, inside their own subjective worlds, each in its own Umwelt. Thus, while modern biology employs the objective term *ecological niche* (that is to say, the set of conditions — in the form of living space, food, temperature, etc. — under which a given species lives), one might say that the Umwelt is the ecological niche as *the animal itself* apprehends it.

One of Uexküll's prime examples was the tick, known to crawl up in branches only to wait, nearly lifelessly, for a warm-blooded animal eventually to pass by below. Only when this happens will the tick let go of the branch and land itself upon the animal, where it quickly burrows itself into a fixed position on the animal's skin. The one signal that awakens the tick is butyric acid, a compound secreted by all mammals, and thus the Umwelt of the tick consists mainly in the presence or absence of butyric acid.

In *Bedeutungslehre* (*The Theory of Meaning*), published in 1940, Uexküll writes, "If we stand before a meadow covered with flowers, full of buzzing bees, fluttering butterflies, darting dragonflies, grasshoppers jumping over blades of grass, mice scurrying and snails crawling about, we would instinctively tend to ask ourselves the question: Does the meadow present the same prospect to the eyes of all those different creatures as it does to ours?" (Uexküll 1982 (1940), 45). And to illustrate why the answer to this question is no, he uses the example of a meadow flower:

1.) A little girl picks the flower and turns it into a decorative object in *her* Umwelt;

2.) An ant climbs up its stalk to reach the petals and turns the flower into a natural ladder in *its* Umwelt;

3.) A larva of the spittlebug bores its way into the stalk to obtain the material for building its "frothy home," thus turning the flower into building material in *its* Umwelt; and

4.) A cow simply chews up the flower and turns it into fodder in *its* Umwelt.

---

2   The term *Umwelt* can be traced back to a Danish poet, Jens Baggesen, who lived in Kiel (which is now part of Germany, but still belonged to Denmark in Baggesen's time). Baggesen wrote in German and translated the Danish term *omverden* (surroundings) to German as *Umwelt* around the year 1800 (Albertsen 1990).

Each of these acts, he says, "imprints its meaning on the meaningless object, thereby turning it into a conveyor of meaning in each respective Umwelt" (ibid., 131).

The species-specific Umwelt of the animal, the model it makes of its immediate surroundings, is for Uexküll the very point of departure for a biological analysis. As the two parts in a duet must be composed in harmony (tone for tone, bar for bar), thus, he says, the organism and its Umwelt must also be composed in a contrapuntal harmony with those objects that enter the animal's life as meaning-carriers (ibid., 68). It is this idea of contrapuntal harmony that lets Uexküll call the flower beelike and the bee flowerlike, or the spider flylike, and the tick mammallike.

Yet, poetic formulations like these probably are much responsible for the rejection of Uexküll's ideas among most biologists and philosophers as being suspiciously *vitalistic*. For as we saw in Chapter 1, the term *vitalism* may cover a range of different conceptions, which makes it a difficult accusation to refute, once levelled. Uexküll, however, never *defined himself* as a vitalist, and whereas Driesch, in his attempt to capture the essence of the life-world, returned to the Aristotelian concept of *entelechy*, Uexküll used the much more commonsensical word *plan* (Kull 1999d). Now, there can be no doubt that the Uexküllian conception of evolution as a sort of overarching regularity (*Planmässigkeit*) or composition of a big symphony, goes against the ontological intuitions of most modern biologists, who see chance mutations as the ultimate source creativity in the organic world.

And true enough, the more nature is seen as a "perfect symphony" the more difficult it becomes to connect von Uexküll's Umwelt theory with the evolutionistic conceptions that hold that either (if everything is perfect now) the world wasn't perfect earlier (when it was different) or conversely, if everything was perfect in earlier times, it cannot be so now. If, however, one tries to fit the Umwelt theory into an evolutionary framework, there emerges, as Frederik Stjernfelt (2001, 88) has observed, an important finding: "There remains, namely, a gestaltist and hence non-irrational account of the organization of the life of an organism." For in describing the behavior of animals as being arranged according to distinctive *qualitative* categories, (that he termed *tones*), Uexküll is on the track of a phenomenon that was later in phonetics and psychology to be christened *categorial perception*. Stjernfelt says,

> The melody — arch-example for the Gestalt theorists from von Ehrenfels, Stumpf, and the early Husserl onwards to the Berlin and Graz school — articulates an organized structure disconnected from the here and now of physics and implying a teleological circle foreseeing the last note already by the intonation of the first. Thus — as Merleau-Ponty (1995 (1968), 233) remarks — this metaphor makes it possible to see the life of the individual as

a realization, a variation of the theme, *requiring no outside vitalist goal* — a variation, we may add, which constitutes the condition of possibility of modification of the animal's system of functional circles and hence the acquiring of new habits, possibly to govern evolutionary selection in Baldwinian evolution, . . . Music may be perfect but it is far from always the case (Stjernfelt 2001, 87–88; italics added).

Interpreting Uexküll's work in this way, we can see that *Planmässigkeit* does not imply a deterministic unfolding of a preordained order. And although the *telos* involved in *Planmässigkeit* is of course very different from Peirce's vision of evolutionary cosmology, it is not necessarily antagonistic either to Peirce or to the modern-day biosemiotic understanding (see Sebeok 1979, chapter 10). Rather, Uexküll's *Planmässigkeit* may be understood in its *purely local* and *situated* context:

> The semiotics of corporeal life in any creature — ourselves included — does take part in the dance of ecosemiotic motifs, the *local Planmässigkeit*, which has been framing the evolutionary processes and has formed the particular form of the Umwelt of each species. The Umwelt must serve to guide the animal's activity in the semiotic niche, i.e., the world of cues around the animal (or species) which the animal must necessarily interpret wisely in order to enjoy life. The semiosphere, as I use the term, i.e., the totality of actual or potential cues in the world, is thus to be understood as an externalistic counterpart to the the totality of Umwelts. Together they form, in the term of Jakob von Uexküll, an unending set of "contrapuntal duets" (Hoffmeyer 2006, 94).

I shall not delve further into the details of Uexküll's ontological positions but simply conclude that, whether the deeper presuppositions that nourished the work of Uexküll[3] are deemed acceptable to a modern scientific sensitivity or not, his Umwelt theory was, in any case, a milestone on the way to the establishment of a biosemiotic understanding of nature.[4]

For a characteristic concept in the work of Uexküll is the word *hinausverlegen* — a word that I, in agreement with Thure von Uexküll (Jakob's son), will translate as "projected to the outside" (Uexküll 1982). What is projected to the outside is precisely the Umwelt:

> No matter what kind of quality it may be, all perceptual signs have always the form of a command or impulse. . . . If I claim that the sky is blue, I am doing so because the perceptual signs projected by myself give the command to the farthest level:

---

3   Stjernfelt has recently returned to an in-depth analysis of these questions, with particular emphasis on the connection between Uexküll's, Husserl's, and Peirce's positions (Stjernfelt 2007: see also Bains 2001).

4   In this computer age of ours, the term *virtual reality* may perhaps express Uexküll's fundamental idea better than the slightly awkward term *Umwelt*.

Be blue!... The sensations of the mind become, during the construction of our worlds, the qualities of the objects, or, as we can put it in other words, the subjective qualities are building up the objective world. If we, instead of sensation or subjective quality, say perceptual sign, we can also say: the perceptual signs of our attention become the perceptual cues (properties) of the world (Uexküll, 1973; quoted in Uexküll 1982 (1940), 14–15).

Animals unconditionally and throughout their lifetimes conjure up internal models of the outer reality that they have to cope with. And these virtual realities apparently may sometimes entail an interactive aspect, too, since it is known that almost all vertebrate animals do on occasion dream. The Umwelt theory of Jakob von Uexküll is presumably the first serious effort ever made to subject virtual reality to scientific investigation (Hoffmeyer 2001c).[5]

The idea that animals possess internally experienced or phenomenal worlds that they then project back upon the outside world, however, has never been well received by mainstream twentieth-century biology. Rather, as John Collier has observed, the "modern synthesis" of the 1930s and 1940s signified a pervasive turn towards *behaviorism* in biology (Collier 2000): Organisms began being treated as black boxes, operated upon by the external forces of mutation and environmental selection. What went on *inside* the black box (morphologically, physiologically, or psychologically) was no longer seen as part of the *generative dynamics of nature*, since only the consequences of such processes — i.e., the actual survival patterns and population differentials — needed to be taken into consideration, it was thought, in order to understand and explain the great scheme of natural selection. In this scheme, the eventual possession by animals of phenomenological worlds was at best considered to be an unnecessary complication — much in the same way that human consciousness for most of the twentieth century was a nonexistent subject in mainstream psychology, and for many of the same reasons. And worse yet, the idea of Umwelt was thought to signal a dangerous return of anthropomorphic or even animistic atavisms in a biological science priding itself on its potential to approach the scientific status of physics.

The automatic rejection by modern science of all theories carrying even the faintest trace of anthropomorphism (a rejection reminding one of the *horror*

---

5   And while animals are *absorbed in* their virtual realities, it is human fate to see through the illusion. The "gift" of speech implied that humans could not escape comparing their individual experiences with those of each other, and thus drawing the logical conclusion: there exists a shared world, a *reality*, which is neither wholly another's, nor mine. Thus, whereas the Umwelt comes to us *for free* (in that experientially, we *start from there*), the notion of a mind-independent reality per se, was from the beginning an intellectual achievement. That we now, finally, have come to understand that this reality, too, is itself (at least partially) a construction, does not make it any less real. (This discussion is found in detail in Hoffmeyer 1996b, chapter 8.)

vacui of an earlier epoch) is itself deserving of critical study (see Favareau 2006). As Karl Popper once remarked, if we are talking about the *nose* of a dog, we are also anthropomorphizing the dog, but we are doing so for good reasons, because the nose of the dog and the nose of the human individual are homologous organs, i.e., their structural and functional similarities are accounted for by the well-established fact of common ancestry. Likewise, claimed Popper, we are well-justified in speaking about *knowledge* in animals to the extent that homology implies that animal brains and human brains are evolutionarily related organs performing related functions (Popper 1990, 30). In fact, any claim that human beings are the only animals to possess Umwelt*s* (or perceptually experienced subjective states) would require additional theories to explain why other mammals should be so fundamentally different from us. No satisfactory theories pertaining to such an effect is known to this author.

Umwelt theory does not, of course, represent an atavistic revival of animism in biology. Quite to the contrary, one might say that modern science, in its obsessional rejection of animism has itself maintained a strange trace of that which it rejects, in that in its very fear of spiritualism, science has closed itself off from vast areas of the world which most of us would take to be very real even if *objectively* immaterial in some modest sense of this term. Or to state this differently (and using these terms as science understands them), materialist science *spirit-ualizes* and, consequently, denies that area of lived experience that is the *virtual reality* of all animal perception.

The realization that the human experience of reality is always a *virtual reality* (though not one to be confused with any supposed *supra-reality* existing *independently* of our human knowledge) has, of course, been known to philosophers for centuries. But in general, the scientific community has been little influenced by the obvious consequences of this insight, namely that scientific reality itself is a human, and therefore humanly limited, construction. Or, to put it in the famous quote from Einstein, "Scientific concepts are free creations of the human mind, and are not, however it may seem, uniquely determined by the external world" (Einstein and Infeld 1938).

Yet considering the heated debates — the so-called science wars — arising in the wake of the Sokal affair (Robbins and Ross 1996, see Brown 2001) it may be necessary to emphasize that the social construction of reality which we label as *scientific* is not, in my view, independent of the genuinely mind-independent reality which it purports to investigate. In fact, I believe that the particular strategy underlying the scientific endeavor assures a probably unequaled dependence of knowledge on reality. And yet, in some sense this project still must be a *construction*, and the virtuality of human understanding cannot be escaped as Jakob von Uexküll clearly saw. Moreover, the epistemological consequences of this fact must be confronted as Uexküll certainly did in his *Theory of Meaning* (Uexküll 1982 (1940)).[6]

Yet the widespread resistance of the scientific community towards an acceptance of this fact explains why scientists in general feel justified to neglect the whole idea of virtuality as anything *real* in the world. But this willful epistemological innocence may now have become challenged in a way which mainstream biology may find difficult to ignore. This challenge comes from research in what is called artificial life as envisioned by, among others, Christopher Langton (1989). For if organisms are not understood in the Uexküllian way as living systems which are inherently and irreducibly suspended in their own phenomenal worlds or Umwelts, then one might easily imagine that the algorithmic kinds of dynamic systems exhibited by computer simulations do, in fact, mirror the abstract fundamental principles of *life* — whereas, *life as we know it* (i.e., organic life) would then be just one particular *instantiation* of this abstract "life form." And this is, in fact, the idea implicit in Christopher Langton's distinction between *A-life* (for artificial life) and *B-life* (for biological life). What this approach presupposes is that A-life theory and biology are equally valid ways to study life, because both are just new kinds of simulacra of some evanescing general life form (see Emmeche 1994). Both are equally virtual realities, and biology cannot claim privileged access to the reality of life.

If biology maintains its rejection of virtuality as a real constituent of life, it is hard to see how one can escape the logic of Langton's approach. If, however, biology adopted a more Uexküllian and thus semiotic approach to the study of life, then virtuality would be seen as built into life from the very beginning, and one might easily dismiss artificial life as it is presently conceived as fundamentally nonliving (regardless of its physical biology, or lack thereof). This would not, however, necessarily exclude the possibility that computers might some day *in the future* be constructed to host semiotic kinds of *true* artificial life. But in our opinion, this would most likely presuppose that ways were devised to solve what is called the *qualia problem* (Searle 1992; Emmeche 2004 — discussed below).

6   Unfortunately, as Stjernfelt (2007, 228) observes, Uexküll has a tendency to fall prey to "a widespread German temptation to naturalize this constitutive subjectivism" with the consequence that "physical laws of nature, for instance, become mere extrapolations and abstractions in the specific human Umwelt." However, as Stjernfelt himself, points out, the acceptance of a naturalized subjectivism, as Uexküll develops it, is not reconcilable with constitutive subjectivism at any rate. (Nor need we follow Uexküll — or any other thinker — in *all* of his conclusions, in order to use the conceptual tools that he developed. Were such a rule required, science could hardly progress at all!)

Self-Organization, Semiosis, and Experience

In her fine analysis of the concept of intentional behavior as a property of complex systems, the American philosopher Alicia Juarrero reminds us that the modern idea of self-organizing systems runs counter to a philosophical tradition leading back to Aristotle and that, all the way through, is based on the assumption that *causes are external to their effects* (Juarrero 1999, 2). Aristotle claimed, writes Juarrero, that nothing can move, cause, or act on itself in the same respect — and this principle has remained unchallenged throughout the history of philosophy. That a chicken develops from an egg is not, in the Aristotelian conception, due to immanent causes in the egg as a substantial thing; it is due rather to formative determinations that characterize hens in general (see Giordano Bruno's critique of Aristotle in Chapter 3).

Likewise, Kant inherited and expanded on this idea of causes as being external to their effects. He certainly seems to have intuited the self-organizing properties of organisms as a characteristic trait of life, but for him this property became a reason for *not* counting life as a field for scientific understanding. Juarrero puts it this way:

> Organisms' purposive behavior resists explanation in terms of Newtonian mechanics and is likewise a major impediment to unifying science under one set of principles. These considerations convinced Kant that natural organisms cannot be understood according to mechanism in general or its version of causality in particular. Since only external forces can cause bodies to change, and since no "external forces" are involved in the self-organization of organisms, Kant reasoned that the self-organization of nature "has nothing analogous to any causality known to us." Kant thus upheld Aristotle: causes are external to their effects; self-cause, and therefore, self-organization, are phenomenally impossible (ibid., 47).

In this elegant but, as seen from the point of view of rationality, strangely powerless way, Kant escapes the obvious antinomy between a Newtonian understanding of nature and life's self-organizing finality (see also Stjernfelt 1999).

For better or worse, natural scientists rarely let themselves be impeded by such philosophical reflections on the permissibility or nonpermissibility of this or that theoretical construction. And by the end of the twentieth century, the idea of self-organization little by little begins to take hold in science, thanks to developments in a range of advanced studies inside physics, biology, cognitive science, economics, and elsewhere. Too, a relative consensus seems to have been reached in viewing complex systems as having dynamic properties that allow for self-organization to occur (Haken 1984; Yates 1987; Kauffman 1993; Kelso 1995; Port and van Gelder 1995). Self-organization is seen here as a process by which energetically open systems of many components "tend to reach a particular state, a set of cycling states, or a small volume of their state space (attractor basins), with no external interference. This attractor behavior is often recognized at a different level of observation as the spontaneous formation of well-organized structures, patterns, or behaviors, from random initial conditions (emergent behavior, order, etc.)" (Rocha 2001, 96).

Juarrero's book is a scientifically well-informed attempt to use the conceptual structure offered by the theory of complex adaptive systems as a resource for the establishment of a "different logic of explanation — one more suitable to all historical, contextually embedded processes, including action" (Juarrero 1999, 5). In complex adaptive systems, there occur such kinds of positive feedback loops whose products are themselves necessary for the process to continue and complete itself, thereby producing a *circular cause* or a *self-cause* (ibid.).[7] Such systems, furthermore, form dynamic wholes that are not just, as science so often assume, *epiphenomena*, but are capable, as systems, of exerting causal power over their own components, and of exhibiting both *formal* as well as *final* kinds of causality. Juarrero furthermore claims that causal connections between different levels in the hierarchical structure of these systems are best described as *constraints* — in the sense of restrictions in the space of possibilities for processes to be able to manifest or realize at any given particular level.

Juarrero's scenario for the formation of complex adaptive systems capable of intentional and meaningful action is an impressive *tour de force* and is, in any case, a decisive contribution to the understanding of the philosophy of self-organization. Nonetheless, her analyses omit the semiotic aspect of selfhood as one of its concerns. It therefore remains unexplained how the element of first-person perspective that necessarily clings to intentionality — i.e., the fact that intentionality always presupposes an intentional *subject* — might possibly have appeared out of sheer complexity. How, in other words, could a self-organizing system that — in principle at least — might be described algorithmically in terms of sequences of ones and zeroes ends up with intentionality in the first-person sense of this term?

Traditionally the argument has been that the reason why evolution — though based on a continuous stream of chance events — can nevertheless create strange phenomena such as people, is that we, in the words of Eugene Yates (1998, 447),

> are the result of a random variation blocked at the statistical "left wall" of simple organisms, by the fact of their minimal complexity. The thus-constrained drift through chance must be toward the right (increased complexity), but it has no special outcome or elaboration. By a concatenation of accidents encountered and

7   See the bladderwort system — analyzed by Ulanowicz (Chapter 3) — that exhibits this same type of causality. Kant already, with some alarm, noted this kind of causality (Stjenrfelt 1999, 2007).

avoided, we are here, along with Venus flytraps, humming birds, and crocodiles. But the *modal* (most frequent, widely distributed, and most totally massive) forms of life are the bacteria.

I concur with this argument as far as the appearance of complex organisms is concerned. But this kind of explanation fails to take seriously the fact that we are not just complex material aggregates, but also *subjects*. Every person is genuinely an "I" phenomenon, whereas complexity in principle can be exhaustively described as an "it" phenomenon. How"it"s can possibly become "I"s is the puzzle that must be explained — and not even dynamic systems theory does yet offer a solution to this puzzle. What is missing, I would argue, is the admission of a semiotic dimension of explanation.

When we are often bothered (or offended in our scientific taste) by the badly hidden anthropomorphisms in Uexküll's writings, it is because it is maintained through his whole work that animals are much more like us than science has so far been willing to accept. And this is exactly because the animals have an Umwelt, an internal model of the relevant parts of their environment (i.e., those parts of the environment that are relevant for them), and that this model has to be included in any fully explanatory analysis of their life.

We need to take care to express things correctly here, and it may be a problem that language simply does not readily provide us with the appropriately subtle words. A tick waiting for butyric acid to reach its sense organs hardly has any experiences (as this term is normally understood). In fact, my guess would be that it is about as interactive as a computer in standby position. But in the moment its receptors catch the signal *butyric acid* in intensities that exceed the lower threshold value, a reflex-like movement occurs in it, immediately causing it to drop down upon (what turns out to be) its prey below. Now, even in this very split second, the state of the tick probably does not rise to the level of what we might call an experience, but here one might perhaps imagine the presence of some glimpse-like state of feeling — a *let go* impulse. On one level, of course, it is pointless to discuss unanswerable questions such as this. I do mention it here, however, because the question of the evolutionary history of *experiential* existence has huge theoretical implications, and raises the natural-science question: What might be the *function* of an experiental world? In other words, what good is the having of experiences in a biological sense?[8]

We shall suggest that experiences quite generally serve as holistic markers, causing the brain machinery to focus its (our) attention upon one single track in the spatio-temporal continuity. In animals that have admittance to the world of experiences, as for instance, the cephalopod (presuming it actually does expe-

8   I shall return to the philosopher David Chalmer's discussion of these problems in Chapter 8 (Chalmers 1996).

rience its demonstrated seeing), the sensoric aparatus continuously processes the changing production of an astronomical number of impulses being sent to different parts of the brain that — equally continuously and in parallel — activate a number of physiological and/or motoric mechanisms. All of this might presumably proceed quite efficiently in the absence of any experiential dimension — without qualia, as the philosophers might say. But there is a reason why a holistic control must interfere, and this reason is that the organism is a unitary agent in its own life. Holistic control, then, is needed in order to track the finality of brain processes in accordance with an organism's ever shifting current needs and intentions.[9] Thus it is through our experiences that the brain becomes a tool for the survival project of the bodily unity. As a tool for such holistic control, the body has at its disposal first its emotional equipment — as when young birds duck their heads at the sight of big-winged objects moving above the nest. Such emotional reactions are accompanied by measurable alterations in the physiological and biochemical preparedness of the body.

Secondly, there must minimally be an ability to build up a favorable correlation (or *ontogenetic optimization*) between the patterns of emotional reactions on the one hand, and the brain's sensoro-motoric coordination schemas on the other. Here we are talking about a kind of correlation — or *calibration* — that is unique to the individual's life history and cannot, for that reason, be encoded in the "innate manual" of the genome. And this is precisely where and why *experience* enters the picture. *Experiences serve to focus brain processes according to bodily finality by the creation of an approximated isomorph or analog virtual reality, a single dominating "lead track" that, as in a computer simulation, extracts an iconic surface out of its deep cerebral activity.*

That experiences appear to us as analog codings of meaningful parts of our surroundings, so that we can, to some extent, justly project them out as Umwelt, is probably due to the fact that such codings establish the simplest possible functional mechanism. Since we are bodily creatures bound to operate by and in a world of space and time, the simplest — or safest — way to organize our calculatory imagination, also is also in time and space, or in other words iconically. Our muscles are not preprogrammed to their functions, but are calibrated in the course of our ongoing interactive life processes, and our muscles and our experienced worlds are tightly reciprocally calibrated. For these reasons, it would not

---

9   The bizarre finding that in Siamese twin salamanders — i.e., salamanders that have developed two independent bodies apart from a shared stomach — each head competes for food intake, although the food will end, anyway, in the shared stomach, illustrates the necessity for such a holistic marker, the absence of which, of course, is the reason why this pointless competition takes place in the poor creature (Hoffmeyer 2006). The German embryologist Hans Spemann has told that the wonder he once felt toward this little creature was the reason why he was originally spurred into a lifelong career in embryology (mentioned in Hamburger 1988).

be unlikely that experiences are iconically coded in all animals that have experiences at all. The holistic control function is an emotionally anchored focusing of our brain processes. It has nothing to do with directly *controlling* the processing of the infinite multiplicity of input that the brain receives, but only deals with *establishing an overarching directional perspective*. The experience is at each moment the superior, immediate, and unconditional *interpretant* in the ongoing *biosemiosis* of the organism.

Then what about an animal whose nervous system is not sophisticated enough to produce such higher-order interpretants in the form of analog-coded models? The need for some primitive version of a holistic marker is probably present in all forms of life, and I imagine a graded series of such markers that in the lowest end consisits in the patterns of attraction and repulsion characteristic to chemotactic behavior in bacteria. In other words, I suggest that the phenomenon of experience has primitive parallels all over the life world.Uexküll distinguished sharply between plants and animals. Only the former had nervous systems and, therefore, Umwelts. Plants instead possessed what he called a *wohnhülle* — a cover of live cells by which they select their stimuli. Like Anderson et al. (1984), I shall prefer to use *Umwelt* as a common concept for the phenomenal worlds of organisms, of whatever kind these might be. Although plants, fungi, and protists do not possess nervous systems, they do have receptors to guide their activities, and they all, in our view, possess some kind of semiotic freedom, however limited it might be.

The experiential component of life, *qualia*, is thus seen as an integral aspect of life as such — an aspect that has had its own evolutionary history from its most primitive forms in prokaryotic life to the sophisticated kinds of Umwelts that we find in big-brained animals. In this respect, our view is in line with the American philosopher Maxine Sheets-Johnstone, who has sketched a natural history of consciousness where especially proprioceptive senses play a central role for what she calls a *somatic consciousness*. The capacity for proprioception seems itself to have evolved in the metazoans, Sheets-Johnston (1998) claims, via an internalization of the simple receptors that were originally localized at the surfaces of our protistan ancestors (discussed below).

Additionally, it turns out that our *holistic marker* hypothesis is also in agreement with the American philosopher John Dewey (1948, 91):

> The true stuff of experience is recognized to be adaptive courses of action, habits, active functions, connections of doing and undergoing sensory-motor coordinations. *Experience carries principles of connection and organization within itself.* . . . These principles are none the worse because they are vital and practical rather than epistemological. Some degree of organization is indispensable to even the lowest grade of life. Even an amoeba must have some continuity in time in its activity and some adaptation to its environment in space. Its life and experience cannot possibly

consist in momentary, atomic, and self-enclosed sensations. *Its activity has reference to its surroundings and to what goes before and what comes after. This organization intrinsic to life renders unnecessary a super-natural and super-empirical synthesis*. It affords the basis and material for a positive evolution of intelligence as an organizing factor within experience (italics added).

The Semiotic Niche

The claim that there is an internal or subjective aspect to biological phenomena, and that this aspect must be taken into account in our theoretical understandings, has been called *internalism* and has lately been advanced in boundary explorations of evolutionary systems (Matsuno 1989; Matsuno and Salthe 1995; Van de Vijver 1996; Van de Vijver, Salthe, and Delpos 1998; Chandler and Van de Vijver 2000). While traditional neo-Darwinism clearly is *externalistic* in this sense, the Uexküllian Umwelt theory potentially takes us directly into the area of internalism.[10] As the Japanese biophysicist Koichiro Matsuno has explained, internalism is concerned with situations where a system finds itself in a state that might be grammatically characterized as its *present progressive tense*: the state of being in the midpoint of action — going towards, changing, recognizing, etc. Science never deals with such states but only with states that belong to the *present* or *past tense*.[11] According to Matsuno (1996), this omission by science of considering the unique properties of states in their present progressive tense — states of *becoming*, rather than *being* — springs from the universalist and externalist ambitions of science.

Honoring such an ambition presupposes the synchronization of all parts under one single measure of time. Since, however, nothing can move faster than the speed of light, synchronization — and thus universalism — can, in principle, never be realized in the *midpoint* of acting, for no matter how small an entity might be, there are always, even inside the atom, distances between parts that must be overcome in shared action. The synchronized perspective only applies after the fact, i.e., not *while* something happens but *when* it happens.[12] This understanding seems to be in deep agreement with Peirce's thinking on continuity, as this is for instance expressed in "The Law of Mind" from *The Monist* 1892 (*CP* 6:102–63).

10   Uexküll did not himself say anything about the subjective and experiential aspect. In this regard he is, as Stjernfelt (in personal communication) has said, a "methodological behaviorist."
11   As Don Favareau (in personal communication) comments, this leaves the scientists in the same position as Zeno of Elea (of the fifth century BC) who could not understand how an arrow can ever be in motion, since at each discrete time interval it must be located in a single place!
12   Zeno again! Here, the well-known Achilles paradox.

Now, I should point out very clearly here that I do not wish to contest the view that the inner side of subjectivity per se is beyond the reach of the objective methods of science. The qualitative differences between the pleasure of looking at paintings of Rembrandt and the pleasure of being on the receiving end of a baby's first smile, is rightly considered a topic for the humanities, and definitely not for science. But even though science might not need to concern itself with examining the inner side of subjectivity, it may and should be concerned with examining the external side of subjectivity, such as the question of how the possession of subjectivity affects the living systems under study. It is not the task of biology to say what animal experiences are like (considered *as experiences*), but it is the task of biology to deal with the fact that at least some animals *have* experiences, and to study how this affects their livelihood.

The most obvious way biology could do this is by directing more attention to what I have previously referred to as *the semiotic niche* (Hoffmeyer 1996b). For the niche concept has a long ancestry in ecology. In 1917, Joseph Grinnell defined the niche as the totality of places where organisms of a given species might live. Ten years later, Charles Elton gave the concept a functional turn — seeing the niche as a description of the ecological role of the species, its way of life, so to say. The resulting duplicity in the understanding of the concept of niche has clung to it to this day: On the one side, the niche is a kind of *address* (Grinnell) on the other hand it is a *profession* (Elton).

In 1957, G. Evelyn Hutchinson gave the niche concept its modern definition, namely as an imaginary *n-dimensional hypervolume*, whose axes would indicate the multiple ecological factors of significance for the welfare of the species (Hutchinson 1957). Thus, the niche of a plant might include the range of temperatures that it can tolerate, the intensity of light required for its photosynthesis, its specific humidity regimes, and the minimum quantities of essential soil nutrients needed for its survival. Hutchinson also in this context introduced the distinction between an organism's *fundamental niche* and its *realized niche*. The fundamental niche of a species includes the total range of environmental conditions that are suitable for existence without the influence of interspecific competition or predation from other species. The realized niche describes that part of the fundamental niche actually occupied by the species.

In the *Oxford Companion to Animal Behavior*, the following more down-to-earth explanation is offered: "Animals are commonly referred to in terms of their feeding habits; terms such as carnivore, herbivore, and insectivore being widely used. The concept of *niche* is simply an extension of this idea. For instance, there is the niche which is filled by birds of prey which eat small mammals, such as shrews (*Soricidea*) and wood and field mice (*Apodemus*). In an oak wood this niche is filled by tawny owls (*Strix aluco*), while in the open grassland it is occupied by the Old-World kestrel (*Falco tinnunculus*)" (McFarland 1987, 411–12).

This latter conception of what constitutes a *niche* has the advantage that with it it becomes possible to pose a series of interesting questions — e.g., How many ecological niches are there in the world? Are there more niches in warm climates than in cold? Were there more (or fewer) ecological niches a hundred million years ago than there are today?

Yet the concept of the *ecological niche* has framed a controversy about whether it is possible for two species with identical ecological niches to coexist, or whether one of the species will always, in the end, outdo the other via *competitive exclusion*. The question is difficult to decide because the n-dimensional character of the ecological niche makes it impossible to definitively clarify whether the ecological niches of the two species are indeed identical in all respects. This problem, of course, will only become so much more insolvable, if one includes the semiotic dimensions of the niche concept, as we are going to do in a moment.

Traditionally, it has been assumed that natural selection would favor those individuals inside the competing populations that evade competition by entering into a partnership of reciprocal specialization in the choice of resources, what is called the strategy of *resource partitioning*. The result of resource partitioning is that niche overlap between different species is minimized. In tropical forests in South and Central America, for example, several hundreds of species of birds, monkeys, and bats all eat fruit as their primary food source — but the enormous diversity of available fruits there has allowed all of these species to specialize such that the overlap between their diets has become very slight. Similarly, in a now classic study, Robert MacArthur found that five species of singer birds with nearly identical niches self-segregated in a surprising way. Not only did they each seek food in different zones of the fir, they also ate insects in different combinations and timed their nest building differently.

Since Hutchinson's niche concept is n-dimensional, it is in principle wide enough to also embrace the semiotic dimensions of an organism's need for a living place.[13] It is plain, nevertheless, that the niche concept — as currently used in ecology — is grounded in a de-semiotized understanding of the interplay between organisms in nature. Behavioral ecology may well have become a fashionable part of ecology, but the methodology of this approach is based upon a selectionist frame of understanding that leaves no space open for a semiotic perspective. It is therefore necessary to introduce a special concept to cover the semiotic dimension of the niche concept, and my suggestion of the term *semiotic niche* was intended to do precisely this.

---

13   Myrdene Anderson hints that Hutchinson may have been acquainted with Uexkull's work and tells us that when Hutchinson was once asked, late in his life in 1991, to indicate "the singular puzzle left us at the end of the twentieth century," he spontaneously replied, "Insides and outsides" (Anderson 1998). And this is of course a basic semiotic theme (see Chapter 2: "Surfaces Within Surfaces").

The idea behind the concept of the semiotic niche was to construct a term that would embrace the totality of signs or cues in the surroundings of an organism — signs that it must be able to meaningfully interpret to ensure its survival and welfare. The semiotic niche includes all of the traditional ecological niche factors, but now the semiotic dimension of these factors is also strongly emphasized. The organism must *distinguish* relevant from irrelevant food items and threats, for example, and it must *identify* the necessary markers of the biotic and abiotic resources it needs: water, shelter, nest-building materials, mating partners, etc. The semiotic niche thus comprises all the *interpretive challenges* that the ecological niche forces upon a species. Here are the words I originally used when introducing this concept: In order to occupy a semiotic niche, an organism or species "has to master a set of signs of a visual, acoustic, olfactory, tactile, and chemical nature, by means of which it can control its survival in the semiosphere" (Hoffmeyer 1996b, 59). To these means of semiosis one ought to add, as I have now learned, ultraviolet, ultrasonic, magnetic, electrical, solar, lunar, and presumably a host of other communicative media (Hediger 1974, cited in Sebeok 2001b, 24).

The semiotic niche in this way may be seen as an *externalistic counterpart* to the Umwelt concept. It makes the Umwelt concept easier to handle in an evolutionary context, since now one may pose the question of whether the Umwelt of a species is up to the challenges posed by the available semiotic-niche conditions. *Magnetotactic* bacteria, for instance, are anaerobic organisms that find their livelihood in the border zone between water and sediments. Because these bacteria do not tolerate oxygen, they must by all means avoid surface water, and evolution solved this problem in an inventive way. Their cytoplasm contains a smart protein-based compass, a magnetosome that tells them, what is up and what is down. However, if by accident these bacteria shifted hemispheres from north to south (or vice versa), they would soon perish, because the magnetosome would lead them to swim to the surface where the oxygen would kill them. In this (admittedly speculative) case, their Umwelt would not fit the semiotic niche available to them.[14]

Semiotic Freedom

The so-called Cambrian explosion that took place in the Cambrian era, half a billion years ago, refers to fossil findings that were interpreted to show a dramatically rapid appearance of new types of animals at this point in life's history. It has been suggested that all (contemporary as well as extinct) major phyla (main

14   This example, by the way, also illustrates the danger or insufficiency of the Uexküllian concept of life as a perfect symphony!

groups) of animals were established in one big "moment" of creativity at that time — with not a single basic architectural form (*bauplan*) having been added since (Gould 1989). (Much controversy has ensued around these ideas, and for our purposes we do not need to take a side in this battle.)

One theory — just for illustration — has it that the explosion was caused by rising oxygen content. Before the Cambrian era, oxygen concentrations were supposedly too low to support life of anything but the smallest animals. Algae in the oceans in these distant times may have produced more oxygen than bacteria and other primitive marine animals could consume. Accordingly, the oxygen content on the whole began increasing, in spite of much turbulence. At the start of the Cambrian era, a critical threshold might have been attained, however, allowing bigger (more oxygen-using) animals to survive. At that point in time, therefore, a whole new competitive parameter was introduced — i.e., that of being big, and this then gave birth to radical evolutionary experiments involving how to make the most of bigness. The Cambrian explosion, in this view, reflects the fact that it took evolution less than seventy million years to invent the approximately thirty-five fundamentally different ways (i.e., basic body plans) to be an efficient big animal (with *big* here meaning more than a few centimeters).[15]

Be this as it may, what interests us here is the question of whether the space of morphological possibilities for constructing animals (which apparently filled up at this time) would have simultaneously caused a filling up of the *possibility space* at the level of ecological niches. Were the fundamental ecological roles already established several hundred million years ago, and has evolution since then mostly been concerned merely with the finer adjustments of these basic settings? (As, for instance, when marsupials spread into many of the niches left open by the extinction of the dinosaurs only to find themselves replaced later, for the most part, by placental mammals.)

The answer that I propose to this question is *no*. And the reason I feel confident in saying so is because the property that we have called *semiotic freedom* (Hoffmeyer 1992) has an inherent tendency to grow, as we shall see. Over time, this has occasioned the formation of a range of new semiotic niches — thereby also, according to Hutchinson's niche concept, a corresponding range of new ecological niches.

Semiotic freedom was defined as "the depth of meaning that an individual or species is capable of communicating" (ibid., 109). The use of the word *depth* in this connection is related to Charles Bennet's concept of *logical depth* — his attempt to supply the concept of *information* with a measure for

---

15   An alternative and perhaps equally likely explanation might be that these thirty-five basic animal architectures simply were the "lucky ones" that came in for the share before all others, effectively blocking the way for newcomers (Gould 1989).

the *meaningfulness* or *complexity* of the information, quantified as the number of calculatory steps spent upon producing it. I have no illusions as to the possibility of transferring this kind of calculation from the world of computers to the reality of nature, but intuitively it seems clear that the meaning of different messages may indeed have different depths.

Thus, the saturation degree of nutrient molecules upon bacterial receptors would be a message with a low depth of meaning, whereas the bird that pretends to have a broken wing in an attempt to lure the predator away from its nest might be said to have considerably more depth of meaning. In talking about semiotic freedom rather than semiotic depth, then, I try to avoid being misunderstood to be claiming that semiotic freedom should possess a quantitative measurability; it does not. But it should also be noted that the term refers to an activity that is indeed *free* in the sense of being underdetermined by the constraints of natural lawfulness. Human speech, for instance, has a very high semiotic freedom in this respect, while the semiotic freedom of a bacterium that chooses to swim away from other bacteria of the same species is of course extremely small.[16] The middle ground between these two extremes is the main arena of biosemiotics.

In biology it has been widely discussed whether evolution might be seen as having optimized certain specific parameters for organisms. A range of parameters have been suggested as candidates for such a role, but none of them have been generally accepted. Probably the most common assumption has been that evolution exhibits a trend towards the increased complexity of organisms. The problem is, however, that it is not exactly clear just what this complexity amounts to. According to the evolutionary biologist Daniel McShea, it is more or less agreed that the morphological complexity of a system is determined by the number of different parts of which it is comprised and the greater or lesser irregularity of their arrangement. A complex system is therefore "heterogeneous, detailed and lacking in any particular patterns" (McShea 1991; Hoffmeyer 1996b, 60–61). Accordingly, McShea (1991) concludes that despite what common knowledge would have us believe, there is hardly any empirical evidence to support the theory that complexity, in the above-mentioned sense, has grown greater in the course of evolution. And apropos of this, he quotes the distinguished paleontologist George Gaylord Simpson (1949, 252): "It would be a brave anatomist who would attempt to prove that Recent Man is more complicated than a Devonian ostracoderm" (the ostracoderm is a species of fish, to which the trunkfish belongs, that was in existence between three and four hundred million years ago).

16   But even a bacterium is a very complex physico-chemical system that is underdetermined by its internal parameters in the sense that its contextual situatedness cannot fully account for cellular controls.

From a biosemiotic point of view, however, the focus of this analysis is mis-directed. I have nothing against the idea that the purely morphological com-plexity of organisms reached its upper limit already in the Devonian period, or even earlier for that matter. But it seems obvious that as evolution little by lit-tle created animals with central nervous systems to be players "in the ecolog-ical theater" (to borrow Hutchinson's famous phrase), the play itself changed character so that increasingly, evolutionary gains would turn upon the develop-ment of efficient mechanisms for social interaction and cooperation — as well as upon such misinformative practices as cheating and faking — and, in short, that evolutionary games would be expected to increasingly concern the acquir-ing of semiotic competence. Therefore, as I originally suggested, "the most pro-nounced feature of organic evolution was . . . not the creation of a multiplicity of amazing morphological structures, but the general expansion of 'semiotic freedom.' . . . The anatomical aspect of evolution may have controlled the earlier phases of life on Earth but my guess is that, little by little, as semiotic freedom grew, the purely anatomical side of development was circumscribed by semi-otic development and was thus forced to obey the boundary conditions placed on it by the semiosphere" (Hoffmeyer 1996b, 61–62). And indeed, as soon as one puts on one's semiotic glasses, the evolutionary trend towards the creation of species with more and more semiotic freedom becomes so obvious that one may wonder how it can be that it was never suggested.[17] The main reason for this may well be that *anthropomorphism* is generally considered such a deadly sin of the first magnitude, that in setting up semiotic freedom, as I do here, as the pivotal point of evolution — at least in its later phases — we almost by def-inition must accord to human beings the status of being the foremost creatures in the natural history of the Earth. Perhaps this is also the reason why science in general is suspicious of the semiotization of nature implied by the biosemiotic approach. It is time to stop this farce[18].

Semethic Interaction

The growth in semiotic freedom through evolution is caused by the possession in living systems of an extreme semiogenic capacity, a capacity based on their ability to *read omens* in the broadest possible sense of this expression — in other

17   We do not claim complete knowledge of the literature, of course, but the Nobel laureate and French molecular biologist François Jacob's statement that "evolution depends on setting up new systems of communication" (Jacob 1974, 308) is the closest case known to us.

18   In the original Danish edition of this book, I said this more succinctly, I suppose, by using the old expression: It is time to call a spade a spade. But apparently this, in the U.S. context, might be read with racist connotations that are absent in the Danish context and were, of course, not intended.

words, to take advantage of any regularities they might come upon as signify-ing vehicles, or signs. And indeed, although the word is not often any longer used this way these days, I must stress at the outset that by the word *omen* I mean nothing at all mysterious or supernatural. Anything is an omen until we understand its true significance. Thus, whether this reading of omens occurs via genetic adjustments down through generations or occurs as an effect of the cog-nitive system of an individual organism, is, in this connection, virtually irrele-vant. What happens in both cases is the same — seen from the standpoint of semiotics — although the time scales of events are, of course, widely different in the two cases. I have called this pattern of interaction *semethic interaction* (from the Greek, *semeion* = sign + *ethos* = habit) (Hoffmeyer 1994a; 1994b). Whenever a regular behavior or habit of an individual or species is interpreted as a sign by some other individuals (conspecific or alter-specific) and is reacted upon through the release of yet other regular behaviors or habits, we have a case of semethic interaction.

The bird that lures the predator away from the nest by pretending it has a bro-ken wing — and then flies away as soon as the predator has been misled a suf-ficiently long way — is an obvious example of a partner in a game of semethic interaction. And, in fact, at least two cases of semethic interaction are involved here: first, the predator has perceived (genetically or by experience) that clumsy behavior signifies an easy catch. The bird's behavior is therefore (mis)interpreted as a sign for an easy catch. Here we have a very simple semiotic process, where a nearly lawlike (and clearly nonsemiotic) relation (of one certain physical state to another — i.e., clumsiness with vulnerability) serves as a signifying regularity, or sign for the predator (Hoffmeyer 2008a).

The bird, however, takes advantage of a much less safe relation, the relation between a sign and its interpretant. By pretending to have a broken wing, the bird can "count on" (and again, this may or may not be a gentically fixed inter-pretant) the predator to misjudge the situation. In other words, the success of this strategy counts on a false interpretive act in the predator. That the preda-tor will misinterpret the bird's behavior may be a safe assumption — seen from our view — but it is hardly a lawlike necessity. Clearly the act of pretending in this case has to be well executed. In this way then, semethic-interaction patterns are built upon other semethic-interaction patterns in chains or webs of increas-ing sophistication.

Among biochemists, there is a rule of thumb saying that whenever nature keeps a store of energy (e.g., food) there will also always be a species that makes a living on consuming it. I shall suggest a quite similar rule of thumb by saying that there never occurs a regularity or a habit in nature that has not *become a sign* for some other organism or species. Admittedly, this rule may be less well investigated (so far!) than the biochemical rule, but it does catch an important

semiotic aspect of the evolutionary process, and that is this: Due to the mechanism of semethic interactions, the species of this world have become woven into
a fine-meshed global web of semiotic relations. And I shall claim, furthermore,
that these semiotic relations, more than anything else, are responsible for the
ongoing stability of Earth's ecological and biogeographical patterns.

Semethic interactions have been at play from the earliest steps of evolution.
An example on this is the invention of light sensitivity in early heterotroph
organisms.[19] Swenson and Turvey (1991, 340) give the following description:

> Photopigments were first used in photosynthesis, and in locating or moving toward
> or away from places where the wavelength of light was suitable or not suitable for
> photochemistry. . . . At some point, when (photosynthesising) cyanobacteria are
> presumed to have constituted a major portion of the biomass on earth, they them-
> selves represented a field potential on which heterotrophs . . . began to feed. The
> heterotrophs used the same photopigments for detecting light, but not to photo-
> synthesise; instead the pigments were used to detect light that was specific to where
> the autotrophs (photosynthezising cyanobacteria) were feeding (on the light).
> Light distributions specifying not light as food itself, but *information about* the
> location of food, was evolutionarily instantiated in its modern sense.

These heterothroph organisms evolved light-sensitive receptors, not because
they needed light, but because their prey needed it.

Some amusing examples of semethic interaction have already been given
earlier in this book. For instance the squid that survives through a mutualistic interaction with light emitting spirochetes, or the fungus that profits from
the regularities inherent to the sexual schemata of the male fly. Semethic interactions are probably involved in most — if not all — interspecific relations.
Both predator and prey must in their opposing projects necessarily be aware of
those signs that tell them about the habits of the opponent. A funny case in this
respect is the hare-fox interaction as described by Anthony Holley (1993). A
brown hare can run almost 50 percent faster than a fox, but when it spots a fox
approaching, it stands bolt upright and signals its presence (with ears erect and
the ventral white fur clearly visible), instead of fleeing. After ten years and five
thousand hours of observation, Holley concluded that this behavior is energy
saving: if a fox knows it has been seen, it will not bother to give chase, so saving
the hare the effort of running. Holley rejects the alternative explanation — that
the hares just want to better monitor the movements of their predators — partly
because the behavior does in fact not help them to see the fox more clearly, and

---

19   Heterotroph organisms, like animals, cannot make organic compounds from inorganic compounds and therefore have to procure such organic compounds by eating them. This is contrary,
of course, to the ways of autotroph organisms, such as plants that survive by photosynthesis — or
bacteria that get energy by taking in and degrading energy-rich inorganic compounds.

partly because they do not react the same way to dogs. While a fox depends on stealth or ambush to catch a hare, the dog can run faster and it would therefore be counterproductive for a hare to signal its presence. . . . The hare "knows" that the fox has the habit of not chasing it if spotted. Thus it develops the habit of showing the fox it has become spotted. Whether this habit has become fixed in the genomic setup of the hare or whether it is based mostly on experience is probably not known, but doesn't matter (Hoffmeyer 1997a).

The amazing semiogenic competence of many animals was perhaps most famously brought to the attention of the scientific community in 1907, when the German psychologist Oskar Pfungst disclosed the trick behind Clever Hans — the horse that surprised audiences all over Europe with its ability to do simple calculations. Recall this famous story: Hans's trainer would pose to it a simple arithmetic problem, such as 3×4, by writing with chalk on a black-board, and Hans would then reply by tapping one foreleg twelve times. In spite of many attempts, nobody was able to disclose any cheating until Pfungst began his studies.

The horse, of course, did not possess any capacity to do mathematics or under-stand writings on a table, but it did an eminent job of reading the wishes of indi-viduals from a foreign species. If the horse could not see the person posing the question, it could not then perform, and the explanation for its artful tapping was shown to reside in the horse's ability to notice an ever so slight — and obvi-ously unconscious — body movement by the trainer, when the correct number of tappings was reached. At the point when the cue showed up, all Clever Hans needed to do was to stop tapping. Now, unfortunately, this famous story has probably contributed more to bring ethology into ridicule than anything else. And, indeed, a series of other clever-animal stories have appeared since then, perhaps the most notorious being the many experiments purporting to show that great apes had been taught to talk (Sebeok and Umiker-Sebeok 1980). Yet in a comment on the Clever Hans phenomenon, the Swiss pioneer in nonverbal communication studies, Hans Hediger (1974, 27–28) writes, "The apparent per-formance of these 'code-tapping' animals is only explainable by the continually repressed fact, that *the animal* — be it horse, monkey or planarian — *is gener-ally more capable of interpreting the signals emanating from humans than is con-verse the case.* In other words, the animal is frequently the considerably better observer of the two, or is more sensitive than man; it can evaluate signals that remain hidden to man" (cited in Sebeok 2001b, 23). Hediger quotes Pfungst for the observation that horses are capable of perceiving movements "less than one-fifth of a millimeter" in the human face (Hediger 1974, 32).

Similarly, a fascinating example of semethic interaction between humans and birds concerns the African Boran people and a bird known as the black throated honey guide, *Indicator indicator* (Sebeok 1979, 14–18). Collecting honey is an ancient human practice as witnessed by 20,000-year-old cave paintings. The honey guide often accompanies the Boran people when they go out to collect honey. *Indicator indicator* guides them from tree to tree by characteristic call-outs. Thanks to this assistance, the time Borans expend finding the bees' nests (which is otherwise approximately three hours) is shortened by one third. The bees are smoked out, the hives are opened, and the honey collected. And while the honey guide birds cannot themselves open the hives, after the Borans have taken their honey, much valuable larvae and wax still remain in the hives for the birds to eat. The species designation *Indicator* bears witness to the spontaneous semiotic intuition that many biologists have upon discovering such interactions.

Too, semethic interactions may, in some cases, be very complex and involve several species. This is, for instance, often the case in *plant signaling*, where plants that have been damaged by insect attacks emit signals that are received by undamaged conspecifics. Undamaged fava beans (*Vicia fabea*), for instance, immediately started attracting aphid parasites (*Aphidus ervi*) after having been grown in a sterilized nutrient medium in which aphid-infected fava beans had previously grown (Bruin and Dicke 2001). The damaged beans thus had managed to signal their predicament through the medium to the undamaged beans, which then immediately started to attract aphid *parasites*, although no aphids were, of course, available for parasites to find.

Perhaps the best studied examples of this mechanism in plants concern cases where the sign vehicle is a volatile airborne compound (but soluble waterborne compounds that spread to neighboring plants through the earth also often function as messengers). The complexity of these relationships is further increased by the intervention of nonconspecific plants that may gain advantage from the density of freely available parasitoids (insects whose larvae lives as parasites that eventually kill their hosts)) and it is therefore conceivable that these nonconspecific plants themselves may develop sensitivity to the volatile signal molecules. Bruin and Dicke, reviewing a series of examples on this kind of communication, also advance the speculation that signals might be transferred by direct contact between the roots of neighboring plants, or even through the fungal bridges (*ectomycorrhiza*) between them.

Parasitic wasps (*Cotesia marginiventris*) that lay their eggs in caterpillars offer another intriguing example of semethic interaction. When a caterpillar munches on the leaves of a corn seedling, a component present in the oral saliva of the larva induces the formation of a signal that spreads to the whole plant. This signal causes the corn seedling to emit a volatile compound, a terpenoid, which is carried off with the wind. Eventually, the terpenoid arrives at the antennae of female wasps and is interpreted as a sign for oviposition, prompting the wasps to fly upstream towards the source of the terpenoid. Upon detecting the caterpillars, the wasps lay their eggs in the young larvae, one egg in each, and a couple of days later the eggs hatch and the parasitoid starts eating up the interior of the caterpillar. Ten days after oviposition, the parasitoid emerges from the caterpillar and spins itself a silky cocoon, leaving the host larva to die.

Seen from outside, what happens here is that the wasp and the corn plant have common, if opposite, interests in the caterpillar and have each worked out a cooperative way of satisfying those interests by actively sharing a small part of the semiosphere. Or, more concretely, a habit (the emission of a terpenoid by the corn plant when leaves are munched upon by caterpillars) has become a sign for the wasp, leading it to a suitable opportunity for oviposition. But should this wasp have any natural enemies, this very same successful oviposition mechanism might yet serve as a perfect habit for that enemy to exploit, building up even more layers of semethic interaction upon semethic interaction.

And in fact, parasitic and mutualistic symbioses are more or less unthinkable without a subtly developed pattern of semethic interaction between the involved organisms — as we have already seen in the case of the squid and light-emitting bacteria, where a multiplicity of signals and signal receptors interact back and forth across the species barrier. In Chapter 4, we discussed several other kinds of symbiotic relations such as lichens, mycorrhiza, and the cultivation of fungi by ants. Further examples include cellulose-degrading microorganisms in the intestinal tract of ruminants, pollination relations between flowering plants and insects, and the close cooperation between coral polyps and algae (usually dinoflagellates). The scheme is nearly inexhaustible, and if sufficiently broadly defined, every organism on Earth does, in some sense or other, enter into mutualistic symbiotic community with other species.

Semethic interaction is often involved in *intraspecific ritualization* — i.e., the development of stereotypic displays with communicative content that are intended for conspecific individuals. An interesting theme in this context is the iconic use of typically feeding-related items for mating purpose. For example, during courtship the water mite (*Neumannia papillator*) male will make a vibrating movement with its front four legs while wandering around the female — a behavior called *courtship trembling*. This trembling behavior is iconically indistinguishable from the vibration in the water surface that discloses the presence of the small animals that the mites feed upon. The water mite strikes a specific attitude while watching the vibrations in the water surface ready to seize the prey with its forelegs, and this is precisely the way the female seizes the male in the initial step in mating. It is, of course, particularly intriguing for the imagination that the hungrier females are more likely to gravitate towards the male and clutch him (Johnstone 1997, 161). Seen from the standpoint of semiotics, what goes on is that the male takes advantage of the female *foraging Umwelt* for the purpose of communicating his mating wish. Students of behavioral ecology aptly call this phenomenon *sensory exploitation* or even *sensory trap*.

And again, the same theme is repeated with many different variants in many different animals. For instance, it is well known that female birds will often, in a late phase of the mating ritual, strike an attitude that is otherwise only seen in very young birds when they are begging for food. This exclusively happens in a late state of the ritual, and the birds are not especially hungry. Wish for feeding has thus developed to become an icon for wish for mating. Sebeok discusses a more cruel case of courtship ritual in that of the balloon flies of the genus *Empididae*, where swarming males bring with them insects caught as "wedding gifts." "The male offers his gift to a female," writes Sebeok, (1979, 18), "which sits peaceably sucking it out while the male inseminates her. As soon as copulation is completed, the female drops her present, but if the empidid bride is still hungry, she may consume her amorous groom next."[20]

The normal case of semethic interaction concerns the interplay between two or more organisms, but abiotic regularities may also be used as a substrate for the semiogenic inventiveness of living systems, as we saw in the case of magnetotactic bacteria. Similar cases are found among the migratory birds that find their way across continents — or between them — by interpreting stellar configurations by night.[21]

Too, semethic interaction is by no means exclusive just to the organismic level but may also take place at levels other than the organismic. Thus, in the case of the eyeless mutant in salamanders, we saw that this very same principle of semethic interaction is also an important principle during embryonic development. The presence of a developing eye at a distinct stage of cerebral development is used as a semethic trigger for those tissues from whence the hypothalamus is supposed to develop — so that, in the absence of the eye, the hypothalamus and the gonadotropic hormones go missing as well. The example of female nest cooing in the ring dove (see Chapter 5, the section "The Semiotics of Reproduction") shows that semethic interaction is used by evolution to connect habits and

20   Sebeok's point in presenting this example is a little more sophisticated than we have felt necessary to show here. For, interestingly enough, this "wedding gift" varies quite a lot from species to species, so that in some species of *empididae* there are no such gifts at all (and there is thus a corresponding risk for the male of succumbing to the cannibalistic propensity of the female), whereas rituals amongst other species form a graded series wherein the insect steadily decreases in size (and hence in food value) while the balloon that the flies construct around the gift increases commensurably in complexity. Finally a stage is reached in evolution where the female receives only the empty balloon. Sebeok (1979, 19) says that at this stage, "from a strictly synchronic point of view, the link between a representamen and the object for which it stands has now become 'arbitrary,' and . . . thus (as well as in other familiar ways) the sign meets every viable definition of a symbol." This succession of evolutionary steps is, in itself, a splendid illustration of evolution's tendency to develop higher and yet still higher levels of semiotic freedom.

21   Alternately, there are dung beetles that forage by reading the polarization patterns of moonlight (*Nature* 424, 33).

signs *across* the traditional levels in the biological hierarchy. In this case, we saw that the physiological timing of ovulation had become scaffolded by habits of vocalization during courtship, since the brain of the dove apparently had evolved to interpret the body's own cooing behavior as a sign for the initiation of ovulation.

The Ecosemiotic Perspective

Traditionally, ecology has had a hard job in trying to map the multiple physical and chemical interactions between organismic populations, as these are reflected through such things as trophic structure and nutrient cycles. But the task of unraveling the semethic-interaction patterns between such populations is, of course, magnitudes more complex. Probably we have only seen the beginning of such studies, and my guess would be that our present knowledge gives us only a small glimpse of a nearly inexhaustible stock of *intelligent* semiotic interaction patterns taking place at all levels of complexity from cells and tissues inside the bodies up to the level of ecosystems.

The situation, in other words, has a matrix-like structure with multiple interdependent relationships binding populations of many different species into a shared interpretive universe or motif. Against this background, it would be reasonable to suggest that evolution may be as much constrained by the existence of these *ecosemiotic interaction structures*, as it is by developmental constraints (Alberch 1982).[22] In an earlier paper (Hoffmeyer 1997a), I suggested the term *ecosemiotic discourse structures* with reference to Michel Foucault's exposition upon the *discourse concept*, which, very briefly stated, refers to the symbolic order relating human subjects to a common world (Foucault 1970, Cooper 1981). However, the term ecosemiotic *interactions* may be preferable to that of *discourse,* since there is no reason to associate this activity to the human sphere of symbolic minds here.

Thus, while most biologists suppose that *symbiotic mutualism* is an exceptional case of no general importance for evolutionary theory, I believe that *semiotic* mutualism involving a delicate balance of interactions between many species is widespread (see Margulis and Sagan 2003). And if this is indeed the

---

22 *Developmental constraints* refer to the limitations that the developmental process puts on the evolutionary construction of phenotypes. One might, for instance, think that it would be simpler for horses to develop hoofs directly, but instead the horse embryo develops through a stage where the embryonic limbs have five digits. The likely explanation for this is that the internal logic of the developmental schemes in tetrapods makes it impossible to skip the five-digit stage. Evolution is thus constrained by structural bindings caused by the historical process whereby the developmental schemes were first established. It is probably for the same reason that no tetrapods have wings although such a feature might well be advantageous in some species.

case, it has significant consequences for our thinking about evolution, for it implies that the relative *fitness* of changed morphological or behavioral traits become dependent on the whole system of existing semiotic relations that the species finds itself a part of. Accordingly, the firm organism-versus-environment borderline will be dissolved, and a new integrative level intermediate between the species and the ecosystem would have to be considered — i.e., the level of the *ecosemiotic interaction structure*. Clearly, this possibility becomes most interesting in cases where experience and learning enters the interaction pattern, as will often be the case in mammals and birds. Such learning might on occasion even subsume the evolutionary process, as is the case in human culture. Conversely, one might wonder if a relatively autonomous ecosemiotic interaction structure is precisely what is needed for learning to evolve in the first place. In this way, eventual increases in semiotic freedom will be prone to feed back into the evolutionary process by strengthening the advantages of possessing semiotic freedom.

Thus, semiotic freedom is an emergent property and should always be analyzed in relation to its proper level. For example, the semiotic freedom of the free-living individual cell must have been severely diminished in the process that transformed unicellular organisms to multicellular organisms. The necessity for single cells to obey the *somatic ecology of the body*, as Buss (1987) termed it, must have constrained the freedom of each individual cell, but these constraints at the level of the cell made possible the enormous gain in semiotic freedom acquired at the higher level of the organism. Through the differentiation of its tissues, the multicellular organism obtained a much greater capacity for processing and communicating knowledge, in the sense that it could deal with larger parts of its environment both in space and time.

We shall thus suggest the term *interpretance* as a measure of the capacity of a system to respond to signs through the formation of *meaningful* interpretants. High interpretance allows a system to "read" many sorts of cues in the surroundings; such high-level interpretance means that the system will form interpretants in response to complex cues that might not be noticed, or even be noticeable, by low-level agents. Thus, a unicellular organism cannot interpret complex patterns such as animal tracks, and in this sense it has a low-level interpretance. Mammalian organisms, on the other hand, are capable of interpreting extremly complex cues — such as the individual behavioral patterns of conspecifics — accordingly, they may be said to have high-level interpretance.

All this indicates that there is an aspect of *play* in the evolutionary process, an aspect which has been more or less overshadowed (virtually to the point of invisibility) by the Cyclopsian focus on selection. For *play*, it is often said, is an activity which carries its purpose in itself. "What is characteristic of 'play,'" writes Gregory Bateson (1979, 139) "is that this is a name for contexts in which the constituent acts have a different sort of relevance or organization from that which they would have had in non-play." Bateson (1979, 151) also suggests the definition of play as "the establishment and exploration of relationship" as opposed to ritual — "the affirmation of relationship." Thus, to the extent that the living world is engaged in an open-ended and nonsettled exploration of relationships between systems at many levels of complexity, it can truly be said that nature does, in fact, exhibit play-like behavior. It therefore will be as legitimate to talk about *natural play* as a force in the evolution of life forms, as it is to talk about *natural selection*. Selection acts to settle things — i.e., to fix behaviors, morphologies, or genetic setups — thereby putting an end to some element of ongoing play in the system while simultaneously providing for the beginning of whole new kinds of play.

Thus it was, for example, that more than fifty million years ago a particular ant species began interacting with a particular kind of fungus and the processes of natural selection eventually settled this as a new ant habit for farming fungi (as discussed above). The counter-processes of natural play, however, continued exploring this newly created semethic-interaction pattern (or *ecosemiotic interaction structure*), since now all two hundred of the existing fungus-growing species have evolved from this single ant species, And with few exceptions they all grow fungi from the same family, *Lepiotaceae*. In fact, the higher forms of ants have now become so specialized that they cannot survive without exactly the right variety of fungus (*New Scientist* 17/12 1994, 15). So here, the long, slow, interactive processes of natural selection may finally have resulted in the total crystallization of the relations from the open form of *play* to the closed form of *ritual* (or as it has sometimes been called, *instinct*).

Obviously, an increase in semiotic freedom will tend to push the influence of selective forces to higher levels. Thus, the more there is of inter-species semiotic interaction, the more will the *selective aspect* of evolution be loosened at that level, and the more dominating will become the *play aspect*. This is because a rich semiotic interaction pattern produces *fitness ambiguity* — for when organisms are bound up in a web of complex semiotic relations, virtually any newly developed property or behavior can potentially be counteracted or integrated in many different ways. Thus, the number of possible solutions for *selection* to scrutinize — and the subtlety of the communicational interactions — will tend to produce a no-win situation. As a result, selection cannot really measure the stakes of single players (individuals, demes, or species) in the game, though it could still influence the *choice of the game* itself. Because ultimately, it is plays, not players, that are selected for. Accordingly, I have suggested that instead of the evolutionarily *derivative* concept of *genetic fitness*, evolutionary biology should try to develop a concept of the evolutionarily prior phenomenon of *semiotic fitness* (see Chapter 4, section on life cycles, and Hoffmeyer 1997a).

The Biosemiotic Core of Evolution

If a morphological or behavioral trait has a relatively unambiguous genetic anchoring, and if, on the average,[23] it conveys an increased advantage, in survival or reproduction, to the organisms carrying it, then one would expect this trait to spread in the population, thanks to natural selection. And, indeed, we have no problems in ascribing such authority to natural selection. What I do question, however, is that this principle can be said to, even approximately, suffice as a comprehenisve explanation for evolution. For the problem is that this principle does not itself explain the establishment of the conditions under which it applies, i.e., under which it both operates and became possible in the first place.

Whether a trait conveys an *advantage* to its carrier or not depends on a complex, self-organizing context of semiotic relations that were gradually established through massively combinatorial trial and error events *at the lived ecosystemic level* and is therefore beyond the reach of genetic prespecification. This especially applies to later stages of evolution, where the *semiotic competences* of species are more unambiguously pronounced. Our implication is not, of course, that selection is no longer very important in later stages of evolution, but only that selection cannot be said to explain the evolutionary process *as such* — since this process to a great extent has been played out on premises given solely by the force of organisms' semiotic context.

The Scottish geneticist Conrad Waddington (1957; 1968–72) fought strongly to get recognition for the idea of embryogenesis as an autonomous factor in evolution. His idea of *developmental canalization* is still an important resource for our understanding of the developmental process. According to Waddington, the ontogenetic process may be seen as analogous to a ball running downhill through a branching system of valleys in an *epigenetic landscape*, the contours of which are determined as the effect of interplay between multiple individual genes (Figure 6.1).

This illustrative idea conveys an immediate understanding of why genes do not usually determine distinct traits, but rather, in a cooperative fashion, maintain the structuring of a developmental course. Even slight changes in the height of the floor of a valley in the epigenetic landscape might force the ball into a deviating route, and if this happened in early stages, it might have dramatic final effects. By supposing, furthermore, that the contextual situation in which development takes place influences the structure of the landscape, we get a picture of a true interactive dynamics involving both genetic *and* environmental influences upon the embryological process. By adding the epigenetic landscape as an interactive layer between genotype and phenotype, Waddington attempted

---

23   Meaning here, measured over a wide range of genetic backgrounds.

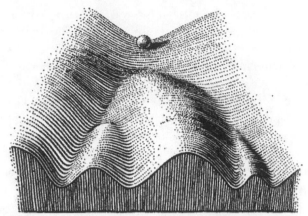

Figure 6.1 The epigenetic landscape. See text (Waddington 1957, 29).

to transcend the behaviorist black-box conception of the role of the organism (i.e., as an evolutionary dead end; see Chapter 4) and to gain some space within which to explore his own new ideas about genetic assimilation.

The crux of this latter idea proceeds from the well-known occurrence of *exogenous adaptation* (e.g., muscles that become thicker and stronger when continually and intensely used). Most organisms are, to some extent, ontogenetically adaptable to changing environmental conditions, which in Waddington's terminology means that the *canalization* of the respective property or trait is only partial, leaving open a range of optional phenotypic end products. There must however be limits to such flexibility: "If for instance, there was no canalization of the growth habit of a plant ecotype, every cold spring would convert the lowland forms into alpine types unable to take advantage of a succeeding warm summer" (Waddington 1957, 168). *Natural selection* would therefore be expected to tend towards some kind of *balance* between flexibility and genetic fixation of adaptable traits. Thus came the idea of *genetic assimilation*:

> It looks as though it must be too difficult for natural selection to produce organisms which *always* respond in a *perfectly adjusted* adaptive manner to fluctuating environmental circumstances, and that *faute de mieux* it tends to fix, by canalization, a type which is *reasonably well adapted* to the situation it will *most* frequently encounter. When this occurs in a population in an environment that remains relatively unchanged for considerable periods, it is the *process* that I have called *canalizing selection*. When it happens to a subpopulation which is carrying out exogenous adaptation to a *new* environment, it converts this into a pseudo-exogenous adaptation, and the "acquired character" becomes genetically *assimilated* (Waddington 1957, 68).

Thus, the overall effect of this mechanism is to create genotypes that reflect the conditions of life of the organisms concerned. In other words, the *actual life* of a population influences its evolutionary future. Waddington was even capable of showing that such a mechanism was, in fact, at work in *Drosophila melanogaster,* as he for many generations subjected the flies to straining conditions such as ether vapor (Waddington 1956).

Like the Baldwin effect that we are going to discuss later in this chapter, Waddinton's theory of genetic assimilation was not well received at the time by leading figures of what was called the new synthesis (between Darwinism and genetics) (see Depew 2003).[24] What is important about Waddington's idea for our biosemiotic understanding is his insistence, via the image of the epigenetic landscape, upon the autonomy of an intermediate zone between the genotype and phenotype — for this is a zone where not only embryological but also semiotic influences are of the essence.

And in fact, perhaps the most crucial single aspect of the embryological process (after the attainment of brute viability) is the development of the Umwelt of an organism. The role of the Umwelt is to regulate behavior (or, in general, organismic activities), for if it happens that the Umwelt of an organism is not well tuned to the semiotic niche, the chances of this organism surviving, much less leaving healthy offspring,will be diminished. Thus, it follows that the establishment of a good fit between the Umwelt of an organism and the semiotic-niche conditions it must cope with, stands as a central theater for natural selection.

But this immediately raises the question of the genetic anchoring of the Umwelt — a problem that once again brings us to the question of canalization. There must be species-specific determinants behind the selection of a potential sign in the surroundings of an organism that the organism can become capable of interpreting with some success. A moth, for instance, is equipped with a totally silent Umwelt, apart from the narrow chink that is kept open for registering the bat's fateful frequencies of approximately 20,000 Hz. When the bat is far away, the moth naturally veers away from the sound, but when the bat comes up close, the moth instead makes sudden and unpredictable movements. The moth, in other words, displays Umwelt-controlled behavior.

Now, it is true that *individual variations* in moths regarding these sensomotoric couplings are, of course, extremely limited, and Waddington's landscape probably does not reveal to us its true value at this level. But as soon as we move on to more advanced behavioral schemata — involving, for example *associative learning,* as in the octopus — it becomes necessary to explain the occurrence of *individual calibrations* to the semiotic surround.

24   As pointed out by Terrence Deacon (lecture at Copenhagen University May 29, 2002), *genetic assimilation* and the Baldwin effect are in fact describing opposite events. The Baldwin effect consists, as we shall see, in the *masking* of genetic weaknesses by the help of social adaptations that *compensate* for the potentially lowered fitness these weaknesses have caused. Genetic assimilation, on the contrary, depends on a *de-masking* whereby a "weak" gene suddenly becomes "visible" to selection through the creation of an extreme situation (see further in Wiles et al. 2005).

The concrete shaping of the nervous system of an individual is, as we have seen, dependent on the sensory inputs that the individual receives, as well as on the brain's own interpretive activities. As a consequence of this, it might clarify matters to provide an extra — largely independent — layer of interpretive processing, that we have called the *Umwelt landscape* (Hoffmeyer 2001c). The canalization process then consists of a ball at the same time tracing a route through both of these landscapes. One might, of course, depict this as an n-dimensional landscape, and the combinatorial output of these two under-determined canalization processes thus creates the individual as a morphological-physiological system (*modus* Waddington) with an Umwelt calibrated to the de facto challenges of the semiotic niche it is supposed to encounter. Taken together, these two coupled canalizations effectively break the determinism generally supposed to rule over the genotype-phenotype transformation.

In creating big-brained animals, natural selection managed to take advantage of the adaptive capacity of brains, but in doing so, it also partiallly subsumed itself under the semiotic determinations that this new kind of adaptive talent opened up the way for. An octopus quickly adapts to changing conditions — and if these changed conditions persist through many generations, it is hard to see how it could be avoided that other new adaptations would not also occur, so that an eventual return to the original conditions would now lead *not* to a *loss* of the original adaptations, but rather, to yet other adaptations.[25] This idea comes close to *genetic assimilation* as Waddington conceived it. The point is that it is *the semiotic competence of animals* that seriously puts (or should put) this *intermediate embryological zone* (the combined epigenetic-Umwelt landscape) at the forefront of the agenda in evolutionary thinking. Genes are assimilated *as support mechanisms* for changes of behavior that are necessitated by changes in the eco-semiotic interactions of organisms.

If we put such genetic assimilations of the semiotic niche into the context of the interspecific semiotic patterns described in the preceeding section (and in general with the widespread occurrence of semethic interaction and niche construction described in Chapter 4), it seems amply substantiated that *selection is a tool for the increasing semiotic refinement* rather than the converse. Selection surely does occur, but it flows down semiotically constructed pathways.

25 Parts of the genome that are not functional under prevailing conditions undergo much faster changes than do functional parts of the genome. As observed by Kalevi Kull, temporary adaptations lead to a changed use of the genomic resources so that formerly functional areas of the genome, under the new adaptation, may become nonfunctional and thus experience a fast accumulation of mutations. This might, in turn, contribute to a blockage for an eventual return to the earlier adaptive strategy. This model, that Kull called "evolution via the forgetting of the unused" is, in fact, a model for nonselective adaptation (Kull 2000; Hoffmeyer and Kull 2003).

Louis Pasteur has been credited for coining the saying that "chance favors only the prepared mind." In essence, this captures the idea being expressed here. Chance mutations are not selected because they are beneficial; they are beneficial because they happen to appear in a relational system which was already well prepared for them. That blind selection should be the sole cause of evolution is one of the mightiest fictions of our time. Selection is never blind; it is always guided by the prior formation of developmental and semiotic integration. Semiotic integration is not exclusive to the level of species, but instead takes place on many levels — e.g., as symbiosis and as more diffuse ecosemiotic interaction structures.

Semiotic Partitioning

One particular aspect of the biosemiotic approach that should be mentioned in connection with this discussion is that of semiotic partitioning (Hoffmeyer 2001c). *Semiotic partitioning* consists in the sympatric isolation of particular segments of a population that happen to share a particular kind of Umwelt (whether by common conditioning, learning, or heritage). By sharing a particular deviation from the prevailing dominant Umwelt, organisms are lead to share in certain aspects of behavior as well, and this in itself might tend to bring them closer together in sub-niches. Such *semiotic partitioning* has a built-in positive feedback mechanism in that individuals that share in similar behavior will also tend to share in similar conditioning or learning outcomes, thereby reinforcing or accentuating the shared deviation. If further reinforced by *genetic assimilation*, semiotic partitioning might eventually lead to genetically based isolation mechanisms and sympatric speciation.[26]

The Baldwin Effect

As already mentioned, Waddington's theory of genetic assimilation did not resonate well with his contemporary neo-Darwinians. The leading figure in paleontology at that time, G. Gaylord Simpson, identified it with a theory that had been advanced half a century earlier by the American child psychologist James Mark Baldwin — a theory that, according to most neo-Darwinists was considered as Lamarckism through the backdoor (Depew 2003).[27] As David Depew

---

26   A special, but by no means exhaustive, case of semiotic partitioning is the specific mate-recognition system as studied by Hugh Paterson (Kull 1992; 1999b; Paterson 1993).

27   A largely identical theory was suggested the very same year by the British psychologist Conwy Lloyd Morgan, but to his historical disadvantage this fact was overlooked by Simpson, who ascribed the mechanism, or effect, to Baldwin alone.

has shown, Simpson's evaluation of the theory was heavily colored by a kind of paradigmatic blindness towards both theories: Waddington's and Badwin's ideas were in some respects related, but they were definitely not identical (Wiles et al. 2005; Longa 2006).

Through the last two decades, however, a change of view has happened in evolutionary thinking that has led to a revival of Baldwinism. It is perhaps noteworthy that this change was not inspired by biological findings, but by computer simulation studies (Hinton and Nowlan 1996 (1987)). Among others, Daniel Dennett was enthusiastic but also — and in a very different sense — was Terrence Deacon (Dennet 1995; Deacon 1997; 2003). Baldwin's theory, quite generally, assumes that learned behavior may *feed back* on both the direction and the rate of the continued evolutionary process. Superficially seen, this does indeed look like Lamarckism in disguise — but, in fact, Baldwin believed that this effect might be caused through wholly Darwinian processes of evolution. As a child psychologist, he was particularly attentive to the phenomenon that we today would call social inheritance: "In the child's personal development, his ontogenesis, his life history, he works out a faithful reproduction of his social conditions. He is, from childhood up, excessively receptive to social suggestions; his entire learning is a process of conforming to social patterns. The essential to this, in his heredity, is very great plasticity, cerebral balance and equilibrium, a readiness to overflow into the new channels which his social environment dictates" (Baldwin 1902, 53).

Baldwin was also a confirmed Darwinian and, in essence, what he suggested was that social inheritance was operational not only in children, but in the animal world at large, if only to a lesser extent. In this way, he could synthesize what he could not deny as the results of his studies in child psychology with his belief in Darwinism. He saw the intellectual *plasticity* of the child — or, in general, the young — as a *trait* for natural selection to work upon. And since social transmission is itself enough to explain the behavioral likenesses bewteen father and his son, there is no need for a Lamarckian theory of *acquired mental characteristics* (Hoffmeyer and Kull 2003). In fact, quite to the contrary, "the only apparent hindrance to the child's learning everything that his life in society requires would be just the thing that the advocates of Lamarckism argue for — the inheritance of acquired characters. For such inheritance would tend so to bind up the child's nervous substance in fixed forms that he would have less or possibly no plastic substance left to learn anything *with*" (ibid., 55).

The decisive point here, of course, is that Baldwin thought social heredity might fascilitate the formation of genetic heredity — for this is where the modern Darwinist has been trained to suspect a Lamarckian "catch." Yet Baldwin explained this phenomenon through a mechanism he called *organic selection* that implied the appearance of developmental adaptations in the lifetime of individual organisms.[28] These adaptations were caused by "the great series of *adaptations secured by conscious agency*, which are all classed broadly under the term 'intelligent,' such as imitation, gregarious influences, maternal instruction, the lessons of pleasure and pain, and of experience generally, and reasoning of means to ends" (Baldwin 1996, 442–43; italics added).

Taken together, these adaptations would mean that individuals might survive even under odd conditions and "thus kept alive, the species has all the time necessary to perfect the variations required by a complete instinct" (Baldwin 1902, 97). *Organic selection* and *natural selection* were thus, as Baldwin saw it, opposing mechanisms, and what organic selection achieved was to give stressed individuals a place where they could breathe freely as a safeguard against the sharp knife of selection. This, he thought, might permit them to survive until genetic adjustments appeared and were fixed through natural selection, offering more permanent support for the new adaptation.

Baldwin thought that his theory on organic selection showed "that the ordinary antithesis between 'nature and nurture,' endowment and education, is largely artificial, since the two are in the main concurrent in direction" (Baldwin 1902, 106). He illustrated this by reference to complex instincts where physical heredity and social transmission are inextricably combined. Thus, in certain instincts, "we find only partial coordinations given ready-made by heredity and the creature actually depending upon some conscious resource (imitation, instruction, etc.) to bring the instinct into actual operation. . . . [In animals,] social heredity serves physical heredity, while in man we find the reverse" (ibid., 107).

Probably the most serious conflict between the Baldwin effect and classical neo-Darwinism concerns the question of the primary causal role of the genes. It is close to dogma to claim, as Simpson (1953) did, that "the ability to acquire a character has, in itself, a genetical base" (Simpson 1953, 116; Depew 2003). According to neo-Darwinian dogma, any population therefore always carries a large reserve of hidden genetic variation, and by implication, variants will nearly always by necessity appear to take advantage of eventual changed conditions.

The Russian-born American geneticist, and one of the great figures behind the *Modern Synthesis*, Theodosius Dobzhansky, proposed a theory of *balancing selection* that reflects this very conception. The disease of sickle cell anemia is the prototypical example of balancing selection at work. In areas of Africa with severe occurrences of malaria, there is also a high frequency of sickle cell anemia — a disease caused by a point mutation in the gene for the beta-chain

---

28   The use of the word *adaptation* here (instead of *adaptive behavior*) is nearly incomprehensible from a neo-Darwinian perspective, where an adaptation is, by definition,"the result of selection" (Depew 2003).

of hemoglobin that seriously impairs the health of persons that are homozygous for the mutation.[29] Only few homozygotes survive to adulthood, whereas heterozygote carriers of the disease gene normally have no problems as long as they are not exposed to low oxygen pressures.

The reason why selection has not eliminated this serious monogenetic disease is that the heterozygote carriers of the disease are much more tolerant to malaria than are the healthy people. In malaria-threatened areas, a balance therefore will often obtain, causing the disease gene to be maintained in the population through its being rewarded by the increased number of healthy offspring left by heterozygous carriers. How, precisely, this balance will be set, is wholly dependent on how serious is the malaria threat.

Terrence Deacon (1997, 323), however, has given a Baldwinistic explanation for the incidence of sickle cell anemia:

> The sickle cell trait spread quite rapidly in Africa in recent prehistory because of human activity.... Probably the critical historical event that catapulted malaria to an epidemic disease was the introduction of agriculture and animal husbandry into Africa between five and ten thousand years ago. This culturally transmitted practice modified the tropical environment to create the perfect breeding ground for mosquitoes.... The human population was thrust into a *context* in which powerful selection favored reproduction of any mutation that conferred some resistance to malaria.

As Depew explains in his analysis of the modern revival of Baldwinism, one need not see any strict opposition between Deacon's position and that of Dobzhansky. Rather, what has happened is that genes are now to a lesser extent seen as deterministically *coding for a trait* but rather as tools recruited to support already established practices — such as, for example, making it attractive for people to eat vitamin-C-rich citrus fruits rather than reestablishing the gene that was lost, by mutation in our distant ancestry, thereby eliminating the human capacity for synthesizing ascorbic acid (vitamin C) (Depew 2003).

From a biosemiotic standpoint, Baldwinism in its modern version is quite unproblematic. Organisms, and the cells and tissues of which they are built, are not just objects but also subjects — in the sense that they are semiotic agents capable of interacting with their surroundings in "intelligent" ways. And the history of how these semiotic interaction patterns have been scaffolded into the myriads of ontogenetically consistent dynamics of this world — i.e., the life cycles of organisms — is what evolution is all about. Genetic fixation, of course, plays a crucial role in such scaffolding — but I believe that there are countless

---

29   Homozygote persons in this case have a mutant gene for the beta-chain on both of their chromosomes, contrary to heterozygotes that carry a mutation on only one of the two homologue chromosomes.

semiotic ways of obtaining a relatively secure scaffolding of intra- and inter-specific interaction patterns (semethic interactions) (see Hoffmeyer 1995). I see no reason to believe that all — or even most — of these semiotic scaffolding mechanisms are unambiguously "coded for in the genomic setup." On the contrary, I think that there are serious reasons to believe they are *not*, since *flexibility* is at the core of such semiotic scaffolding (Bateson 1963; Hoffmeyer and Kull 2003).

Intelligence and Semiosis

The extent to which different animals *possess intelligence* has been highly disputed. It has often been overlooked, however, that intelligence is not just something one has between the ears, but is very much a social skill, an ability to use physical marks as well as social relations to scaffold and organize one's knowledge and behavior.

From the very beginning, nerves were developed as tools for movement. Their task was to facilitate long-distance communication between cells in different parts of a moving animal. But the presence of fast moving animals implied the creation of fast moving environments (e.g., the co-presence of fast moving prey — or predators!) and brains developed to allow certain animals to cope with this situation in new ways. The combinatorial possibilities of moving in a moving world are enormous of course, and from the beginning, the task of brains was to help the animal make proximal decisions which might be assisted by learning, but which could not possibly be deterministically based on genetic anticipation. Brains were means for nurturing nature.

One aspect of brain action which may deserve special emphasis is *proprioception* — i.e., the awareness of one's own movement and position. Even the simplest movement presupposes a continous feedback from proprioceptive organs in the body measuring muscle tensions and displacements of cell layers including the sense of gravitational orientation. The American philosopher Maxine Sheets-Johnstone (1998, 284) has recently suggested that the proprioceptive sense serves as a *corporeal consciousness*: "Any creature that *moves itself*, i.e., that is not sessile, senses itself moving.[30] By the same token, it likewise senses itself when it is still. Distinguishing movement from stillness, motion from rest, is indeed a fundamental natural discrimination of living creatures that is vital to survival."

---

30   Unicellular organisms who can move themselves (e.g., using flagella) may not always be able to distinguish between the moving of their own bodies and the changes or movements in their surrounding. In that respect, Sheets-Johnston's case may be just slightly overstated. However, the principal meaning of her statement is clearly correct.

It was the French philosopher Maurice Merlaeu-Ponty (2002 (1945), 160) who observed that "originally, consciousness is not an 'I think that' but an 'I can.'" Sheets-Johnstone (1998, 285) echoes this insight when she writes that "a creature's initiation of movement is coincident with its kinesthetic motivation, its dispositions to do this or that — turn, pause, crouch, freeze, run, or constrict; its kinestethic motivations fall within the range of its species-specific movement possibilities . . . [which] are the basis of its particular repertoire of 'I can's . . . [and thus] any item within its repertoire of 'I can's is undergirded proprioceptively (kinesthetically) by a sense of agency."

It is a well-known fact that animals can and do dream. This implies that mental states may sometimes be uncoupled from bodily action. But the extreme *extent* of *uncoupling* between behavior and mental activity that characterizes the human mind is probably unique among animals. The uncoupling has made philosophers wonder how it can be that mental states are always *about* something. But seen from the perspective of biology, this is no surprise at all, since *mental aboutness*, (human intentionality) grew out of a *bodily aboutness* (Hoffmeyer 1996 a). Whatever an organism senses also mean something to it — e.g., food, escape, sexual reproduction. This is one of the major insights brought to light through the work of Jakob von Uexküll (1982 (1940), 31): "Every action, therefore, that consists of *perception and operation* imprints its meaning on the meaningless object and thereby *makes it into a subject-related meaning-carrier* in the respective Umwelt."

Seeing "I can" as the center around which mental processes are organized by evolution implies a blurring of the mind-body dichotomy. The acts of thought and the acts of body are not totally separate categories, but are essentialy connected via the intentionality of the animal that instantiated them — and therefore mental activity is just a particularly sophisticated extension of traditional animal behavior. It follows from this understanding that we do not have to operate with two quite different categories such as *phenotypic flexibility* and *learning*. Learning is just an especially smart form of phenotypic flexibility.

The Ghost of Lamarckism

The French naturalist Jean Baptiste Lamarck — who in 1809 (fifty years before Darwin) suggested the first scientific theory of evolution in the history of the world — is a sad figure in the history of biology, outmaneuvered and overruled by his contemporaries, scorned and misunderstood by posterity (Burkhardt 1977). Lamarck's misdemeanor, seen with modern eyes, was that he believed that properties acquired by plants or animals in the course of their lifetimes could become inherited by their offspring. It is this, to the best of our knowledge, false conception that nowadays is called *Lamarckism*. For instance, as the wading bird delicately set out to feed at still deeper water, Lamarck posited that its stilted legs would become incrementally prolonged in the process, and this, he claimed, would prove of use to the offspring as manifested in an ever so little prolongation in the length of the legs already from birth. Lamarck felt that through many generations, this process might lead to the substantial kinds of change that we can observe when comparing present species with fossilized specimens. By suggesting this (intuitively quite reasonable) connection, Lamarck in one bold stroke broke down the millennia-old wall inherited from both the Bible and from Plato and Aristotle that guarded the *static* image of the composition of the natural world. And yet, this world-changing figure is nowadays remembered mostly as the defender of a wrong theory that was successfully replaced by Darwin's.

The idea of acquired properties as inheritable was, in fact, common sense in Lamarck's own time and to identify his theory with this simplistic idea is to blind oneself to his real achievement.[31] Nobody at the time had the faintest idea about the existence of genes, and there was therefore no good reason to distinguish so sharply between biologically *innate* and biologically *acquired* properties. For Lamarck it was, in fact, something very different that seemed central — namely, that *habits create forms*. When circumstances change, organisms will have to change their patterns of activity accordingly — or, in other words, they must take up new *habits*. But new habits will usually make anatomical, physiological, or behavioral innovations desirable, and Lamarck thought that the "inner feeling" (*la sentiment interieur*) of the species imperceptibly guided the appearance of innovations that satisfied just these needs. This, of course, required a huge number of generations, and it was therefore necessary that the small improvements acquired in each generation were heritable, so that they would be added to the accumulated result of the efforts of the preceding generations.

The idea that a species could possess an *inner feeling* is of course a stumbling block for the modern scientific mind. Lamarck himself speculated that this inner *feeling* was caused by so-called *subtle fluids* (an expression that is not likely, either, to meet acceptance by the sharp scientific minds of of the twenty-first century). But these immeasurable subtle fluids were, in fact, the only explanatory tool eighteenth-century science had at its disposal for explaining strange phenomena such as electricity, magnetism, or even wickedness.[32] Lamarck's

---

31   One might just as rightly (and wrongly) conflate alchemy with Newtonism, for Newton spent the last thirty years of his life doing alchemical studies in his search for the deeper causes behind the mathematical connections he had discovered.

32   Our modern parallel to "subtle fluids" would be the (still far too common) misconceptions of "gay genes" or "genes for morally decent behavior." Here too, we are concerned with speculative entities that none has ever seen nor measured directly, and yet the existence of which is accepted

own time, of course, lies right at the border of the nineteenth century where subtle fluids were no longer looked upon with much sympathy. But perhaps for the same reason, nineteenth-century science no longer pondered the kind of "big" questions that were the focus for Lamarck's work.

Lamarck, however, further had the sad misfortune, long after his death in 1829, to have his name drawn into the heated controversies surrounding Darwinism. This would last for three quarters of a century until, in the 1930s, the neo-Darwinian synthesis finally seemed to extinguish the last hopes for a Lamarckian kind of evolution. And, although many neo-Lamarckian biologists could and did adduce quite weighty arguments in defense of their opposition toward Darwinism, Lamarck's thinking unavoidably became vanquished by the thinking of the more victorious theory.

It didn't help either, of course, that neo-Lamarckism had increasingly become an asylum for religious, antiquated, or nostalgic elements in the debates. That *the inheritance of acquired characteristics* should be misunderstood as the central core of Lamarckism is precisely what might be concluded when the theory is evaluated through Darwinian glasses. For Darwinism sees evolution as a product of the *differential reproduction* between individuals — and in this light it becomes fatal, of course, that Lamarck's theory poses an *instructivistic* concept of change.

Offspring (in Lamarckism) are instructed to perform better; they do not (as in Darwinism) perform better because they happened to have inherited winning properties.

Yet the linking of *change* exclusively to the hereditary mechanisms is a Darwinian bias — and seen from this bias, the *essence* of Lamarckism dwindles away to be caricaturistically replaced with Lamarck's (admittedly poor) understanding of how inter-generational heredity works. Seen with a Lamarckian bias, on the other hand, Darwinism is a narrow-minded exegesis of an absurdly *mechanical* philosophy, and it never achieves an explanation for what it ought to explain, i.e., how it is that the perfectly adapted decends from the less perfectly adapted. With our twenty-first-century eyes, Lamarck's error is not difficult to see, but do we yet see Darwin's?

In this book I have pointed to a diversity of epigenetic-heredity forms with inherently instructivist potentials — and yet, I basically agree with Darwinism that evolution can not be explained through instructivist *heredity*. The core of the Lamarckian theory, however, does not so much depend on his theory of heredity, for evolution, in his eyes, was a process that operated on species, not on individuals. Not mystical, but rather, *biological* "inner propensities" *at work*

---

by many researchers because they might explain human personality traits without any need for psychosocial theories (which are looked upon with skepticism).

in a species was the real causal agent in Lamarck's scheme. I am inclined to think that Lamarck in this respect had discovered an important point, and that the Darwinian focus on hereditary mechanisms has tended to distort our understanding of evolution.

It is interesting in this connection that Baldwinian ideas are Lamarckian in the broad sense that "something learned" *influences* evolution, but they are not Lamarckian in the narrow sense that Darwinists have attributed to his name. Neither Baldwin, nor Waddington, believed in the direct inheritance of acquired properties in a genetic sense. But it seems that the justified rejection of Lamarckism in its narrow sense is confused with a never-justified rejection of Lamarckism in its broad sense. The implication of this blatant ambiguity is that a highly legitimate discussion of the eventual *influence of the organism upon evolution* is relegated to a dim no-man's-land. Susan Oyama (2003, 172) observed,

> Once Lamarck was firmly identified with the inheritance of acquired characters, and once the inheritance of acquired characters was set in place as the defining contrast to Darwinian natural selection, all sorts of other things followed. Whether Lamarck's heresy was ruled out of bounds altogether . . . or safely confined to the "transmission of culture," anyone wishing to explore the evolutionary roles of organismic activity, phenotypic plasticity in general, or learning in particular was obliged to engage in some theoretical acrobatics to do so. These might involve opening up a separate informational "channel" relying on hidden genetic variation, or hoping for fortunate mutations, but there seemed to be a need for fancy footwork to avoid the dreaded charge of Lamarckism.

Lamarck believed that evolution was not just a process of change, but also a process of progression. He even suggested a new term, *biology*, as a designation for the study of this phenomenon of perfection that characterized the two kingdoms of animals and plants in contrast with the kingdom of minerals. And thus his idea of the *inner feeling* was needed as a means to justify *la marche de la nature* in this sense — i.e., as a progression (Burkhardt 1977).

The twentieth century's landmark discoveries in thermodynamics and complexity research imply that we no longer need explanations *à la* mysterious subtle fluids in order to explain evolution as a directional process. The modern scientific version of subtle fluids is called *self-organization* and is generally considered quite legitimate (although, as we saw earlier in this chapter, the notion of self-organization implies some rather heavy philosophical or ontological problems). If in place of *inner feelings* we put *the processes of self-organization* at work in a species, the Lamarckian scheme does, in fact, approach the most modern conceptions of the ways of nature.

Compared to this, Darwinians generally are obstinately opposed to the conception of organic evolution as obeying a deeper "directedness" of *any* sort. For modern Darwinists, the flow of *chance mutations* coupled to *competition among conspecific organisms* is all we need to explain not only the multiple forms of life on Earth, but also the superordinate ecological and behavioral patterns that have appeared among these entities. And one may be allowed to suspect that the popularity of Darwinian explanations does not suffer damage by being so close an analogue to the dominating "economic realities" of Western societies. The idea that an "invisible hand" behind the back of the endlessly *competing* creatures has — all by itself — assured a healthy evolution of nature, has shown itself to possess an overwhelming appeal to the modern mind.

In this chapter, I have suggested that the agency of organisms has an experience-like component, and I have sketched evolution as a perpetual increase in semiotic freedom produced through the semiogenic interactions of organisms. To call this *perfectioning* (*modus* Lamarck) is, of course, to apply a very anthropocentric perspective. But it feels hard — and this is no superficial feeling — not to think that the string quartets of Beethoven or the songs of John Lennon surpass the cries of macaque monkeys, or that the songs of birds are more exciting than courtship trembling in the water mite. The fact that creatures and interactive patterns expressing high levels of semiotic freedom make stronger appeal to our sensitivities than do the more law-based activities of simpler animals may, as we shall see, have an anchoring in the natural history of human origins. We shall return to this question in Chapter 9, but here it will be adequate to note that this propensity apparently brings us into harmony with the internal dynamic course of the universe — which should not surprise us too much, since the universe has itself created us.

7
Endosemiotics

Innenwelt

Endosemiotics is a term for "sign processes within the body" that Thomas A. Sebeok[1] introduced in about 1976 (Sebeok 1985 (1976), 3). Its use has become customary, although in denoting this *within*, the term could be said to be somewhat paradoxical. For the processes of life are, as we saw in Chapter 2, played out *across* multiple surfaces, one enveloping another — and therefore no single surface or location can reasonably be appointed the *essential* one. In Chapter 2, the example of the Norwegian physician who suffered an attack of Guillan-Barrés syndrome showed that although the skin is indeed a borderline serving to keep the world away from us in a rough physical sense, at the same time it opens us up to the world in a psychological sense, allowing us to belong to it and vice versa.

The skin, then, mediates contact with the surrounding world via its manifold of surfaces, on the physical, biological, psychological, and social levels. Among these, let us just mention the invisible surface that Sebeok (1979, 45) referred to as the *Hediger bubble*:[2] "a variably shaped zone of personal space that admits no trespass by strangers and is defended when penetrated without permission." By this very definition alone, we can draw the more general conclusion that the prefix *endo* in endosemiotics has no obvious or clearly-demarcated physical referent. Moreover, a sign process is itself necessarily *always* a border-crossing process in which an interpretant is called forth by something else.

In habitual usage, however, it is the body *individuating* an organism that most conveniently defines an endo-exo *asymmetry* of causal relations — and consequently, endosemiotics has become the pragmatic designation for sign processes taking place in the interior of an organism. Yet since the concept of the *individual organism* is not always as clear and unambiguous as it appears in the order of vertebrates (see Chapter 4), endosemiotics sometimes also deals with

---

1   Sebeok, himself, writes that he used the expression "cavalierly" (Sebeok 1979, 22).

2   The designation is in reference to the Swiss animal psychologist Heini Hediger, whose life-long interest in the training of animals in the context of circus performances made clear to him that the animal trainer must always attempt to determine, and to respect, such invisible physical "borderlines" around the animal.

semiotic relations across the limits of the individual — as in cases of mutualistic symbiosis such as between squid and light emitting spirochetes. Such endosemiotic processes in damaged corn seedlings, for example, cause the excretion of a *terpenoid* that eventually becomes integrated into the semiotic activity of parasitic wasps (as mentioned in Chapter 6) illustrating that the *endo* versus *exo* distinction is primarily an analytical tool. At the same time, such examples point to the impossibility of upholding any sharp dividing line between biochemical and semiotic activity, however reasonable such a borderline may once have appeared.

For example, even we *rational* humans emit pheromones that unconsciously influence our emotional responses towards one another — and in so doing, become indelibly embedded in the complicated web of human sign use, or anthroposemiotics (see e.g., Russell 1971; Schleidt, Hold, and Attili 1981).[3] Nowhere in this web of semiotic processes is there any sharp border to be found that might justify the distinction between an area of Peircean Secondness and an area of Thirdness (i.e., an ontological distinction between *the body* — as traditionally conceived as distinct from the mind — and *the mind*). And while it could be reasonably argued that endosemiotic sign systems are much more tightly anchored in the determinism of natural lawfulness (i.e., they have less semiotic freedom) than the kind of human mental sign systems traditionally studied in anthroposemiotics, still no absolute distinction between a semiotic, nonbiological domain and a nonsemiotic, biochemical domain can be justified.

Analyzing animal endosemiotics, Thure von Uexküll, Werner Geigges, and Jörg Hermann (1993) drew a parallel between endosemiotics and Jakob von Uexküll's concepts of *Gegenwelt* and *Innenwelt*. These two concepts are constitutively connected with the Umwelt concept because they refer to the internal layer of processes that brings forth a common *inner world* in the animal. In the words of Uexküll, Geigges, and Hermann (1993, 41) — in a section that focuses primarily on the human psyche, but which may be broadened to include nonhuman semiosis simply by substituting the more general Umwelt for the species-specific human *consciousness*,

> Paradoxically, we are "outsiders" to all "inner" (i.e., endosemiotic) sign processes, whether they occur in the nervous system or in the immune system. We can illuminate them only by complicated methods, and "understand" their messages only through biosemiotic interpretation. Whereas we are "insiders" of all sign processes by which the outside world and our body are presented to us as conscious realities. The "inner" sign processes of the brain are, like all "inner" or endosemiotic events,

---

3   *Anthroposemiotics*, as the name suggests, is the study of human uses of signs, whether verbal or nonverbal.

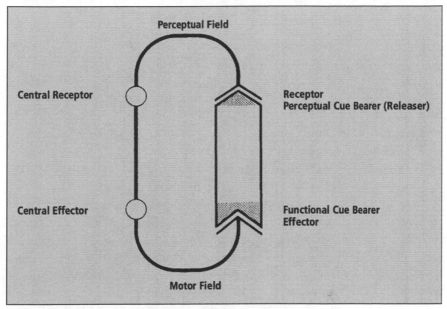

Figure 7.1 Uexküll's functional circle (from Sebeok 1979, 10).

inaccessible to our conscious experience; but, unlike all other endosemiosis, they participate directly in its formation. That is, all consciously experienced "outside" realities are translations of "inner" sign processes which occur in our brain and are inaccessible to our experience and understanding. Here it should be recalled that originally "inside" and "outside" are semiotic concepts which take on spatial connotations only when we look at our motoricity.

These formulations may be understood as a modern interpretation of Jakob von Uexküll's famous functional circle (Figure 7.1) where the Umwelt-Innenwelt pairing closes together, so to speak, around the animal's immediate object of attention.

Thure Uexküll and his co-authors introduce a distinction between *Gegenwelt* and *Innenwelt* in that *Gegenwelt* is the designation for endosemiotic systems that — like the nervous system or the immune system — each have their own internal logic (or *programs*) for representing states of the world. *Innenwelt*, on the other hand, designates a superordinate level of integration where such counter-worlds have become woven together to function as one unitary entity via circular processes: "Such an 'inner world' contains templates, as it were, of the sectors of the environment that are significant to the living being. These templates are recorded in signs exchanged between cells and between organs. Therefore, although they are closely related to the organism's environment, we have to call them *endo*semiotic" (Uexküll, Geigges, and Hermann (1993, 6).

The Metabolic Code

The first modern endosemiotic analysis was proposed by the biochemist and bio-physician Gordon M. Tomkins, who in 1975 published a paper in *Science* entitled "The Metabolic Code" (Tomkins 1975). This paper focused on a group of signal compounds in bacteria that seem to function as warning signals for stress, and thus go by the name of *alarmones*. One of these compounds is a substance that we have already met in Chapter 5: cyclic AMP (cAMP) — a molecule that accu-mulates in cells lacking glucose (which for many bacterial species is the preferred nutrient sugar). A high concentration of cAMP causes the cell to start producing enzymes that are specialized for metabolizing nonglucose sugars.

Another alarmone, *Guanosine tetraphosphate* (ppGpp), accumulates in cells lacking amino acids and will cause a radical decline in the rate that the cell makes mRNA. This is biologically appropriate, since amino acids are substrates for the synthesis of proteins. A lack of amino acids means that proteins cannot be made anymore, and that therefore further transcription of genes to mRNA would be useless — or worse, a waste of metabolic energy. The interesting thing about these regulatory mechanisms is that there is no necessary physical molec-ular relationship between the *signal* (alarmone) and the process that the signal controls — which is why Tomkins called the alarmones *symbols*, defining a sym-bol as "a specific intracellular effector molecule which accumulates when a cell is exposed to a particular environment" (Tomkins 1975, 761).

Tomkins emphasized that alarmones are *metabolically unstable,* meaning that they will quickly disappear if or when prevailing conditions become better, and he also defined a *domain of action* as comprising all the metabolic processes that are controlled by a given alarmone (or *symbol*).

As such, alarmones probably constitute the most primitive example of a dig-itization of analog coded processes (e.g., cAMP concentration as an indexical code for glucose starvation), as discussed in Chapter 4 in the section on the use and nature of analog codes. Tomkins even suggested a possible scenario for the evolutionary digitizing of the alarmone concentration, starting with the enzyme *glucose kinase* that catalyzes the first reaction in a chain of processes whereby glu-cose is normally metabolized in cells. This first reaction consists in the attach-ment of a phosphate group to the glucose molecule, rendering the molecule prone for further degradative processes. This phosphate group is furnished by the compound that cells normally use for carrying around energy in the cell. This compound is called adenosine triphosphate (ATP), so named because it consists of adenosine plus three phosphate groups. In the process of glucose phosphorylation, one of these phosphate groups is transferred from the ATP to the sugar, leaving behind a partly dephosphorylated ATP molecule called ADP (adenosine with two phosphate groups attached).

Tomkins now suggests that glucose starvation originally may have led to the accumulation of cAMP, due to a chance side reaction that would not under normal conditions play a significant role. For example, a primitive version of the enzyme *glucose kinase* might have catalyzed a minor side reaction consisting in the conversion of ATP to cAMP (by splitting off two phosphate molecules, as pyrophosphate, from the ATP). If the cell was starved for glucose, the main reaction of the glucose kinase would then lack its substrate (glucose), and catalysis would quickly be brought to a stop. Conversely, the side reaction would make available a lot more of its substrate, ATP. Glucose starvation would thus lead to an accumulation of the product of the *side reaction*, cAMP.

But since the shape of the cAMP molecule has many traits in common with the shape of the ATP molecule (after all, each consists mainly of adenosine), cAMP might easily bind to a range of enzymes at the exact same positions where ATP would normally bind. A high concentration of cAMP would therefore cause a *displacement* of ATP from those enzymes (*ATPases*) that normally degrade it. In this way, cAMP serves to *block* the cell's energy consumption (i.e., the degradation of phosphate bonds in ATP) whenever the supply of energy is diminished, as for example, by glucose starvation.

Seen from above, what has happened is that the cAMP-concentration has become a "switch" locating the cell in one of two possible situations: either energy metabolism goes on at full speed (low cAMP), or it is blocked (high cAMP). In the course of evolution, bacteria have developed means to utilize cAMP in a way that is even more different from the original biochemical function of this molecule — i.e., as a release mechanism for specific transcription processes whereby mRNA is formed from the genes of precisely those enzymes that are necessary for the cell to switch to other energy sources (for example, to switch on the synthesis of enzymes that specialize in the burning of lactose).

Corresponding scenarios may be suggested for the other *alarmones* (Kilstrup 1998). This case shows that cAMP in bacteria at one and the same time may function as an *icon* for ATP (i.e., as a specific conformation that enzymes may take to be ATP), as an *index* for increased ATP concentrations, and as a *symbol* for glucose starvation. Danish molecular biologist Mogens Kilstrup (2000) writes that, in post-genomic science,

> the quantity of information about regulatory pathways will be both enormous and unstructured, and new ways of extracting information will become necessary. "Regulation" is *information* transmission per se and a [purely] biochemical description of regulatory networks is both too complex and too unfocused. By focusing on *the representation of information* rather than on biochemical reactions, interesting questions will appear, such as (1) How is the particular stress factor represented? (2) How can the system be fooled (e.g., cross-talk, alternative pathways)?

and (3) Are there alternative representations of the same stress factor? And if so, are the signs *synonymous,* or do they convey different information?

Attempting to work out a sign-model that might fit the needs of a formalization of the semiotic character of intracellular biochemical networks better than the traditional two-dimensional Peircean graphic of a tripod, Kilstrup developed a new form of notation for sign relations. An important concern behind the development of this new notation system was that it should be applicable to a chain of representations: A is represented by B, which is represented by C, and so on, where Kilstrup uses the symbol :-- to represent the relation "is represented by" and notes that "the sigh *chain* must be a one-dimensional path through a multidimensional network in such a way that all sign elements may be allowed to represent, and be represented by, many different sign elements," writes Kilstrup (1998, 107). In its basic form, this notation has the following structure:

Object :-- Representamen :-- Interpretant [R:O]

Here the Object is *represented* by a Representamen that again is *represented* by the Interpretant, but in such a way that the Interpretant refers to *both* the Representamen and the Object *and also to the relation between them* (ibid.). With a deep-drawn sigh, Kilstrup notes that this one-dimensional notation, of course, does not immediately convey the true triadic nature of the relations. But this loss of direct appeal is, in the eyes of Kilstrup, the necessary payment for a notation that is capable of specifying all the elements of the sign. For comparison, he says, the equation of H_2O as equivalent to water does not bring forth "the same associations, to a beautiful lake in a forest or a cool glass of water that we might get by hearing the word *water*. But few are those that would *doubt* water to be water, just because its formula consists of two hydrogen atoms and one oxygen atom. One should not, therefore, by the mere change of notational technique, think that the linear atomized account of the sign would indicate that the sign consists of only two dyadic representations in a row" (ibid.).

Kilstrup offers an immunological example of how the new notation may be applied in the case of a physician examining the blood sample containing HIV antibodies:

[HIV virus] :-- [HIV antibodies] :-- Mental representation[HIV antibodies]: [HIV virus]

This should be read to indicate that an antibody towards HIV is a sign (most properly, a *sign vehicle*) for the presence of the HIV virus. The functional sign, the *representamen,* is *not* the antibody as such. Rather, it is the concentration of that antibody that in chemical notation is indicated by placing the chemical compound in angular parentheses, so that the concentration of a compound,

X, is written [X]. The (immediate) object of the sign, as given in the notation above, is one or several distinctive traits pertaining to the *real* object of the sign (the dynamic object, in Peircean terms), which in this case would be a distinctive trait of the HIV virus itself, namely its concentration — or, more accurately, the concentration of an *epitope* on the surface of the virus (see below). The *interpretant* consists of "an *induced change* of the interpreting system, that thus appears as a *new* distinctive trait and therefore a new potential sign; in this case, the neurophysiological change that a physician experiences by learning that a patient has HIV virus in his blood" (ibid., 108). The two *links* (:-/-:) are icons for the relationships that *taken together* are responsible for the establishment of the representation. The reference link, or the *interlink* ([HIV antibodies]:[HIV virus]) refers to the knowledge that the physician has of the relationship between antibody concentration and virus concentration. The reference function thus always presupposes learning and memory, either as mental learning in the traditional sense or in the form of evolutionary (or gene-based) adaptations.

An important point in Kilstrup's notation is that it makes it possible to distinguish between the Peircean sign-categories simply by assigning to each of them a distinct notation: (:-/-:) for *recognition links*, (:-:) for *causal links*, and (:->) for law-based or *habitual links*, respectively. Kilstrup's notational system is a promising point of departure for explorative endosemiotic analysis, but unfortunately we are still awaiting his publication of a fully worked-through version of the system.

Dictyostelium discoideum

The *problem of multicellularity* — i.e., the difficulty of establishing a stable reproductive coexistence between, on the one hand, mobile cell lines lacking reproductive potential and, on the other hand, cell lines with reproductive potential — has, as already mentioned, been solved only a few times in evolution. However, there are a number of intermediate forms between unicellular and multicellular organisms, and examining these may help us get a better grasp of the role that semiosis plays in organizing unicellular life into multicellular life. The best studied among such intermediate forms is the slime mold *Dictyostelium discoideum*. *D. discoideum* alternates between a phase of free unicellular life and a phase wherein hundreds or thousands of single cells crowd together and come to work in concert as one, as they form a superorganism called a *plasmodium* — a slug-like structure that is capable of moving around and searching out the suitable conditions whereby the plasmodium may form a stalk with fruiting bodies from which spores bearing the next generations of unicellulars eventually burst forth.

Here once again, cAMP is at center stage for semiotic control — and whether we choose to call it *endosemiotic* or *exosemiotic* becomes a matter of concern only to the analyst — the life cycle of *Dictyostelium discoideum* itself, as we shall see, makes no such distinctions. There, just as in the bacterial system, cAMP functions as a stress code. So long as there is plenty of food and moisture, the amoeboid cells move freely around eating bacteria, living independent of one another and largely indifferent to each other's presence. But as soon as a few cells start experiencing a deficit in one or both of these two life essentials, they start producing cAMP and excreting it into the shared medium. The presence of this cAMP then attracts neighboring amoeboid cells and induces them to excrete cAMP, too.

At the same time, however, the cells also excrete an enzyme, a *phosphodiesterase*, that degrades cAMP — and in doing so, creates a pulse of cAMP/non-cAMP *waves* spreading through the population. Experiments have shown that it is this pulse that attracts cells, whereas a fixed concentration of cAMP has no effect. Yet in order for this wave to work, an additional refinement is needed to stop the wave from running backwards (Goodwin 1995, 48). For when cAMP is excreted from a cell, it will normally diffuse away equally well in all directions — but this would spoil the wave structure. This does not happen in the case of *D. discoideum* because the amoeba can produce cAMP only in short bursts, each of which is followed by a *refractional* period, where it is insensitive to the signal while it reconstitutes itself. In this way, the possibility of the wave traveling backwards is avoided.

As the cAMP pulse moves forward, cells increasingly begin to stream up against the biochemical tide. A differentiation and asymmetric distribution of the formerly homogenous cell population now sets in, leading to the formation of a series of definite architectures, where the destiny of cells depend on their position relative to the pattern of the gradient — some cells ending up as the stalk, others as the fruiting bodies (see Figure 7. 2). Yates (1992, 478) has described it in the following way: "The same signal molecule, cAMP, according to the ambient context, can be interpreted as a command to aggregate, to differentiate, or to disaggregate. . . . This single signal molecule can induce different sets of genes (informational structures) at different times through different receptors. Here is chemistry in a very functional guise, giving commands: 'Aggregate'; 'Migrate'; 'Differentiate.'"[4]

All in all, this chemical exchange of commands between cells entails a collaborative effort aimed at the plasmodium's reaching up just a few centimeters

---

4   As we saw in Chapter 5, in eukaryotic organisms, cAMP has been accorded far more complex and inarguably biosemiotic tasks, including that of "second messenger" in intracellular signal-transduction processes.

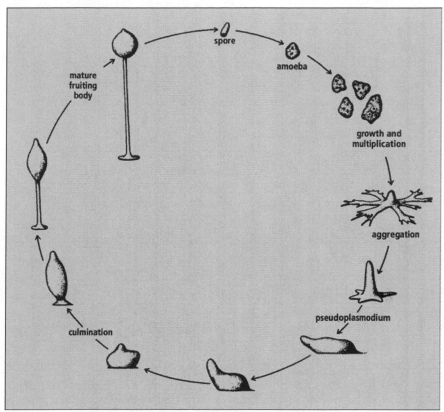

Figure 7.2 The life cycle of the slime mold, *Dictyostelium discoideum* (Goodwin 1995, 46).

above the ground to a position where the wind may get hold of its spores and spread them to more favorable locations, where unicellular life can begin again.

The Multicellular Self

As mentioned in Chapter 5, evolution has succeeded only three times in crossing the threshold from unicellular to multicellular life with any real success (if by this commonly misused expression, we refer to the capacity of a species to leave offspring in the sense of a large quantity of taxonomic groups that include many species). In these three cases — of *fungi, plants,* and *animals,* respectively — the problem of multicellularity was solved by the use of three very different principles. Most fundamentally, the differences turn upon how to establish a multicellular self. The *unicellular* self was said to be due to the establishment of a stable integration of a system for other-reference into a system of self-reference in the earliest prokaryotes (see Chapter 2). But how, the Darwinian must ask, could *clones* of cells descending from one and the same zygote find ways to collaborate on the task of creating a *communal* self, where single cells work for the common good, rather than for their own best interests?

The difficulty of maintaining slime mold cultures in the laboratory offers an interesting illustration of just this very problem. Routinely, such cultures are maintained by the collecting of spores (for example, by flooding the petri dish). But cultures that are propagated in this way have a tendency to grow shorter and shorter stalks, says Buss (1987, 127). This is presumably because the washing technique does not select against eventual mutations that increase the probability of cells to end up as spores, rather than carrying out help functions such as the forming of the stalk. Over the course of successive generations, one will therefore witness a falling ability of the progeny to form high stalks, for there will remain fewer cells available to carry out such necessary supportive functions. In the wild, this condition is prevented by sexual reproduction, whereby a recombination of the genetic material will take place so that individuals without the degenerated mutations will appear among the recombinants. These *genetically healthy* individuals may then found new cultures where a sort of multicellular self can be maintained — at least long enough for the relevant purposes.

The task of protecting such a multicellular self against mutational deterioration has manifested in widely different strategies in animals, plants, and fungi, respectively — and these differences apparently go all the way back to the three very different types of protists from which these three kingdoms were originally derived. Two essential traits are of special importance here: (1) possession of rigid cell walls and (2) possession of a cell-based structure. Plants have both, whereas animals and fungi must do with only one of them. Fungi have rigid cell walls but no unambiguous separation of cells; animals consist of cells lacking rigid cell walls. As we shall see, these fundamental differences each pose a very different set of challenges.

The most vulnerable point of departure seems to have been that of the animals — because the free mobility of animal cells (due to their lack of rigid cell walls) means that the animal multicellular self is in imminent danger of becoming deranged by anarchistic mutant cells. These are cells that are supposed to give rise to important somatic functions but instead — due to mutation — migrate to positions in the embryo reserved for the sex cell lines.[5] Our remote Metazoan ancestors solved this problem hundreds of millions of years ago by instituting very early differentiation of the embryonic cells. This implies a loss of totipotentiality

5   The reader should not misunderstand the apparent anthropocentric way of thinking here. Through the spans of millions of years, all kinds of chance mutations are statistically bound to appear — among them, some that threaten the multicellular self through inappropriate migratory propensities.

(the ability to become any cell in the body) for all cells except those few cell lines destined to become sex cells. This does not prevent the possibility that variant cells might appear and migrate to more favorable positions in the embryo, but it does prevent any asexual *reproduction* of such an odd individual — and therefore it protects the multicellular self against mutational degeneration.

And here we find the root of Weismann's cut, i.e., the strict separation of the germ cell lines from the somatic tissues. The result is that an animal's development is differentiated at the cellular level long before the differentiation is unfolded spatially as the adult form becomes established through a complex pattern of relative growth rates in different tissue areas combined with cellular migrations. One may say that an animal's self is bound to have a high degree of predictability as far as its basic form (*bauplan*) is concerned — but that this rigidity is richly compensated for by a very low degree of predictability as to the actual location of the animal, since animals are mobile creatures.

In plants, there is no corresponding separation of the temporal and the spatial aspects of differentiation, simply because plant cells cannot move. The plant cannot start differentiating its cells at an early time in anticipation of events to occur in later stages, for it is bound to leave the cells behind at the exact location where they were first formed. Plants therefore do *not* have the kind of *preformationistic* (using this word in its modern, non-naïve sense) scheme of development that is characteristic of animals, but are free to develop as the wind and the weather invites them to. For contrary to the animal cell method of development, plants maintain throughout life a mitotically *totipotent* cell line that takes care of somatic functions (as stem cell lines) as well as the formation of gametes. If regressive cell lines should appear, they will not normally pose a serious threat to the multicellular self of the plant, due to the lack of motility in the regressive cells. This difference between plants and animals is the reason why tumors in plants are relatively benign and clearly localized, whereas in animals they are often malignant and metastatic. Considered as a disease, cancer is precisely an attack on the coherent biological self by a subset of its constituents — one might even call it a *cellular insurrection* of sorts.

The vegetative, multicellular self of a plant is thus well protected against dissolving mutations solely through the rigidity of its cells. The load of endosemiotic controls in the plant kingdom is therefore considerably smaller than in the animal kingdom, where the more rigidly determined developmental scheme requires accuracy in terms of temporal and spatial coordination through semiotic exchanges between cells, tissues, and organs. Conversely, plants — unable to move away from their place of origin — have had to rely on flexibility in growth and structure and the development of a whole battery of exosemiotic tools for influencing surrounding life forms and patterns of interaction in order to survive in the living world.

In the kingdom of fungi, the problem of rigid cell walls, consisting here of chitin rather than cellulose — chitin exhibits far more resistance to attacks by microorganisms — was solved by compensating for the lack of *external* mobility through increased *internal* mobility. The fungi have achieved this by dismantling the automatic coupling of cellular division and division of the nucleus. Instead, most fungi consist of long branching threads called *hyphae* that form a tangled mass called the *mycelium*. The weblike structure that one may occasionally find on the surface of old bread may be the outer part of such a mycelium (though most of it grows invisibly inside the bread). In many cases, these fungal hyphae are *coenocytic* — meaning that they form lengthy, gigantic cells with multiple nuclei that are formed by division of the nuclei unaccompanied by any division of the cell. This method prevents nuclei from being locked up within the prison of the rigid cell walls.

Morphologically, fungi are considerably more underdetermined than even plants. Hyphae spread away from each other when there is plenty of food, and then re-assemble to form fruiting bodies (e.g., mushrooms) when such stores of nutrients are exhausted. Because of their "open contextual borders" (as the British fungal expert Alan Rayner has called it), fungi are involved in numerous symbiotic interactions. In many natural environments, fungi provide the hidden energy-distributing infrastructure — like the communication pipelines and cables beneath a city — that connects the lives of plants and animals in countless and often surprising ways (Rayner 1997). And because of these flexible borders, fungi may continue to live for a very long time. Thus, in a Canadian forest an individual *rhizomorphic* fungus has been observed to have a mycelium that extends through an area of fifteen hectare and has an estimated weight of one hundred tons. This individual is supposed to be fifteen hundred years old (ibid.).

In a dynamical biological system of such age, many somatic mutations are bound to have accumulated — and if we consider it to be one *individual* (as indeed, there is no biological reason why we shouldn't) we must accept that individuals may possess significant suborganismic variation, *even at the genetic level*. Adding to the suborganismic variation in fungi, mycelia often fuse with other mycelia, and often even with mycelia that exhibit difference markers. Such chimeric mycelias may sometimes prosper under conditions where none of the two components from which the chimera was derived would have been able to survive on their own. This is because different nuclei move to different localities inside the mycelium, depending on where their particular synthetic competences (as concerns enzymatic resources) could be of use.

This surprising openness, however, is also a major risk factor. Variant cell nuclei that might further their own rate of division at the expense of their contribution to the somatic tasks of the fungus will threaten to spoil the fungal self. That this is indeed a threat, writes Buss (1987, 134), is illustrated by the observation that "naturally occurring mycelia of the devastating rice pathogen, *Fusarium fujikuroi*, harbor a substantial proportion of nuclei which lack the capacity to invade rice." In order to defend themselves against this threat, these fungi have developed a host of endosemiotic control mechanisms, such as distinct markers, that must be accepted before mycelia can fuse with each other, and have established synchronization controls for the division of nuclei. Certain fungi have developed *septa* — partial partition walls that subdivide the hyphae into semi-cells — and these septa can be capable of distinguishing normal nuclei from parasitic nuclei, allowing only the former nuclei to pass.

We can conclude, then, that in all three taxonomic kingdoms where multicellularity has been established, the emergence of a coherent multicellular individual (or *self*) has required the development of numerous endosemiotic control mechanisms to safeguard the multicellular (or holistic) self against the ever-present risk of mutant anarchistic cell lines that would catastrophically multiply themselves at the expense of the multicellular individual's life. But only in the animal kingdom do we find that peculiar individualization of the self as a preformationistically designed entity, the animal, with well-defined borderlines in the physical, ecological, and social space.

Compared to animals, plants and fungi have much more freedom to profit from the vicissitudes of life — either through their own growth processes or through asexual forms of reproduction. But because the early segregation of the sex cell lines makes such possibilities virtually unavailable as a life-resource for animals, the animal kingdom has had to invent compensating mechanisms. The solution was the construction of sophisticated senso-motoric systems coupled to a corresponding finely tuned regulation of a *milieu interieur*[6] that could safeguard the stability necessary for reliable performance. The task of integration that this has required is truly impressive, and I will spend the rest of this chapter in illuminating the most biosemiotically relevant aspects of it.

Cognition and Semiotic Emergence

That *life* and *cognition* may be seen as related phenomena has been observed by several thinkers (Bateson 1979; Maturana and Varela 1980; Uexküll 1986), and this conception gains in credibility with the decreasing confidence in the idea of *thought* as a purely logical-rational activity and the understanding of its irreducible bodily embeddedness (see Sheets-Johnstone 1990; Damasio 1994; Kelso 1995; Port and van Gelder 1999; van Gelder and Port 1995; Deacon 1997; Lakoff and Johnson 1999, just to name a few). *Thought seems to swarm out of the body*: "The brain is, as we have seen, immersed in the immune system's floating

6   The term was coined by nineteenth-century physiologist Claude Bernard.

morass of physicality, and the cognitive scientists' search for the brain's supreme center — or 'central processor' — has proved futile. There do not appear to be any such centers or processors. Rather than the brain being pre-programmed to produce intelligence, intelligence seems to swarm out of it" (Hoffmeyer 1996 b, 114).

The idea that thoughts swarm out of the body implies their entanglement in human life processes. Although children are often thought to be able to spontaneously think both rationally and logically, such claims often underplay the rich world of rationalistic social interaction into which most such children have been plunged since birth, and as any parent can attest, the development of truly rational thinking in children requires much practice and incitement to develop. Conversely, narrative thinking, it has been claimed, comes to us for free and usually feels much more natural to us. The psychologist Jerome Bruner (1990, 43) has characterized such *narrative thinking* as fundamentally temporally organized: "Perhaps its principal property is *its inherent sequentiality*: A narrative is composed of a unique sequence of events, mental states, and happenings involving human beings as characters or actors. These are its constituents. But these constituents do not, as it were, have a life or meaning of their own. *Their meaning is given by their place in the overall configuration of the sequence as a whole* — its plot or *fabula*."

The predilection of the human brain for narrative thinking agrees well with the discussion we have had in the preceding chapter that argued that mental life is organized around a central series of "I can"s (Merleau-Ponty 2002 (1945); Sheets-Johnstone 1998). Bruner refers to the French philosopher Paul Ricœur's demonstration of a reciprocal interdependence between "being *in* history" and "telling *about* history" (Bruner 1990, 45) — and in this connection, one might also recall Merlin Donald's elegant theory of the origin of language in the mimetic culture of our quite smart but clearly nonlinguistic *Homo erectus* ancestors (Donald 1991). If human speech was indeed a further development of mimed communication, it would be understandable that our linguistic thinking from the very beginning has been inscribed in the senso-motoric field.

And yet when one comes to think about it, it is nearly unbelievable that for so long we have been able to maintain this basic confidence in the ability of thought to put itself beyond time and place — in spite of the fact that it all occurs in just a few cubic centimeters of nerve tissue at the inner side of the skull. Anyone having witnessed the effect of a cerebral hemorrhage at close range will recognize that this bit of brain tissue is all we have to connect us with the world.

And yet, we automatically project our thoughts upon the outside world. Or, as Carol Feldman has suggested, we tend to endow the conclusions of our cognitive reckonings with a special *external* ontological status. Our thoughts are *in here*, but the conclusions are *out there* — *ontic dumping*, as Feldman called this human naïveté (cited in Bruner 1990, 24). However, it only deserves to be called naive if one also believes that our *conclusions* are themselves beyond time and place.

For in reality, the brain is there to give us a track to act upon, it is not there to give us the truth (as this concept has traditionally been understood).[7] In Peircean pragmatism, *truth* is precisely seen as a track we may follow in our efforts — and not as a position or locale to be reached in a lifetime.

The inextricable relationship between *mental recognition* and *bodily senso-motoric activity* is evident most obviously at the level of our senses, as for instance, in the case of vision. For contrary to the way that we *finally* experience it, the human eye does not see a full picture of the visible field in the sense that all parts of the field are represented at the retina. The full picture is, rather, the work of the brain alone.

The eye focuses at each one-third of a second on less than 0.01 percent of the total visual field — but this small fragment, in turn, is perceived in very high resolution. Unconsciously, the muscles connected to our eyeballs make small movements, *micronystagmus* or *saccades*, whereby the focus is shifted some three times per second. As far back as in 1967, cognitive scientist Alfred Yarbus showed that these saccades are intelligent in the sense that they reflect the *context* of the visual field as it relates to the immediate goals of the perceiver — as opposed to how the field is *in itself* (Yarbus 1967).

In one of Yarbus's experiments, subjects were shown a picture of a room with some people in it and were asked to guess either the age of the persons, or the kind of activity they were engaged in. The experiments showed that eye movements exhibited different patterns depending on which of the two tasks that the subjects were asked to do. People use saccades to attend to the positions in space that are of *interest* to them, and it is a big question if they have any picture of the rest, whatsoever.[8] And indeed, of what use would it be to have such a picture? In doubtful situations, one might just take another look — and as far as memory is concerned, we likewise tend to record exactly those things most relevant to our goals in any given situation, and there is no use for the rest to be encoded into memory. Most likely, claim many cognitive scientists, our brains fill in the gaps according to prior experience and expectations — an everyday phenomenon called *visual closure* (see Ramachandran and Blakeslee 1998).

---

7   Our desire for knowing the truth is, of course, both useful and commendable. But it risks becoming exaggerated beyond our natural limits — as one may sometimes see in those philosophers of the analytical school who maintain a confidence in the scope of reason that may itself be somewhat unreasonable.

8   Recent studies in the widespread phenomena of *change blindness* — the inability to perceive otherwise obvious changes in a visual field if such changes are not being actively looked for — also lends support to this hypothesis (Grimes 1996).

The philosopher Andy Clark, working in the field of robotics that is devoted to the creation of intelligent autonomous agents, has suggested that a general principle for cognition in humans — as well as in modern mobile robots (called *mobots*) — is one that he calls the *007-principle*, wherein the mobots (or the human mind), like the famous British movie hero, use whatever resource they might first come upon to advance their performance. According to this principle, an evolving system always proceeds by *using the world as its own best model*: "In general, evolved creatures will neither store nor process information in costly ways when they can use t*he structure of the environment* and *their operations upon it* as a convenient stand-in for the information-processing operations concerned. [The operative principle seems to be this:] Know only as much as you need to know to get the job done" (ibid., 46; italics added).

The 007-principle is derived from work in new robotics or *situated robotics* — the attempt to construct robots that can cope with the messy and unpredictable environments of the real world. One of the main lessons from this work is the dethronement of the *central planner* model:

> The New Robotics revolution rejects a fundamental part of the classical image of mind. It rejects the image of a central planner that is privy to all the information available anywhere in the system and dedicated to the discovery of possible behavioral sequences that will satisfy particular goals. The trouble with the central planner is that it is profoundly impractical. It introduces what Rodney Brooks aptly termed a "representational bottleneck" blocking fast real-time response [in (Brooks 1991)]. The reason is that the incoming sensory information must be converted into a single symbolic code so that such a planner can deal with it. And the planners' output will itself have to be converted from its propriety code into the various formats needed to control various types of motor response. These steps of translation are time-consuming and expensive (Clark 2002, 21).

Instead of being based on the central planner model, mobots are built according to a principle that robot researcher Horst Hendriks-Jansen has called *interactive emergence* — but which might as well, or better, be termed *semiotic emergence* (Hoffmeyer 1997b). By semiotic emergence, I mean the establishment of higher-level patterns scaffolded by a situated exchange of signs between components.

For illustration, let me borrow an example from computer scientist Martin Resnick (1994). Resnick tells us, with barely concealed delight, how his computer simulations of traffic jams evoked head-shaking distrust in some of his colleagues because the simulation showed that a traffic jam moved *backwards* relative to the driving direction of the cars. This result was easily confirmed from the air, and the reason is simple: the traffic jam does not consist of cars, which of course move forward; it consists of a *relation* between cars. The traffic jam is a *pattern* that appears because of the semiotic interactions between the drivers of the individual cars who, moment for moment, must observe (interpret) the movements of other cars — as well as such constraints on driving that might stem from other factors, e.g., the borders of the lanes, roadwork, a nearby accident, or harsh weather conditions. We have here a kind of self-organization that cannot be explained through any generative causal law but instead demands a *historical* explanation of the phenomenon as an effect of the conjunction of circumstances that produced it.[9] This is the kind of interactive pattern that I call *semiotically situated interaction.*

The theoretical necessity for respecting the reality of historical idiosyncrasies forms a central theme in situated robotics, for such occurrences can not be escaped during the learning processes of animals (or of intelligent robots). A key concept in this connection is what Clark calls *soft assembly.* Soft assembly must be understood as opposite to *hard assembly,* as we know this latter, for example, from a traditional robot arm with a preprogrammed repertoire of movements, whose functionality requires accurate pre-design, manufacture, and installation. As an example of *soft assembly*, Clark mentions, by contrast, the phenomenon of human walking — a recursively adaptive feedback process that gracefully compensates for nearly all kinds of irregularities. Slippery sidewalks, high heels, blisters, or any of a number of various and unpredictable factors may well influence the gait, as well as the pattern of muscle contractions, during human walking — but the overall movement toward the goal usually continues unaffected.

Clark illustrates these ideas with reference to studies of how small children concretely learn simple "reaching-out towards" movements (Thelen and Smith 1994).[10] Although the end result of the efforts of the children to master such movements is quite universal, children exhibit very different styles in how they arrive at the desired result. It behooves us to follow Clark's (1997, 45) description of this process carefully, because the key to the phenomenon in question resides in the details:

> One infant, Gabriel, was very active by nature, generating fast flapping motions with his arms. For him the task was to convert the flapping motions into directed reaching. To do so, he needed to learn to contract muscles once the arm was in the vicinity of a target so as to dampen the flapping and allow proper contact. Hannah, in contrast, was motorically quiescent. Such movements as she did produce

9   Alternatively, one might enumerate a set of conditions whereby potential traffic jams are most statistically likely to occur. But every single case of an *actual* traffic jam will exhibit its own unique pattern and demand its own separate explanation to be genuinely informative.

10   As Clark (1997, 45) notes, this conception fits into a long tradition in developmental psychology: "Theorists such as Jean Piaget, James Gibson, Lev Vygotsky, and Jerome Bruner, although differing widely in their approaches, actively anticipated many of the more radical-sounding ideas now being pursued in situated robotics (Piaget 1952; 1976; Bruner 1968; Gibson 1979; Vygotsky 1986)."

exhibited low hand speeds and low torque. Her problem was not to control flapping, but to generate enough lift to overcome gravity.

Other infants present other mixtures of intrinsic dynamics, but in all cases, *the basic problem is one of learning to control some intrinsic dynamics* (whose nature, as we have seen, can vary quite considerably) *so as to achieve a goal.* To do so, the central nervous system must assemble a solution that takes into account a wide variety of factors, including energy, temperament, and muscle tone. One promising proposal is that in doing so the CNS is treating the overall system as something like a set of springs and masses.[11] It is thus concerned, *not with generating inner models* of reaching trajectories and the like, but with learning how to modulate such factors as limb stiffness so that imparted energy will combine with intrinsic spring-like dynamics to yield an oscillation whose resting point is some desired target. *That is, the CNS is treated as a control system for a body whose intrinsic dynamics plays a crucial role in determining behavior* (italics added).

Here, I want to suggest, is where we find the sources for psychosomatic integration. The central nervous system is not engaged in producing *solutions* to abstract *problems* but is, on the contrary, from the earliest moments of childhood incessantly occupied with the modulation of bodily parameters that show up for us in our attempts to overcome the obstacles posed by the body and the environment. By the expression *show up for us* I am referring not only to the external registration of success or failure in solving concrete tasks, but also to the internal feeling of the movements (e.g., the mechanical resistance that we experience as *exertion* as well as the muscle fatigue that we likewise label *stiffness*). Proprioceptive sense receptors are distributed over the muscles, sinews, and joints throughout the body. Through proprioception, it becomes clear that *cognition* (knowing something to be the case) is as much connected to the registration of movements and to the play of muscles, as it is to the brain and to symbolic reflection — in fact, these two *aspects* of cognition cannot exist apart.

As observed by Uexküll, Geigges, and Hermann, this is dramatically illustrated by the symmetrical set of illnesses called *phantom pains* and *scotoma limbs.*[12] In phantom pains, the brain is stimulated by proprioceptive signals from nerves once connected to a limb that is no longer there, while the proprioceptive signals emerging from a present but undetected *scotoma limb* are

11   Clark (1997, 45) refers to Polit and Bizzi (1978), Hogan et al. (1987) Jordan (1994), and Thelen and Smith (1994).

12   *Scotoma* is the medical designation for the visual blind spot that each of us has at the point where the optic nerve occupies the region where photoreceptors would otherwise be on the optic disc. Like the scotoma-limb patients, what is most curious not that there is an ever-present hole in our perceptual field, but that we do not *detect* it! This is another example of the kind of *sensory closure* discussed above.

never perceived as such, since they do not reach the brain. In the first case, one experiences the presence of a limb one doesn't have, while in the second case one is convinced of missing a limb that one actually does have (see overview in Ramachandran and Blakeslee 1998).

The neurologist Oliver Sacks experienced this phenomenon first-hand, after having suffered a grave accident and a subsequent operation, and has described the disconcerting experience of utter confusion that seized him after the bandages were removed and he could once again see — but not *feel* — his own leg:

> Yes, it was there. Indisputably there! . . . A leg — and yet, not a leg: there was something all wrong. . . . It was indeed "there" in a sort of formal, factual sense: visually there, but not livingly, substantially, or "really" there. It wasn't a real leg, not a real thing at all, but a mere semblance which lay there before me. . . . It was clear that I had a leg which looked anatomically perfect, and which had been expertly repaired, and healed without complications, but it looked and felt uncannily alien — a lifeless replica attached to my body (Sacks 1986, 91).

Uexküll, Geigges, and Hermann (1993, 43–44) have these comments:

> This impressively illustrates that our living body, which we experience as the center of our reality, is the product of a "neural counterbody" (*Gegenwelt*) which is continually shaped and reshaped by the ceaseless flow of proprioceptive signs from the muscles, joints, and tendons of our limbs to the brain, This "neural counterbody" is the center of a "neural counterworld" which our brain constructs and reconstructs from the likewise permanent flow of signs from the sense organs. Counterbody and counterworld are amalgamated to form an indissoluble unity because all the events we perceive in the environment are counteraffordances — that is, they are related to actual or potential affordances of our motor system and combine with these to form the spatial grid by which we orient ourselves.

As we have seen, active movement has been the key to animal life from the very beginning, and senso-motoric integration has subsequently been a major theme in the evolution of the animal kingdom. Step by step, animal species have managed to optimize internal and external coordination tasks, as such coordination was called forth by the imperative to move. And, given the way that evolution works, the achievement of such coordination (and the development of coordinative controls) must have been reminiscent of problem-solving strategies proposed by the research coming out of the work on autonomous agents.

In short, the semiotic interaction between embryonic cells, tissues, and organs must have, at various intervals in evolution, given rise to the unforeseen appearance of organized high-level *patterns* that could stabilize (or *canalize*) the evolution of suitable new developmental strategies or serve as stepping stones to reach yet more overarching organizational structures. Such patterns might consist of morphological elements (as, for example, changes in the relative distribution of embryonic tissues that would open the door for new functionalities) *or* dynamic elements (e.g., the establishment of new semiotic control loops), but in either case the principle would be the same.

And by terming this phenomenon *semiotic emergence,* I wish to bring out as forcefully as possible the common underlying principle exhibited in these cases, i.e., the principle that *the emergence of higher-level patterns is the result of semiotic — and not just physical — interactions between entities at the lower level.* Semiotic emergence will, of course, build on patterns of semethic interactions and it is perhaps a question of terminology when — or from which level of the process — one should talk about *real* emergence.

The appearance of a uniquely flexible kind of consciousness in big-brained animals is perhaps the most obvious case of semiotic emergence. Integrating the ideas put forward by Sheets-Johnstone with those of the robot scientists may give us a hint as to how the emergence of such consciousness might have taken place. Neurobiologists have proved experimentally that proprioceptive information is more important than vision when the task is to direct a correct movement such as reaching out for something: "The remarkable trajectory errors by patients with large-fiber sensory neuropathy indicate that proprioceptive information is critical if accuracy is to be achieved," writes Ghez et al. (1995, 561).

More remarkable, perhaps, is that the task of proprioceptive information does not consist in any simple or direct correction of errors. Rather, sense inputs from proprioceptive sense receptors supply data that the brain must use to calibrate the organism's ongoing action in the environment. As Ghez et al. (ibid.) explain,, proprioceptive information "operates by generating and recalibrating internal models of the mechanical properties of the limbs. . . . It appears that the motor systems predict interaction torques and control their effects so as to achieve the kinematic results required by the behavioral task. In such a system, internal models would provide the means for predicting the unfolding scenario of goal-directed movements."

It is fascinating to speculate (as we have noted earlier — Hoffmeyer 2002a) that these internal-model schemas, which are generated and calibrated by proprioceptive information, might perform the same trick in the moving body as do the *motor emulators* in industrial robot systems. An emulator is a piece of onboard circuitry that replicates certain aspects of the temporal dynamics of the larger system, and its role is to *predict* — i.e., determine probabilistically — what the feedback from the sensory periphery is likely to be. If the device is reliable — i.e., if it is properly calibrated — these predictions can be used in lieu of actual sensory signals, so as to generate faster error-correcting activity. According to Andy Clark (1997, 22), the human body is in need of such a mechanism: "Proprioceptive signals must travel back from bodily peripheries to the brain, and this takes time — too much time, in fact, for the signal to be used to generate very smooth reaching movements." The purpose of the motor emulator is to overcome this problem. A motor emulator "facilitates real-time success by providing a kind of 'virtual feedback' that outruns the feedback from the real sensory peripheries. . . . It models salient aspects of the agents' bodily dynamics, and *it can even be deployed in the absence of the usual sensory inputs*" (ibid.; italics added). The evolutionary advantage of such an emulation device would apply not only to human bodies, but to all animal bodies above a certain size. And brains capable of developing such virtual action-coordination controls must have been created by evolution long before the arrival of humans. The point is that this *substitute-interpreter,* as we might call it, would seem to be a perfect early precursor for human consciousness.

Its primary (survival) function is to facilitate smoothly coordinated goal-directed movement, but it does so by making a construct which is not as such a *direct reflection* of the regularities of sensory input in the naïve inner-model sense (for this would require an infinite regress of homunculi to view the models). Rather this construct is based upon a particular set of sensori-motor schema that are "selected" from the organism's entire behavioral repertoire, so that it orients the organism for the immediate next action. While still probabilistic, this selection of which schema to activate is both evolutionarily and ontogenetically informed. It thus functions as a qualified guess that will be continually re-calibrated by the *actual* incoming inputs from the proprioceptive senses for fine-tuned appropriateness and accuracy. In other words, and most essentially, this systemic substitute or virtual interpreter has already attained a degree of *autonomy* relative to the real world. In the natural history of human consciousness, these properties are exactly what should be expected from an eventual early precursor for what much later would become the *virtual world* of human consciousness, as we know it from our own lives. And this corroborates Sheets-Johnstone's conclusion (1998, 290) that the internalization of *sensing surfaces* to produce proprioceptive organs is an "*epistemological gateway,* one that, by descent with modification, may clearly be elaborated both affectively and cognitively."

We may conclude, that the deepest sources for the *cognizing self* lay inscribed in the very basic senso-motoric *unity* of animal multicellular life. And in the next chapter, I will discuss in detail the appearance of the peculiar abstractive talents that came to characterize the uniquely *human* form of such cognition. It is important at the outset, however, that one understands the basic phenomenon of cognition as a quite general product of semiotic emergence in multicellular organisms. Nothing could be more backwards, in other words, than to understand *human* cognition as the primary or most definitional form of cognition. Human cognition should be understood more modestly as an extraordinarily interesting — but obviously species-specific — development of the cognitive capacity that quite generally characterizes all moving, living, adaptive semiotic systems.

The Internal Surroundings

The realization that organisms do not have just an external environment, but also an internal environment — a *milieu interieur* — was the seminal idea of the French physiologist Claude Bernard (1813–78). Bernard saw that a constant internal milieu was a prerequisite for the efficiency of complex organisms, and he expressed this in saying, "The fixity of the internal environment is the condition for free life" (Olmstead 1938, 254). Inspired by Bernard's ideas, the American physiologist Walther Cannon published a famous book entitled *The Wisdom of the Body* in 1932. It is in this book that Cannon first introduces the term *homeostasis* (from the Greek roots for "the same" (*homo*) and "state" (*stasis*)) as a designation for the surprising ability of all animal bodies to maintain a stable internal state concerning parameters such as temperature and salt balance, even under conditions of severe stress. Science has learned since then that the phenomenon of *homeostasis* depends on the coupling between the saturation degree of specific receptors and a range of response mechanisms in certain cells. If the concentration of an important component falls outside of the usual range, compensating reactions or feedback reactions will occur. For example, the blood sugar levels in a normal person having awakened after a good night's sleep will lie at approximately 90 milligrams per 100 milliliters. If the person then has a breakfast containing a good deal of sugar or carbohydrates, these compounds will successively become degraded into glucose and enter into the circulatory system — accordingly, blood sugar levels will go up. The increase will be registered by specialized cells in the pancreas, and these will command the pancreas to excrete insulin to the blood. Insulin stimulates body cells to absorb glucose from the blood. It especially stimulates liver cells and muscle cells to store glucose as *polysaccharide* (glycogen) — and through this process, blood sugar levels are brought back down to normal.

When, a couple of hours later, this same person's blood sugar level is again approaching a low (due to metabolic consumption of the available glucose) the pancreas is once again in charge of rectifying the condition — this time by excreting the hormone *glucagon* that induces muscle and liver cells to degrade glycogen back to free glucose. The constancy of blood sugar concentrations is thus at all times in a situation of touch and go, with insulin and glucagon as the enforcement officers and pancreas as the executive officer. And, of course, these hormonal mechanisms do not operate in a vacuum, but are tightly interconnected with brain processes. Neurons in the area of the brain called the hypothalamus are continuously monitoring blood sugar concentrations and when the level is too low, the hypothalamus induces a feeling of hunger — whereas, when the level is too high, a feeling of satiety will ensue.

In general, a subtle and highly complicated interplay goes on between the nervous system (and in particular, the hypothalamus, acting in concert with the rest of the limbic system) and the endocrine system (i.e., the hormone-producing glands that physically maintain the homeostasis). Homeostasis, however, is far from a static *set point* as one might find programmed into a thermostat. Rather, some parameters run through well-controlled cyclic oscillations (e.g., those coupled to the menstrual cycle), while others change in ways correspondent with the ongoing phases of life, or in connection to abrupt changes in the conditions of life. In addition, a range of more or less programmed changes of the homeostasis can be initiated by the manifestation of recurrent emotional patterns.

A hormone can be taken as the prototype molecule used to function as a messenger between different locations in the system of the body. Originally, *hormone* was the term used to describe those substances excreted into the blood by a diversity of glands and carried away with the blood to exert their effects in such tissues as were ready to interpret them (i.e., to recognize these substances and to respond accordingly). But the definition has had to be expanded in recent years to include a variety of signal molecules that are emitted not from glands, but from specialized cells in other tissues and organs. Well-known is for example *histamine,* a substance many readers may be familiar with from its role in allergic reactions, but which may also be formed in the event of tissue damage or infection.

Another important group of signal molecules are called *prostaglandins.* These I have discussed previously, in Chapter 2, with regard to their role in the lowering of the threshold for the feeling of pain in the skin. Yet prostaglandins are also involved in enabling many other internal control mechanisms in the body: they dilate passages in the bronchia, they inhibit the secretions of the intestinal tract, they stimulate the uterus, they influence nerve function, they cause fever, and they facilitate coagulation of the blood.

A related group of signal molecules that should be mentioned here are the *growth factors* that stimulate cell divisions and normal development. All these latter signal molecules operate by *parachrin regulation* — i.e., they are excreted to the interstitial fluids (the liquid that fills empty space between cells); they are generally short-lived; and their effects, therefore, are mostly local — which are exactly the qualities one wants in a control mechanism that is required to play its role at a specific moment in the trajectory of development, and then be gone. Yet another particularly interesting group of hormones are the *neurohormones* or *neuropeptides,* peptides that are transported along nerve threads (axons) and released into the interstitial fluids. Neuropeptides play a central role in the communication between the nervous system and the immune system — I will be having much more to say about this critically important communication shortly.

Finally, in Chapter 5, we saw how some hormones (those called *steroid hormones*) may act directly upon the transcription of specific genes, as these hormones are targeted by the relevant cells and become attached to specific proteins. Most hormones, however, operate more indirectly (and more obviously biosemiotically) through the processes of signal transduction. We also saw in Chapter 5 that the same hormone may have quite different effects upon different cells, since the effect depends upon the protein profile of the cell — i.e., on how the actual proteins present in a given cell will interact with the secondary signal evoked at the membrane's inner side by the binding of hormone molecules to surface receptors. Here, as everywhere in living systems, we see the essential principle at work that shows us that *signals do not contain their own message*. Rather, when the concentration of a hormone surpasses a certain threshold value (i.e., when the degree of saturation of specific receptors at the receiving system attains its critical value), the hormone acts as a *sign* that *induces the formation* by the system of an *interpretant*. And the nature of this interpretant depends *exclusively* on the contextual situation of the receiving cell.

The Self as Iconic Absence

An important aspect of endosemiotic control is the systemic property of the body that Leo Buss aptly called its *somatic ecology*. This is the incessant corrective effort exerted by white blood corpuscles (lymphocytes, macrophages, etc.) all over the body, settling potential conflicts between cellular and bodily interests.

Traditionally, the immune system has been conceived of as a "defense department" against infections — a defense that was based on the dynamic cooperation between a multiplicity of white corpuscles that continually patrol the tissues and body fluids of an animal.[13] *Immunological identity*, however, does not consist simply in a demarcation of the *biochemical self* as a stable, unambiguous entity to be protected from the (equally unambiguous) invasions from the outside. Rather, the term refers to a far more complex and ongoing series of *events* — what Francisco Varela (1991, 88–89) has characterized as "a self-referential, positive assertion of coherent unity — a 'somatic ecology' mediated through free immunoglobins and cellular markers in a dynamic exchange." For to act upon some currently residing element within the body as *alien to* the multicellular system of the self and to act upon some other currently residing element within the body as *constituent of* the multicellular system of the self are, logically, two sides of the same behavioral coin. Expressed differently, *other-*

13   In the human body the number of white blood corpuscles has been estimated as 10^{12} (one trillion).

reference and *self-reference* are *inseparable* aspects of the dynamics of the living (see Chapter 2).

The immune system identifies potentially foreign molecules that might have penetrated the body through the processes of exposing their iconic interaction with the body's own lymphocytes. Such indexicality — the clue pointing out the identity of the perpetrator, in this case — paradoxically appears as *a breach of an expected iconic absence*. In other words, the foreign molecules are disclosed because they engage in a kind of iconic semiosis that should have been out of reach to them — as, indeed, it is to all authentic molecular members of the bodily self. Uexküll, Geigges, and Hermann (1993, 38–39) express it this way: "To identify the unknown means to recognize something familiar: nonself is a potential variant of self."

I shall not attempt to present a complete semiotic analysis of the numerous refined mechanisms of the immune system here, for such an analysis would easily take several volumes the length of the present one.[14] Here, I shall focus on just one such mechanism, which I believe will give the reader insight into the general principle under discussion — for this mechanism concerns a sophisticated cellular plot whereby foreign antigens are lured into an iconic trap that then serves as an index to the body.

Few things bear clearer witness to the communicative nature of the body's inner cellular life than does the fact that each cell, at its surface, carries a "tag" — a "Hello, my name is _____" — that identifies its origin. The mechanism seems eerily identical to the one used by travel agencies when they equip tourists at foreign destinations with characteristic hats or scarfs, or other easily recognizable markers for identifying one another. The character of these cellular tags is infinitely more complex and interesting, however, in that they consist of degradation products from the cell's own proteins. These are peptide fragments with a length of –ten to twenty amino acids, that — borrowing Kilstrup's expression — "are stretched out and laid like a sausage in a hot dog bun, between the protein flaps of the MHC molecule" (MHC=Major Histocompatibility Complex) (Kilstrup 1998, 110). The "whole hot dog" (amino acids plus MHC molecule) is then transported out of the cell and presented to the surroundings by being placed at the surface of the cell. (And the term *presented to* is actually a quite precise technical designation in this context, as we shall see below.)

14 Interesting contributions to such an analysis can be found in Sercarz et al. (1988). Unfortunately, none of these contributions seems aware of the possibilities opened by the Peircean tradition in semiotics. A fine discussion of this latter issue may be found in Neuman (2005) and in Kilstrup (1998, in Danish). By the time this book went to press, a biosemiotically informed analysis of the relationship between hosts and parasites was published by Elling Ulvestad (2007). See also Kawade (1992).

MHC is a class of protein molecules that exhibits extraordinary variation from individual to individual. This variation is based, however, on one common molecular theme in the form of a *cleft* (or *hollow*) in the three-dimensional structure "intended" for the binding of cellular peptide fragments. Such wide variation is enabled by the fact that every individual organism has its own absolutely specific MHC molecules — and this is why, for example, transplantations of organismic tissue from one individual to another is likely to provoke the formation of defensive *antibodies* in the new host. Conversely, this individual *molecular signature* means that the body's own cells may be assured that if the MHC molecules at the surface of a cell are correct — i.e., match their own — this cell will indeed be "one of us". Another characteristic trait of the "MHC hot dog" worth emphasizing is that the peptide fragments confined inside are positioned in a stretched (linear) form, rather than in its original three-dimensional structure, indicating that the *amino acid sequence* of the relevant peptide fragment is *displayed*. The significance of such display will become clear presently.

For it is not an overstatement to say that the *key* to the immune system's ability to distinguish between *self* and *nonself* resides in the action of the *T cells* — lymphocytes thus named because, after having been formed in the bone marrow, they get their definitive "tag" through a maturation process in the thymus.

T-lymphocytes carry special receptors on their surfaces (T-cell receptors, abbreviated TCR) that specifically recognize and bind to the stretched-out peptide fragments (as presented via "MHC hot dogs") bearing the signature mark of the individual. Each TCR specifically binds to one, and only one, MHC-presented peptide fragment, iconically corresponding to the empty place on that particular receptor.[15] As long as the immature lymphocyte stays in the thymus, it will only encounter peptide fragments degraded from the body's own repertoire of proteins — and in the rare case that it attempts to bind to one of these, it is immediately eliminated. This ensures that the surviving mature lymphocytes (the ones that leave the thymus in order to take part in the immune defense) will never possess T-cell receptors shaped for the iconic recognition of peptide fragments from their own body. The *immunological self* thus is based on an iconic *absence* — the absence, under normal conditions, of iconic representations of the "empty place" on T-cell receptors.

Moreover, when an organism is infected by a virus or a bacterium, the infected cells will degrade parts of the attacker's proteins to small peptide fragments to

15   Properly speaking, it takes an *interpretant* to establish the iconicity of a given relation. What I mean by calling this correspondence *iconic* is that the body, thanks to its evolutionary ancestry, has constructed this system of *TCR + HCM-peptide binding relations* in the *expectation* (in the sense of genetically fixed past experiences extrapolated to the future through inheritance) that it will serve to (mis)direct foreign pathogens into a binding trap based on likeness of shape between the TCR-receptor and the degraded fragment of a foreign protein.

be *presented* by the "MHC hot dogs" on the cell surface as warning signals. This partial degradation does *not* prevent the cell from being destroyed by the attack, nor does it prevent a new generation of virus particles from being released onto the lymphatic vessels, but it does effectively summon the omnivorous white corpuscles, the *macrophages*, and these will swallow up the prey (vira or bacteria) by *phagocytosis* — a process whereby the cell wall is turned inside out to engulf the prey and to successively digest it through a battery of enzymes released from organelles (the *lysozomes*) inside the *macrophage*.

The next link in the process is bifurcated and consists firstly in the presentation, by the *macrophages*, of fragments of the degraded virus proteins at *their* surfaces. Parallel to this, a very specific reaction takes place whereby the B lymphocytes recognize and bind to certain domains of protein molecules (called *antigens*) that belong either to the surface of as yet unphagocytized vira or bacteria, or to free proteins in the blood or lymph. This binding is made possible by the presence on the surfaces of the B lymphocytes of a kind of receptor that *also* functions as an antibody against the antigen in question. The binding causes an activation of the B lymphocyte that now starts breaking down the infectant and *presenting* the antigens *at its own surface*, in the same way the macrophage did.

To really appreciate the subtlety of this arrangement, it is necessary to consider the numeric relations involved. There are an estimated 10^6 to 10^9 different receptor *types* (each of them corresponding to *one* specific antigen) on the surfaces of the population of lymphocytes in the human body. Each single lymphocyte has one, and only one, of these receptor types, and the enormous diversity depends — as in the case of T lymphocytes — on a chance-generating principle.[16] In practice, this means that there will always be present in circulation just one B lymphocyte to make the single antibody effective towards precisely one corresponding foreign protein that might have entered the body. Conversely, each receptor at the B lymphocytes is specific enough that it will only recognize one single protein among the millions of different kinds of protein molecules present in nature. The next link in the defense process is the formation (through cell division) of *clones* of those B lymphocytes that recognized the foreign protein molecules, and thus must have carried receptors with antibody activity against them. This is where the T lymphocytes enter the picture, and specifically, the kind of T lymphocytes called T helper cells. The receptors on T helper cells are geared to recognize (and to bind to) those exact antigens that

16   The principle is based on *crossing-over events* under the formation of each single clone of lymphocytes. For example, the immunoglobulin G (IgG) of two light polypeptide chains, couple to two heavy chains. Since every cell line, through such crossing-over events, will have attained just one out of three thousand possible light chains, and one out of five thousand possible heavy chains, this arrangement implies that the complete IgG molecule will represent one of 1.5×10^6 (one-and-a-half million) molecular possibilities.

are presented as "MHC hot dogs" on macrophage and B lymphocyte surfaces. Here — as in the other recognition processes we have discussed in this section — recognition is based on iconicity in the sense that significant domains on the relevant antigens have shapes and electrostatic properties that are iconic to the potential (but in the body, nonexistent) protein domain, that "fits" the specific receptor of the T helper cell.[17]

We may now see why biochemists have chosen the (from a scientific standpoint) rather strange (and, in fact, *illogical*) verb, *to present*, for the behavior of the macrophages and the B lymphocytes. For the immunological effect of these processes is precisely to make antigens visible to the T helper cells. When T helper cells successively register and recognize the presented antigens (which are being presented to it via a *correct* MHC complex and at the same time as being *non*self), they respond by releasing the production of yet another class of signal molecules, the *cytokines (interleukins, tumor necrosis factor, lymphokines, interferons)* that cause the cells to engage in a process of reciprocal activation under the formation of clones containing large amounts of those exact B and T lymphocytes that are involved in the inactivation of the infection.

Many of the B lymphocytes in the clone successively develop further into *plasma cells* that release large amounts of specific antibody. Such antibodies will bind to antigens at the surface of intact bacteria (or vira), thereby forming complex aggregates that will further attract bacteria-eating macrophages and also induce an *aftercare* (via the help of the complement system) that assures that the antigen-antibody complexes remain dissolved until the lymph or blood has carried them to locations of intensified *phagocytosis* (lymph nodes, spleen, or liver). There they will be taken up and degraded. Some of the activated B and T lymphocytes will survive in the organism for years and function effectively as *memory cells* that will ensure a quick mobilization of the immune defense system should that same kind of infection appear again in the future.

The interplay between the astronomical number of cells involved in the immunological reaction is, of course, far more complex than this brief description has been able to even hint at. But my only aim here was to show how the immune system discloses foreignness as an iconic *presence* only because of the equally constructed background of a presumed *absence*. The whole system contains the further finesse that while B lymphocytes exclusively form antibodies against antigens (i.e., against certain domains at the surface of proteins called *epitopes*), T lymphocytes will only initiate cytochemical activation when the sequence of amino acids in the epitope — as stretched out in one dimension — is *recognized* as *nonself*. The safety system thus is double-stringed and requires B lymphocytes

---

17   Iconic recognition as described here exemplifies the phenomenon Stjernfelt (1992) has called *categorial perception*.

to recognize the three-dimensional surface structure as foreign at the same time that the T lymphocytes recognize the one-dimensional sequence of the epitope to be foreign.

Seen in a larger view, the immune reaction is based on a special selection process called *clonal selection*. As in so many other cases in the world of the living, the effectual principle here consists in the gross overproduction of entities — followed by the selection for survival of exactly those entities (here, clones of lymphocytes) that have shown themselves most useful. In the nervous system, as we saw, all superfluous or inactive neurons soon become eliminated, but in the immune system the mechanism is slightly different in that an incredibly rich *lymphocyte diversity* is maintained throughout life. *Selection* here is thus rather concerned with the allowance of certain lymphocytes to form clones containing numerous identical cells. The mechanism, in fact, resembles that of *gene regulation* where, similarly, genes passively await the time that cellular machinery eventually recognizes the need to activate the protein resource specified by the gene in question.

In sum, we may say that the recognition of molecular foreignness is obtained through the development of a *maximal iconic openness*, followed by the *elimination* of those openings that allow the self to be recognized. This causes a *digitization* — i.e., a discontinuous distribution — of protein domains into two classes, self and nonself, where nonself is distinguished through the occurrence of iconic recognition processes available exclusively for nonself molecules.

The biosemiotic problem that the immune system has solved in this intricate but elegant way consists in the fact that foreignness is a negative property, a property that cannot be understood in any other way than as nonself. But *non* is a phenomenon meaning "x plus its denial," and as such it is not accessible for immediate iconic representation. By making self negative (in the form of an absence) however, it became possible to make nonself positive in the form of a multitude of potential icons for nonself — i.e., a multitude of T lymphocytes capable of recognizing all conceivable peptide fragments *other* than those that form parts of the body's own proteins. The negative property *non*self was converted to a positive property, *an index for foreignness*.

Complex as it seems here, this very same principle is familiar to us from our experience in the anthroposemiotic area, where numerous cases of the detection of nonself (or *the other*, as the term is used in the social sciences) is based upon the *absence* of a presumed, constructed, and otherwise invisible *uniformity*. Examples might include the tattoos and insignias on the backs of "outlaw bikers," the veils of Muslim women in the West, the Bibles of Christians leaving a church on Sunday morning, all sorts of sports team attire, and — not the least — the various uniforms of soldiers. Employing this strategy also runs the inherent risk that the system can be fooled by the adoption of a false uniform as in "The Captain from Köpenick." Conversely, the police may perhaps just as easily place an informer in a group that *they* might want to infiltrate, simply by having that person satisfy the said group's criteria for uniformity — be it physical or psychological.

This strategy of *faking*, based on the strong organizing power of iconic relations, runs all through the biosemiotic repertoire of living systems. For example, HIV vira have become specialized for infecting T helper lymphocytes and these vira are surprisingly skilled in avoiding detection. Here the *hiding* strategy is made possible by an extraordinarily high error frequency in the translation of virus RNA to virus DNA that precedes integration of the virus into the host chromosome. Thus, the virus simply mutates into any conformation other than its present one *as soon as it has become recognized*. It changes its uniform, one might say — and as it is always the uniform and not the wearer per se that is present to inspection — the virus thus avoids the full effect of the immunological response. In this respect, the HIV virus resembles the influenza virus — and in both cases (and for the same reason), it has proved very difficult to produce vaccines that are not outdated before they reach the marketplace. (In this respect, the HIV virus is approximately sixty times more *iconically deceitful* than the influenza virus.)

The Bodily Psyche

In discussing the internal milieu, I noted how the hormones *insulin* and *glucagon* cooperate to keep our blood sugar level inside an optimal range. But we also saw that mere chemical brute force was not in itself sufficient for true *homeostatic regulation* since it had to be augmented with brain-based supervision — executed by neurons from the hypothalamus that signal hunger when the sugar level goes too low and a feeling of satiety when the sugar level goes too high. It is no longer a secret, then, that our psychological life is influenced by our internal chemistry — and vice versa. We tacitly understand that people get irritable when hungry, or accept, as a matter of course, that unexpected behavior may be caused by PMS (premenstrual syndrome). This is also why we fear for the possible side effects of hormonal treatments and antidepressant drugs, and why a controlled and balanced mental life is important in the world of sports and entertainment — even professional chess players, it has been shown, do better if they undertake decent body training.

In other words, not many people will nowadays seriously deny that body and mind are two deeply integrated aspects of human life. And since one can hardly quarrel with one's wife or play chess while unconscious, it is difficult also to deny that this whole business of what we call human consciousness is effectively part and parcel of the slimy dynamics of the body's incessant corporeal functioning. The question remains, though: *How can this possibly be?*

The Australian-American philosopher David Chalmers has introduced a distinction between *the soft problems* and *the hard problem* that need to be explained in accounting for this consciousness — and his distinction has become a standard one in the cognitive sciences and in the philosophy of mind. The soft problems concern the question of the mechanics of how the brain manages to integrate and process its electro-chemical data. Finding the answers to these problems may take decades, but will not, according to Chalmers, force us to break with any already fully accepted explanatory principles. However, solving the hard problem will, for the hard problem consists in explaining why brain processes are so often accompanied by an *experienced* inner life. Here we are confronted with a problem that, as Chalmers (1996, xiv) carefully reasons, cannot be attacked by the usual scientific methods: "I argue that reductive explanations of consciousness are impossible."[18]

Other philosophers have argued, persuasively I think, that science cannot possibly describe conscious experiences in a language that in principle operates exclusively inside the limits of grammatical third-person descriptions. An "I" cannot, John Searle (1992) claimed, be described by an uninterrupted sequence of "it" phenomena, no matter how complicated such sequences might be. Articulating the same problem a bit more poetically, philosopher of mind Thomas Nagel (1986, 13) once famously asked, "How can it be the case that one of the people in the world is *me*?"

The error of confusing first-person experience and third-person experience is widespread in much "scientific" thinking, as was illustrated by a newspaper story that appeared some years ago under this heading: SCIENTISTS WILL SOON BE ABLE TO SEE CONSCIOUSNESS. The impetus behind this alarming title turned out to be a quote from an expert in mathematical modeling working with brain-scanning technology. "I am pretty sure that one day we will have a picture on our scanner of the activity patterns constituting consciousness," he told the newspaper. But will he?

> Let us imagine him scanning my brain while I — living in the dark winter of
> Denmark — have an experience of longing for summer. Here I personally have no
> difficulty in believing that this experience might somehow have been evoked by
> a brain activity which can be visualized on the scanner. For the sake of the argu-
> ment, let us now further assume (though I take it to be not very likely) that one day

---

18   Chalmers does not doubt that consciousness is a natural phenomenon or that it should be possible to construct a scientific theory of it. But he thinks such a theory would necessarily contain kinds of lawfulness very different from the kind we are used to from physics. In this sense, at least, biosemiotics and Chalmers share the same perspective.

our expert would be able to tell his colleagues while scanning me, that at this time Jesper Hoffmeyer had an experience of longing for summer — and also that I did, in fact, have such an experience at exactly this time. Even then, of course the expert did not "see" *my longing*, for it cannot *be* "seen," it can only be *felt* — and only by *me* (Hoffmeyer 1999b).

Philosophers have used the term *qualia* as a designation for this aspect of our world, the inner feel of lived experience — e.g., the feel of redness when you look at a red tomato or the joy of moving in the rhythm of dance. *Qualia* is thus the essentially *subjective* dimension of consciousness (as this term is used in its everyday — not its Deelian semiotic — sense). And as such it constitutes the deepest challenge to cognitive science. For while all other aspects of our psychological life, at least in principle, might be explained within the ontology of natural physical law, doing so seems impossible as concerns the phenomenon of *qualia*.

And this presumably is the reason why materialistically inclined cognitive scientists go to such efforts to argue that the whole notion of a felt, lived, personal *qualia* is based on some kind of misguided illusion — neuronal brain states exist and are real, but beliefs, feelings, desires, and thoughts, are misdescriptions of these brain states. To believe in them as real *as such* is to fall prey to a widespread, but scientifically illiterate illusion (Churchland 1986, Dennett 2001). But, one may ask, can you really be mistaken about anything if you are not someone? Can pure things have *illusions*? The answer, it seems, has to be no — for if things were pestered by illusions, the *illusion-having* itself would constitute a genuine kind of subjectivity. Our feeling of being a subject can paradoxically only be *illusory* if it is *true*. To call subjectivity an illusion is simply self-contradictory.

One strategy for circumventing the *qualia* problem has been to reject the set of conceptions that — with a certain, and unfortunately not rare, scientific condescension — has been called *folk psychology* (Churchland 1984). By this term is meant the whole set of everyday conceptions about life and the world that we have all grown up with and take to be self-evident — such as the belief that other people have mental states like ourselves, or that maybe even animals have some kind of *animal equivalent* of such, and that neither they nor we are mere unthinking biological machines, as the detractors of this folk psychology insist that we, in reality, are.

One often discussed example of such folk psychology is the idea that the cognitive philosopher Daniel Dennett calls *the Cartesian theater* — i.e., the conception that in some well-hidden locality "in the obscure 'center' of the mind/brain, there is a Cartesian theater, a place where it all comes together and consciousness happens" (Dennett 1991, 39). It is, I suppose, true enough that we all have this immediate experience of a world *presented to us* by the senses — a continuous and coherent world in both time and space, and a world that exists *outside of us* in more or less the same way it is *presented to us* inside our skull. It does not cost Dennett a lot of effort to show that this folk-psychological fundament about our own understanding is not reconcilable with the neurobiological facts concerning perception (nor, for that matter, with the conception of the nature of experience as arrived at by analytical philosophy!).

In this chapter, I have already discussed the "illusion" of visual perception as a spatio-temporal continuum. We saw that the fast movements of the eyeballs, the saccades, could not possibly give us information about anything but a small, if accurately selected, sector of the whole visual field — and that the experience of seeing before our eyes a "filled out space" where none, perceptually, exists, apparently must also be the construction of the brain. Nonetheless, I decline to accept without further ado Dennett's claim that we are all suffering a *folk-psychological illusion* concerning visual perception and that *qualia* belongs in this same drawer of imaginary brain constructs. For it may be objected that the illusion of the Cartesian theater is only an illusion relative to the concrete mechanics of visual perception, not relative to what is seen.

What Dennett has proven is in reality only, as the philosopher Alva Noë (2002, 3) has elegantly expressed it, that "it turns out we are mistaken in our assessment of how things seem to us to be." Our illusion does not concern what we think we see, but only how we *experience* that which we think we see. Vision is experienced as discontinuous, Dennett claims, but we think it is experienced as continuous: "One of the most striking features about consciousness is its discontinuity — as revealed by the blind spot, and saccadic gaps, to take the simplest examples. The discontinuity of consciousness is striking because of the *apparent* continuity of consciousness" (Dennett 1991, 356). But it is far from obvious what should be understood by an "apparent continuity of consciousness." Apparent *to whom*? Dennett's attempts to identify consciousness with the neural processes that are substrate for its appearance seem in this case to lead us to logical monsters that won't be conjured away simply by calling them *apparent*. Don Favareau (in a personal communication) has expressed it thus: "If the argument of the eliminative materialists is that *consciousness* is an *illusion* brought on by neuronal brain activity, this only begs the further question of exactly *how* and *why* this illusion appears to me in the way that it does. What *explains* the experience of the illusion? In other words, this idea — that what we *think* is our mental experience is a fallacy — is *itself* a self-contradicting fallacy."

Neuropsychological research indicates that the cognitive structure that is responsible for vision in human beings may be divided into two cooperative, but partly independent, visual brain systems — one of which (the oldest in an evolutionary context) is specialized in the control of vision-based here-and-now motoric activity, whereas the other (more recent innovation) controls the knowledge- and memory- based selection of well-prepared planned actions referred to as "insight, hindsight, and foresight about the visual field" (Milner and Goodale 1998).[19] For illustration, consider a series of experiments discussed in Milner and Goodale (1995). In these experiments, subjects were given the task of following a visual track with their eyes and, at the same time, required to indicate the path of the track manually. The experience was performed in such a way that the track at unexpected times suddenly was a little bit displaced. Under these circumstances, subjects without difficulty traced the displacements (as disclosed by measuring the saccades) without consciously recognizing that a displacement had taken place.

Successively, it was shown that if subjects that unconsciously had followed the track correctly, in spite of the displacement, were afterwards asked to indicate where the goal (now removed) was last seen, they would place it at the location at which it would have been had no displacement whatsoever taken place. The experiments suggest that *memory* is integrated with *conscious* visual experience while the (largely unconscious) motoric operation is controlled by a separate and more sensitive dynamic that is very much in the here-and-now.

On the basis of a series of experiments like the one described above, Andy Clark has suggested an *experience-based selection model* for visual perception that is based on the principle that "conscious visual experience presents the world to a subject *in a form appropriate for* the reason-and-memory-based *selection of actions*" (Clark 2002, 197; italics added). Clark here takes his point of departure from *skill theory* as developed by O'Regan and Noë (2001). Skill theory, briefly sketched, rejects Dennett's objections regarding visual perception by claiming that visual experience is *nothing other than* the control by the visual brain of immediately relevant senso-motoric challenges. We see exactly what we *need* to see for the accomplishment of any given action — and since the world (in the sense of an object for our actions) lies open to our eyes, there is no reason to also have it *represented* on the inside of the skull. Should anyone be in doubt of this, the skill theorist maintains that such people simply need to take a look — and what they need to see is exactly what their unconscious saccades choose to tell them.

Skill theory, however, does not really take into account that the visual-information processing in the brain is segregated into two tracks (or pathways) — one of which draws upon available knowledge and memory resources: " What matters, as far as conscious seeing is concerned, is that the object/event is 'one of

19   The first of these visual "brains" is identified as "a dorsal visual-processing stream leading to the posterior parietal lobule, and the latter with the ventral stream projecting to the inferotemporal cortex" (Clark 2002, 195).

those' (i.e., falls into such-and-such a class or category) and that a certain range of actions (not movements but actions such as grasping-to-throw, grasping-to-drink, etc.) is potentially available. Both 'visual brains' I am suggesting, represent by activating implicit knowledge of some set of possible actions and results" (Clark 2002, 199). Clark (ibid., 201) concludes his analysis by an outright rejection of the *folk-illusion* kind of theorizing: "Our visual experience is not itself misleading. The scene before us is indeed rich in color, depth and detail, just as we take it to be. And we have access to this depth and detail as easily as we have access to facts stored in biological long-term memory. It is just that in the case of the visual scene, retrieval is via visual saccades and exploratory action. Our daily experience only becomes misleading in the context of a host of unwise theoretical moves and commitments."

In other words, the brain's two-tracked visual processing technique gives us a visual experience that both emphasizes what is relevant in the context of our immediate motivated dealings with the world, *and* positions this processed information in a coherent spatio-temporal visual context reflecting our general memory and knowledge.

In effect, the folk-psychological conception of vision as a Cartesian theater is only an illusion if we, for ideological reasons, identify visual consciousness with the naked sense inputs that reach retina here and now via the saccades — or, in other words, if we insist on identifying *consciousness* with the unprocessed neurobiological basis for sensual experience. Such an identification would, of course, free us from explaining the phenomenon of *qualia*, but it would also, I strongly believe, "free us" from the ambition of ever reaching an understanding of what kind of a thing *mental life* actually is.

What I want to suggest here, however, is that rather than disposing of the *qualia* problem we ought to recognize the absurdity of the concept of a psychological life without *qualia*, and therefore to pose ourselves the challenging question of the evolutionary origin of *qualia*. Seen from a biosemiotic standpoint, *qualia* — and thus the experiential dimension of existence — is a kind of Firstness in the Peircean sense (see Chapter 3), an aspect of semiosis as such, and therefore also of life and of cognition in general. The phenomenon of *qualia* in living systems cannot therefore surprise us, although the concrete manifestations it takes in human psychological life calls, like everything else in science, for an analysis of its historical origins as well as the mechanics of its enabling neurobiology. Thus, while Clark's hypothesis of *experience-based selection* offers an interesting proximal explanation for how visual *qualia* is neurologically based in modern homo sapiens, the American philosopher Maxine Sheets-Johnstone's seminal ideas about bodily consciousness, mentioned earlier in this chapter, provide us with a more distal (evolution-centered) understanding of the *qualia* phenomenon (Sheets-Johnstone 1990; 1998).

Sheets-Johnstone's analysis takes its starting point in the Socratic "know your self" — which is also, and to the highest degree, a *biological* dictum — as I have hoped to show in the preceding discussion regarding the immune system, as well as in Chapter 2. Remarkably, Daniel Dennett also endorses this bio-Socratic point of departure when he states that simple organisms may not have a lot of self-knowledge but must nevertheless possess some rudiments such as "when hungry, don't eat yourself!" (Dennett 1991, 427). But, as Sheets-Johnstone (1998) observes, Dennett's and many other cognitive scientists' theories about the self are in a very fundamental sense *unbodily* — not paying attention at all to the body's feeling of its own movement, and still very much Cartesian at heart, despite all loud protestations to the contrary. For example, Dennett appeals to the visual metaphor when he suggests rules for how the organism orients itself: "Do something, and look to see what moves" (ibid., 427). Yet Sheets-Johnstone (1998, 272) remarks that "were we to examine Dennett's theory of human agency with respect to infants, one would straightaway discover its errors," and explains,

> We humans learn "which thing we are" by moving and listening to our own movement. We sense our bodies. Indeed, we humans, along with many other primates, must *learn* to move ourselves. We do so not by *looking* or *seeing* what we're moving; we do so by attending to our bodily *feelings of movement*, which include a bodily felt sense of the direction of our movement, its speed, its range, its tension, and so on. Our bodily feelings of movement have a certain dynamic. We feel, for example, the swiftness of our movement, its constrictedness or openness, its tensional tightness or looseness, and more. In short, we perceive the *qualia* of our own movement; our bodily feelings of movement have a certain *qualitative* character.

Sheets-Johnstone's own theory of consciousness and *qualia*, as fundamentally rooted in the needs of animals to know themselves as *moving* creatures, may appear provocative to many cognition researchers due to its underlying assumption that our mental dynamics are spun out of the body's biological swarm of processes. The reductive conception of the body as ruled by processes that may, in principle, be explained from the bottom-up by sheer biochemistry (or in algorithmic terms) does not, in the nature of the case, offer any help in explaining how such *knowledge* could appear in a moving animal. The biosemiotic model, on the contrary, has no difficulties with this connection, since semiotic emergence inherently implies the potential for the creation of systems with increasingly high degrees of semiotic freedom.

Biosemiotician Claus Emmeche has stressed that there is always a phenomenal aspect to semiosis — an aspect that in the Peircean sign conception is contained in the concept of the *qualisign*. In our discussion of alarmones, for example, we saw how the *symbolic* function of cAMP was dependent on — though not fully reducible to — its lower-level *indexical* and *iconic* aspects. And, in fact, traces of such Secondness and Firstness can always be unveiled even in complex signs, such as in symbols or in human propositional arguments. Emmeche (2001, 679–80) points out that

> A single sign may be a *token* of some general *type* (e.g., a perceived pattern may be recognized by the organism as being of a certain dangerous kind, say, a predator), but it has always also an aspect of being a *tone* — i.e., being *qualitatively felt* in some way. The type/token/tone is a genuine triad, where the Firstness property of the tone is always hidden, so to speak, within the "objective" or more external property of that sign's belonging to a type. This corresponds to the first trichotomy of signs in Peirce (that according to the character of the *representamen* itself), where every *legisign* is always realized by a particular *sinsign*, and every concrete *sinsign* includes a *qualisign*. What is the qualisign? Only phenomenologically can we approach a clear idea of the *qualisign*; it is of an experiential character, it is, as Peirce says, "any quality insofar as it is a sign," "e.g., a feeling of 'red.'" Thus, the umwelt, as a semiotic phenomenon, includes *qualisigns* with very sensuous "tonal" qualities.

The Psychological Body

Although most of us spend our lives as if our body was a tool for the satisfaction of our psychological needs, it takes only a moment's reflection to see that the body is the *precondition* for the evolutionarily much later psyche, rather than vice versa. The body was there first, and only in the course of evolutionary time did bodies develop brains sophisticated enough to support what we would today recognize as psychological life. It follows that bodies must somehow have benefited by the possession of a psychological tool. In its most primitive versions, psychological life probably served to assure a coupling between, on the one hand, negative and positive emotions and, on the other hand, relevant patterns of action. The function of the development of felt emotional reactions must have been to establish an effective set of semiotic controls upon the operation of the senso-motoric functions. Over the course of many millennia, these couplings have become more and more sophisticated — and gradually they were integrated into the anticipatory and memory-based steering function of embodied brains.

In his important 1994 book, *Descartes' Error*, the American neurobiologist Antonio Damasio (1994, 230), suggests that the cognitive and the emotional aspect of our lives have grown through a kind of co-evolution:

> If ensuring survival of the body proper is what the brain first evolved for, then, when minded brains appeared, they began by minding the body. And to ensure body survival as effectively as possible, nature, I suggest, stumbled on a highly effective solution: representing the outside world in terms of the modifications it causes

in the body proper, that is, representing the environment by modifying the primordial representations of the body proper whenever an interaction between organism and environment takes place.

This reinterpretation of the standard neurobiological account was largely based on Damasio's observations on a range of brain-damaged patients that simultaneously suffered reduced reasoning/decision-making and reduced emotion/feeling. To his surprise, Damasio found that many of these patients were unable to make rational choices for the one and only reason that their brain damage had interfered with their ability to generate appropriate *emotional* responses. Damasio (1994, 128) concluded from these and similar studies that "the apparatus of rationality, traditionally presumed to be neocortical, does not seem to work without that of biological regulation, traditionally thought to be subcortical. Nature appears to have built the apparatus of rationality not just on top of the apparatus of biological regulation, but also from it and with it."

Thus, *feelings* are as fully *cognitive* as any other perceptual image, claims Damasio (1994, 128), "But because of their inextricable ties to the body, they came first in development and retain a primacy that subtly pervades our mental life. Because the brain is the body's captive audience, feelings are winners among equals. And since what comes first constitutes a frame of reference for what comes after, feelings have a say on how the rest of the brain and cognition go about their business. Their influence is immense."

The distinction between *emotions* and *feelings* is central here. The emotional reactions are spontaneous, in the sense that they occur without any interference from consciousness and are released by the limbic system in the subcortical region of the brain, primarily via the structure called the *amygdala* (LeDoux 1996). Through these emotions are established characteristic functional states of the body, or rather, *kinds of readiness* that are connected to basic survival functions such as defense against dangers, reproduction, foraging, or aggression. Emotions are manifested either endogenously, or as the result of inputs from the surroundings. The emotions may therefore be seen as *bodily interpretants* that immediately release subsequent interpretants in the form of characteristic kinds of behavior.

Too, the neural organization responsible for the primary emotional types of behavior is rather uniform for all vertebrates (fishes, amphibians, reptiles, birds and mammals, including humans). This does not, as underlined by LeDoux, imply that the brains of all these animals are identical. But it does imply that our understanding of what a human person is in the biological constitution as an emotional being must take into account this fact that we are so like other animals with whom we ultimately share a common ancestry (LeDoux 1996, 17). Nor did Charles Darwin himself ever doubt this: "The fact that the lower animals are excited by the same emotions as ourselves is so well established that it will not be necessary to weary the reader by many details. Terror acts in the same manner on them as on us, causing the muscles to tremble, the heart to palpitate, the sphincters to be relaxed, and the hair to stand on end. Suspicion, the offspring of fear, is eminently characteristic of most wild animals" (Darwin 1981 (1871, chapter 3).

But while the *emotional* response, to a high degree, works much the same in all vertebrate animals, *feelings*, in Damasio's terms, are far more specific for humans — for feelings consist in the *experience of emotions*. The experience of *fear*, for instance, appears as "the *conscious* recognition of the *emotional* response to a dangerous situation."[20] As such, feelings are much more varied and subtle than emotions, since they integrate the emotions *into* the rich repertoire of nuances available for conscious awareness — the danger may be imminent or more distant, it may be life-threatening or just inconvenient.

In human beings, claims Damasio, this leads to the formation of *secondary emotions,* which arise "once we begin experiencing feelings and forming *systematic connections between categories of objects and situations, on the one hand, and primary emotions, on the other*" (Damasio 1994, 134). The formation of these secondary emotions is no longer dependent only on the limbic system, but now also involves the pre-frontal and somato-sensoric brain cortexes.

The separation of the *emotional response* from the *experienced feeling* opens up towards a decisive insight in that it frees the understanding of the emotional response from its mysterious (and quite incorrect) anchoring in a conscious brain. It thus becomes possible to understand the emotional response without any need to necessarily solve the intractable mind/body problem that immediately presents itself when we talk about conscious emotional experiences. In fact, claims LeDoux, *emotions* as such are as absent from our awareness as is most of what else goes on in our brain. Biologically understood, the reason for this is simple. When an organism encounters a threat, the first command is for the body to discover the source and to react to the danger — and this, precisely, is what the *emotional response* makes happen. Only at the next higher level do we need the *fear*, the strongly inconvenient *feeling* that can only be extinguished by confronting the threat through appropriate action.

Consciousness about bodily states and feelings may therefore be conceived as a fundamental *carrier wave* running through our entire psychological life, always ready to absorb and to carry forward the stream of interpretations and

---

20   In this way, William James's famous claim that "we do not start running because we are afraid of the bear, but, on the contrary, we are afraid of the bear because we start running" is confirmed to the extent that one is talking about *feelings* and not — as James did — about primary *emotions*.

re-interpretations that occur incessantly in both body and brain. Almost implicit in this understanding is the acknowledgement that these emotional body states are *themselves* subjected to intervention by *feelings* — and thus, to cortical modification, giving rise to what Damasio called *secondary emotions*. The amygdala and the rest of the limbic system (and in particular, the hippocampus, hypothalamus, and nucleus accumbens) not only feed nerve impulses to the cortex but are themselves fed by impulses from the brain cortex (some even consider the orbitofrontal cortex as *part* of the limbic system now, largely as the result of Damasio's work). The secondary emotions are *bodily interpretants* that are continually re-patterned, reflecting the psychological reality that the person finds himself in, and these *emotional interpretants* then themselves enter into further semiotic loops of *feeling-based experiences* in an uninterrupted chain (or network of chains) constituting the psychosomatic reality of a human being. By far — and like almost all other biological functioning — the major part of this swarming semiotic control activity remains unconscious to the person within whom it is incessantly taking place. Yet it tacitly takes care of the necessary survival-orientated integration by which body and mind are (or *the body/mind* is) established, moment by moment.

Another important tool for the accomplishment of these operations is the *autonomous nervous system* that governs the central components of the body's internal life (e.g., breathing, blood flow, heart beat, and digestion). But it has become increasingly clear that *both* the endocrine *and* the immunological system are involved in the semiotic activity by which the psychological situation of the organism feeds back into its somatic readiness potential. These connections are the theme for a new field of research in medical science that has developed strongly through the last few decades under the name of *psychoneuroimmunology* (PNI) — which, in spite of the name, also includes the endocrine system as part of its subject matter.

In a pioneering work, Robert Ader and Nicholas Cohen (1975) showed that changes in the immune defense in mice might be conditioned in much the same way that dogs, in Pavlov's famous experiments from the beginning of the twentieth century, had been conditioned to salivate when hearing the sound of a certain bell that had previously repeatedly been rung at the time of feeding. In Ader's and Cohen's experiments, the compound *cyclophosphamide* (a compound that has frequently been used in chemotherapy) was injected into mice at the same time as they were fed a solution of the sweetener *saccharin*. Cyclophosphamide induces nausea, and it was thus not unexpected that the mice responded by developing an aversion to the taste of the saccharin solution. In earlier experiments, Ader and Cohen had noticed that some of the experimental animals who were conditioned this way quite unexpectedly died upon drinking the saccharin solution, even in the absence of *cyclophosphamide*. And since cyclophosphamide is also a strong immunodepressant — i.e., a substance that weakens the immunological response potential — Robert Ader suggested that the mice were in reality dying because the saccharin solution had worked as a *conditioned reflex* (like the bell in Pavlov's experiments) and had thus *elicited reduced immunological readiness* in the mice. This would have weakened the mice's general resistance towards infections and some of the animals might then have succumbed to fatal infections.

In Ader's and Cohen's experiments from 1975, this hypothesis was essentially confirmed. The intake of saccharin in the absence of any injection of cyclophosphamide by the conditioned mice evoked a significant inhibition of the immune system, as measured by reduced antibody reaction in the mice toward the injection of red blood cells from sheep. The intake of the quite innocent compound saccharin was thus — in the *conditioned* animals — capable of influencing the immune system of the animals. This and many of the subsequent studies are discussed in further detail in Ader and Cohen (1993).

A recurrent objection to the psycho-immunological interpretation of these and other experiments of the kind has been that it may not necessarily have been the taste aversion per se that conditioned a reduced immune response. Instead, the reduction might have been caused by the general situation of stress inflicted upon the mice by the cyclophosphamide-saccharin regimen. It is well-known that stress causes immunodepression, and due to the conditioning process itself the mice may have experienced the saccharin solution as a *stress factor*. The effect of saccharin intake on the immune system need not have been caused by cortical impulses (e.g., taste recognition) — rather, stress signals from the subcortical regions may have sufficed.

It has proven extremely difficult to fully rule out this possibility, so long as animal experiments are our only source of information and the corresponding experiments on human persons are, for good reasons, excluded. In the wake of the original experiments by Ader and Cohen, however, an almost exponential growth in PNI research has taken place, and retrograde attempts to explain away the persistent indications for psychological constituents in the endocrinological and immunological systems appear increasingly forced.

The advertising industry, for one, is in no doubt of the efficiency of conditioned reflexes in influencing our choices of consumption. The stream of delicacies presented in commercials are certainly not intended for the appeal to reason — in fact, reason would be much too risky to depend on. If, however, our endocrine response is successfully stimulated, we become unconscious victims of the suggestive ads.

An important channel for delivering messages from the cerebral cortex to the immuno-endocrine control systems follows the *hypothalamic-pituitary-adrenal axis* (HPA axis) where, depending on conditions, the hypothalamus secretes *neuro-hormones* that significantly influence the activity of the neighboring pituitary gland, and thus the release of pituitary hormones to the blood. Via the blood, pituitary hormones reach the adrenal glands, where they control the production of *adreno-cortical hormones* such as *cortisol*. The following, from an online resource (http://en.wikipedia.org/wiki/Hypothalamic-pituitary-adrenal axis), sums up the relationship nicely: "The fine, homeostatic interactions between these three organs constitute the HPA axis, a major part of the neuroendocrine system that controls reactions to stress and regulates various body processes including digestion, the immune system, mood and sexuality, and energy usage. Species from humans to the most ancient organisms share components of the HPA axis. It is the mechanism for a set of interactions among glands, hormones and parts of the mid-brain that mediate a general adaptation syndrome."[21]

Let us now have a more detailed look at just one small part of this homeostatic interaction system, and this will concern the role played by *opioids,* so named because of their molecular similarity to opium. Many different substances, and among them opioids, are produced by cells from the immune system and subsequently recognized by specific receptors at the surface of nerve cells. *Endogenous opioids* are formed naturally in the body and contribute to the pleasant — sometimes nearly euphoric — feeling experienced upon successfully executed physical or mental work. And this experience is most likely the reason for the strong attraction that many people feel towards artificial opium derivates such as morphine or heroin. Even animals have been shown to greedily pursue these substances and to become addicted to them. This implies that the direct effect of the opioids is subcortical or, in other words, that they exert their influence on one's mood through brain processes that do not depend on prefrontal involvement (Panksepp 2001). On the other hand, we know that the experience of positive feelings can also lead to the release of opioids, suggesting that relations between mood and opioids work both ways. And with this acknowledgement, a broad gate is opened for the intrusion of psychological forces into the regulatory machinery of the body. Or, as Damasio reminds us in the quoted passage above, the operation of the cerebral cortex "does not seem to work without that of biological regulation, traditionally thought to be subcortical. Nature appears to have built the apparatus of rationality not just on top of the apparatus of biological regulation, but also from it and with it" (Damasio 1994, 128).

One animal experiment that most persuasively suggests the existence of a two-way interconnection between the brain and the immune system was based on measurements of the activity of a class of immune cells called natural killer cells (NK cells) in mice. The experiment showed that stress-provoked reduction

21   Wikipedia November 1, 2006.

in NK-cell activity may be counteracted if, simultaneously with the stress factor, animals are fed specific substances that bind chemically to the brain's normal binding sites for endogenous opioids. If, on the other hand, animals are fed an opium derivative that is unable to penetrate into and influence the brain, no effect on NK-cell activity was observed (Shavit et al. 1986). Shavit's and coworkers' experiment thus unambiguously showed that the stress-provoked effect upon NK-cell activity is mediated through opioid receptors in the brain, and is thus broadly open to psychological influence.

All this said, it should be noted that, as of the time of this writing, PNI research still finds itself in an explorative and somewhat immature phase. Numerous *binding relations* between a diversity of body/brain parameters are now being scrutinized, yet it is still too early to claim the emergence of any clearcut pattern. Seen from a semiotic point of view, one may worry that the narrow biochemical approach that still counts as the prevailing paradigm for PNI research may disallow the introduction of theoretical tools that might help PNI to transcend the tyranny of trifles.

The vast quantity of potentially relevant substances, the multiplicity of cell types, and the incalculable number of combinatorial possibilities that must be considered if the contextual outcomes of antecedent events are included in the analysis — as of course they should be — quickly leads the researcher into an overwhelming jungle of possible parameters to keep track of. As an alternative, I would suggest that PNI researchers consider whether insights into the semiotic logic of the communicative dynamics of the system might make it possible to select and focus the analysis upon a relatively minor fraction of nodal points. At the present time this, of course, is pure speculation, but it might turn out to be just the theoretical tool that PNI research needs in order to be able to approach the *data* of simultaneously cascading cellular interaction with anything resembling the efficient *logic* of the cell itself.

Beyond doubt, PNI research is a much needed attempt to establish a scientific understanding of the deeper nature of the psychosomatic interplay — an interplay whose significance for human health and disease is unmistakable, but whose precise choreography so far remains dimly lighted.

The Global Organism — Ecological Endosemiotics

Our discussion on the phenomena of *endosemiotics* would not be complete without mention of that world of prokaryotic communication that led the microbiologist Sorin Sonea (1991; 1992) to talk about a *global organism*. Compared to the early prokaryotes (essentially the bacteria), the development of eukaryotic organisms may be seen as a privatization of the hereditary material. Bacteria

easily engage in widespread reciprocal exchange of DNA fragments across species barriers, and they thereby effectuate the rapid spread, around the globe, of
any DNA-written "talent" that can potentially be interpreted by prokaryotic
machinery. Since bacteria are surrounded by cell walls, uptake of DNA fragments from the surrounding milieu is not normally a passive process, but must
be actively facilitated by the bacterium's own activity. In the process of *conjugation*, bacteria take up linear DNA fragments transmitted from other bacteria,
and in the process of *transduction*, bacteria take up a plasmid (or *phage*), which
is a little bit of self-replicating DNA that is successively integrated into the chromosome of the host cell — becoming a *prophage*.

This transduction process has interesting semiotic consequences, for prophages
often contain genes that are useful for the bacterium. Pathogenic bacteria are
notorious for developing resistance toward antibiotics — and this resistance,
more often than not, is caused by prophages hiding in the bacterial chromosome and possessing genetic instructions that furnish the means for the bacteria to resist the fatal effects of the antibiotics. Normally, prophages will remain
stably integrated into the chromosome, but if growth is inhibited (for instance,
because nutrients are missing), prophages are released into the cytoplasm and
immediately start using the bacterial machinery to multiply themselves at the
expense of the bacteria. The multiplication of the plasmids will, like any other
normal biological process, proceed with a certain margin of error, so that occasionally a bacterial gene will become incorporated into plasmid DNA.

Now, considering the astronomical numbers of bacteria in the world (latest count, 5×10^{30} — five nonillion — a magnitude that we have encountered
nowhere else in our discussions so far!), even a low frequency of chance errors
will imply that, in the long run, not a single bacterial gene will escape ending up
as part of plasmid DNA. And upon release from their dead host cells, these plasmids will spread via wind and water, and through infections. When this process is seen from a global perspective (for bacteria are practically everywhere
on Earth) it becomes clear that virtually all bacterial genes ever constructed
will be present in plasmid-form somewhere on Earth — and probably in many
locations.

Seen from the point of view of a starving or otherwise stressed population
of bacteria, this state of affairs represents a "hope" for being rescued. For eventually the stressed bacterial population might be fortunate enough to become
infected by a phage that possesses a gene for an enzyme that can be used to
overcome the stress condition. And while, for any single population, the probability of such a "lucky" outcome is, of course, infinitesimal — considering the
astronomical numbers of bacteria present everywhere, a successful result is virtually guaranteed to occur *somewhere*. Thus the global existence of the bacterial species is rarely, or never, threatened. "Whenever a new type of bacteria

successfully joins a mixed bacterial community," writes Sonea (1992, 380), "it also brings its own genes, including the plasmid and/or prophages it carries. These newly arrived small replicons will, in time, be freely offered to any other member of the community."

In sum, we may say that prokaryotic life on our planet has managed to make use of the law of great numbers to establish a kind of global prokaryote semiosphere — a semiotically based coordination of prokaryotic life into one big worldwide swarm. And as Sonea suggests, because of this holistic integration, we should not look upon bacteria as individual organisms but rather as short-lived units in a huge global organism, consisting of the worldwide network of prokaryotes. The exchange of DNA-coded signs between all these entities may then be seen as a kind of *global endosemiotics.*[22]

Compared to the extreme openness toward foreign DNA-messages exhibited all over the prokaryote world, eukaryotic organisms appear to be very isolated beings. Too, it should be noted that in spite of the statistical character of the semiotic interactions in bacteria, these interactions are nevertheless highly controlled and rely upon the well-regulated occurrence of specific receptors at the surfaces of cells, as well as the induction in the cells of specific enzymes suitable to support this kind of communication. With the appearance of eukaryotic life forms, these *horizontal* mechanisms for DNA-communication disappeared, and DNA-communication became exclusive to the "family" line. Thus, in eukaryotic cells, DNA-transmission is strictly *vertical* or *temporal*, limited to the events of cell divisions. The one major exception to this, of course, is the fusion of genomes that takes place during the course of sexual reproduction.

A deep difference in semiotic logic between prokaryotic and eukaryotic life is buried here. According to current theory, eukaryotic cells arose through endosymbiosis from prokaryotic life forms (see Chapter 5). As we saw present day mitochondria, chloroplasts, and microtubules (the intracellular organelles concerned with movements) are all, according to Magulis' suggestion, descendents

---

22   The richness of the collaborative patterns and communicative mechanisms to be found in the bacterial world is only now beginning to be realized (see Ben-Jacob et al. 2004). For in order to get reproducible results, microbiologists have almost always had to study bacteria while they are in the *exponential growth phase*, where they are close to optimally supplied with nutrients. Under natural conditions, however, this artificial laboratory situation is probably not typical at all — and it has now been shown that under more poor growth conditions, all kinds of interesting cooperative strategies arise between bacteria — patterns that are never observed under exponential growth conditions. This has inspired some researchers to talk about a *bacterial linguistic communication* and about its networked *social intelligence*. These metaphors are hardly satisfying because they tend, once again, to neglect the code-dual nature of bacterial semiotics, stressing the digital (linguistic) aspect unduly. But it is, of course, encouraging to see that microbiologists now are tempted to engage in clear-cut *semiotic* terminology in order to better grasp the nature of their subject matter.

from individual bacterial species that, long before their present relations of symbiotic cooperation appeared, were adapted to niche conditions that had prepared them for their later function in the eukaryotic cell (Margulis 1970; Margulis and Fester 1991).

Gradually, many of the genes that originally belonged to individual endo-symbionts were transferred to the shared pool in the cellular nucleus (that in this process had its genetic material doubled many times — relative to the typi-cal prokaryotic content of genes). Thus the eukaryotic cell engaged in a new sur-vival strategy. Rather than depending on the reception of visiting genes through plasmid infection, eukaryotic cells relied on genetic self-sufficiency. Yet this strategy had the consequence that eukaryotic organisms became genetically segregated from one another, and increasingly more reliant upon *endosemiotic resources* than on *exosemiotic interactions* for both their survival and for the con-tinuation of their line.

But what eukaryotic life forms lost in capacity for horizontal genetic commu-nication, they copiously gained through the development of sophisticated kinds of communication based on a diversity of nondigital biochemical and behav-ioral signs. For while the evolution of eukaryotic cells implied a strong restric-tion in the channels suitable for digitized communication, it opened the way to the development of life forms that possessed far more architectonic multiplicity and behavioral degrees of freedom than prokaryotic organisms could ever have obtained. The transition from prokaryotic to eukaryotic life forms thus exem-plifies a general principle pertaining to emergent processes — i.e., that *in emer-gent processes, freedom of possibility will always be constrained at the simpler level in order to allow an altogether new kind of freedom to appear and unfold at a more complex level*. The emergence of multicellular life and of social life are but two more examples of this fundamental dynamic principle.

The idea of the superorganism (such as the plasmodium phase of the slime mold that we discussed earlier) has a long pre-history in biology, especially as concerns social insects. Both the ant hill and the bee hive have been suggested as cases of superorganisms, with the implication that the individual insects were seen as just subunits — *mobile cells* in the superorganism. But as we have seen earlier, the delimitation of the individual is not always as clear-cut as it may seem in the world of vertebrate animals. Yet I have no wish to choose sides in this quarrel.[23] From a semiotic point of view, one might perhaps suggest the cri-terion that, if a system's semiotic interaction with its environment presupposes a

---

23   In my book *Signs of Meaning in the Universe* (Hoffmeyer1996b), I did suggest a swarm-semi-otic model of the brain functioning involved in creating a mental life. And this proposal does in some sense accord ontological reality to swarm-dynamic systems of the kind we find in social insects.

finely elaborated internal semiotic activity (a *proto-endosemiotics*), then the system deserves to be counted as an organism.

Deborah Gordon's laborious and highly rewarding work with ants of the species *Pogonomyrmax barbatus* (who live in a harsh zone bordering the deserts between Arizona and New Mexico) has revealed a much more sophisticated pattern of semiotic interactions between individual ants than had been expected — and the survival of colonies of this species are so dependent on this protoendosemiosic regulation that the nomination of ant colonies to the status of superorganisms feels reasonable in this case (Gordon 1995; 1999). Gordon does not, herself, draw this conclusion, however. For (and most remarkably, in this connection) she found that ant-colony behavior is not absolutely deterministic. Rather, a particularly important element in the colony's growth process is what Gordon calls *job allocation* — and she shows that although this task does indeed rely on a quite schematic interaction pattern between different groups of ants, an element of unpredictability persists: "An ant does not respond the same way every time to the same stimulus; nor do colonies. Some events influence the probabilities that certain ants will perform certain tasks, and this regularity leads to predictable tendencies rather than perfectly deterministic outcomes" (Gordon 1999, 139).

Gordon's experiments in this area may also be seen as a response to experiments performed by the founder of sociobiology, Edmund O. Wilson, that were claimed to show a full-blown determinism in the response pattern of ants to chemical signals (e.g., *oleic acid*, Wilson 1975). Gordon's experiments, on the contrary, showed that "just as the same word can have different meanings in different situations . . . so the same chemical cue can elicit different responses in different social situations" (Gordon 1999, 97). Physiological, social, and ecological processes are simultaneously at work, says Gordon, and none of them are more basic than the others: "Linking levels of organization are central to any study of social behavior. For humans and other social animals, an individual's behavior is always embedded in a social world" (Gordon 1999, 96).

The semiotic competence of subunits, then — whether these subunits are human individuals in a society, plants in an ecosystem, cells in a multicellular organism, or ants in an ant colony — is the medium through which the behavior and integrity of the higher-level entity is maintained. To the extent that such a system's endosemiotic relations perceive and utilize cues and signs that indicate (are indices for) the state of the holistic unit and its needs, it seems justified to talk about these processes as genuinely *endosemiotic* — consequently, the holistic system itself deserves to be ascribed a status as an autonomous unit: a superorganism.

The evolutionary formation of this kind of autonomous macro-entity is the quintessence of what is called *emergence* (Figure 7.3). Figure 7.3 claims a

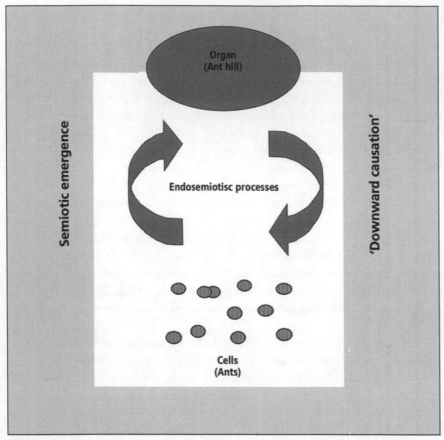

Figure 7.3 The connection between semiotic emergence and downward causation. Downward causation operates through indexical sign relations, i.e., the values of system parameters are interpreted by lower-level agents as indexical signs. But this state of affairs in itself depends on the formation of a large-scale pattern with a behavior that stabilizes the semiotic interaction between parts. Semiotic emergence and downward causation may thus be seen as two sides of the same coin.

connection between *semiotic emergence* (whereby macro-entities or stable large-scale patterns are established through semiotic interactions among small-scale entities) and what has been called *downward causation* (the influence of large-scale patterns upon small-scale interactions) (Bickhard and Campbell 1999; El-Hani and Pereira 2000; Emmeche, Køppe, and Stjernfelt. 2000), and it is suggested that this connection is taken as constitutive for both phenomena.

The semiotic relations between subunits that collectively account for the stability of the large-scale or holistic system (for instance, the ant hill, the multicellular organism, or perhaps the symbiotic system of bobtail squids and light-emitting *vibrio* bacteria) must necessarily be geared to respond to changes in the environment in ways that do not threaten the integrity of the large-scale system. Subunits, for their part, must receive messages telling them how to uphold the macro-system, and probably the easiest way to do this is to distribute the needs of the macro-system via *indexical* signs — as we saw in the case of N-acyl-homoserin lactone in the symbiotic squid-vibrio system. When night approaches, the squid makes sure that the bacterial density in its mantle cavity (and thus the homoserin-lactone concentration) is high enough for the bacteria to respond by allowing for the transcription of *lux-operon* genes — and thus for light emission — to start. The point is that the semiotic emergence whereby this system was gradually established in the course of evolution necessarily also demanded the invention of semiotic means for assuring the stability of the system, and these semiotic means are precisely what we understand by *downward causation*.

Downward causation and semiotic emergence are thus two interwoven, but not identical, aspects of the same process. That this mechanism, based on indexical semiosis, is indeed coupled to the circadian rhythms of the squid, was confirmed by the finding that the squid has means at its disposal to fine-tune light emission. It may, for instance, change the wavelength of the emitted light by help of a yellow filter, or it may weaken the intensity of light by opening a bag of ink (Bruni 2002). Presumably, the indexical sign process itself is too slow in its effects to compensate for momentary variations in background light (as occurs under cloudy conditions), and evolution therefore had to provide the macro-system (the squid) with a number of additional "screws" for fine-tuning.

PART 3
BIOSEMIOTICS
AND THE HUMAN BEING

8
From Animal to Human

Facing Reality

From the biosemiotic point of view, semiotics (the study of sign processes) does not belong primarily, much less exclusively, to the humanities; rather, human beings should be considered as a species with unusually pronounced semiotic talents. Such talents may justify the claim that semiotics, in its point of departure, had to be particularly concerned with human significative and communicative practices — but it cannot justify the privileging of human being as the only semiotic animal on Earth. The human talent for semiosis is not miraculous and cannot be absolutely unique in the world of living organisms, for such talent must necessarily have arisen through an evolutionary process involving prior life forms. To speak most accurately, then, the branches of semiotic inquiry that are concerned with the practices of the human species should therefore be delimited as *anthroposemiotics* — the study of the particularly human forms of sign use, especially as these might differ from other naturally existing sign systems. This, however, immediately raises the question to be discussed in the present chapter: How is such anthroposemiosis evolutionarily connected to a more general biosemiosis?

In posing this question this way, however, it can be seen that I have already turned upside down the hegemonic theoretical dualism that is at the heart of both the humanities and the natural sciences.[1] The somewhat unusual way that the present book has dealt with the concept of *semiosis* through the last four chapters — where purely biological interactions were the main theme — may have irritated some readers, if it has not already chased them away. Little can be

---

1 Charles Darwin would probably have been open to this question, as may indirectly be concluded from the positions that he argues in *The Descent of Man* (Darwin 1981 (1871)). But his contemporary successors have by and large approached human evolution in a reductive — i.e., asemiotic — way. Concerning the humanities, it must be noted that the recourse to ontological positions, like those of Heidegger or Sartre, that characterize human beings in ahistoric and absolute terms, necessarily will have to be transcended in order for the kinds of questions being posed here to be addressed at all in any meaningful (i.e., not in a trivial or reductionistic *evolutionary psychology*) sense.

done to ameliorate this, for such discussion was indeed necessary, as will be evident from the arguments to unfold in the present chapter (see p. 281ff). Human semiotic competence, as exhibited for instance in language and art, are, as we shall see, fundamentally anchored in human corporeal nature. Thus, at the root of these phenomena, we find ordinary organic life processes — and, first and foremost, movement (Sheets-Johnstone 1990). A suitable way to embark upon a journey into the theoretical quicksand dealing with the connection between animals and humans might be to depart from an aphorism put forward by the American philosopher John Deely when discussing the remarkable fact that philosophy, in its earliest expressions, was always poetically elaborated in the form of verse rather than directly written in prose: "Reality was not merely discovered, we might say, it was celebrated," writes Deely (2001, 5).

To celebrate something is, in the nature of the case, to stage it — and an aesthetic staging of reality is certainly one way to protect oneself against the implicit threat in the idea of an extra-personal reality: the realization that the personal *I* must necessarily be decentered, and in some cases, virtually put aside in order to make space conceptually for a reality that is independent of the *I*.

This, however, is a very modern way of stating the matter. What happened to the people who, in our distant prehistory, gradually wrested the idea of a self-subsisting reality from its omnipresent embeddedness in their own experienced worlds, may perhaps rather be compared to the awakening from a dream. Our magic and unquestioned belonging in the world was challenged and made fragile by the unavoidable separation of speech from what is spoken about. The idea, made possible by spoken language, that a reality — "the world" — exists whether or not you yourself exist, must in the long run, have consumed the simple-minded confidence of an unproblematic belonging. Animism, myths, heroic legends, and songs may well have helped to veil or contain the reality of the human thinker's divided existence. But by the time of Greek antiquity, a philosophy appears that has the audacity to express itself in prose rather than in lyrics, and in so doing, starts meeting reality at eye-level, rather than celebrating it behind the talisman of a poem. It is generally supposed that Pherecydes of Syros, in the sixth century BC, was the first philosopher to write in prose (Deely 2001, 4) and Deely endorses the conception (first suggested in West 1971) that "it cannot be an accident that the three oldest prose books that have survived — Pherecydes, Anaximander, and Anaximenes — were all expositions of the *origin and nature* of the world" (Deely 2001, 4). The following citation from John Addington Symonds (1893) illustrates this connection: "It was a great epoch in the history of European culture when men ceased to produce their thoughts in the fixed cadences of verse, and consigned them to the more elastic periods of prose. Heraclitus of Ephesus was the first who achieved a notable success in this new and difficult art. He for his pains received the title 'the obscure'; so strange and novel did the language of science (i.e., knowledge try-ing to achieve exact expression) seem to minds accustomed hitherto to noth-ing but meters" (1920 reprint of third edition, 11–12, cited in Deely 2001). It is (still according to Deely) precisely such linguistic competence that allows (and perhaps even forces) human beings to transcend the simple Umwelt that each human individual, just like all other animals, constructs and projects onto its surroundings. The possession of language brings us into a linguistic *Lebenswelt*[2] where it becomes possible to distinguish between the reality of self-subsisting *things* and the more immediately known and equally real *objects* of our experi-ence. In short, the term *thing*, in Deely's terminology, simply refers to what exists as such in the world, whether or not anybody knows about it.[3] An *object*, on the contrary, refers to what "exists as known" (Deely 2001, 6). Objects, according to this understanding, may well exist as things in the real world at the same time as they are objects in our cognition (the print on this page may be an example), but lots of objects are only objects (i.e., not things), since they refer to entities that exist exclusively in our cognition — as for instance, a letter that you never wrote, or Denmark's King Christian the Twenty-Third.[4]

Deely's terminology may at first seem foreign since it implies that, for instance, the Umwelt of an animal equals the animal's *objective* world, in the sense of the world as it is recognized or modeled *exclusively* as an object of the animal's expe-rience. *Objective* in Deely's terminology precisely means "to exist as known"

2   *Lebenswelt* was Husserl's term for the sensual world as we unconditionally recognize it, and the *Lebenswelt* may thus be seen as constitutive for the world conjured up by science — though this is rarely recognized among scientists, who for the most part are still following in the dualistic tradi-tion of Bacon and Descartes. Yet, since the *Lebenswelt* is a necessary presupposition for all experi-encing, it may be said to be a *transcendental* condition that, according to phenomenology, comes before the world as analyzed by science. Unfortunately, the word *transcendental*, too, now carries all kinds of spiritual and supernatural overtones (for scientists) that are in no way implied by my use of the term as explanatorily presupposed here. Having to always stop and clarify such things shows how wide the gap between science and the humanities has become — and much of the work in biosemiotics right now is devoted to clearing up such incapacitating misunderstandings.

3   What is at stake here is simply that the thing concerned possesses a dimension or an aspect that is not part of the relation through which the thing is recognized as an object. In other words, there is no allegation of a Kantian *Ding-an-Sich* — the idea of some ontologically mysterious *an-Sich* reaching beyond our capacity for cognition — an idea that strangely has managed to mystify Western thinking ever since it was first put forward in 1781 (see *CP* 5:452).

4   I use this example with equanimity, since the occurrence of the name Christian in the Danish list of kings presently ends with Christian the 10th, and since modesty prevents me from consid-ering the possibility that the present book might be in use when, or if, Christian the 23rd even-tually should attain royal status. In either event, the object of our thoughts right now, Christian the 23rd, can fully serve as the object of much logical, reasonable, and predictive analysis — with-out having to exist as an actual thing in the world at this time. Such ontological oddness lies at the heart of our uniquely human semiotic competence (see Hoffmeyer 1996b, chapter 10).

(ibid.) — and to an animal, unlike to us, that which is known immediately and that which exists mutually exhaust each other. The *subjective*, on the other hand, is a mode of existence that has nothing to do with cognition — though, again, on first glance this terminology is the inverse of contemporary usage.[5] My liver thus is an object as soon as I or my doctor think about it — but it is also a subject to the extent that its status as a locus of physical interactions with the rest of reality go on quite independent of my or any other's reflections upon it.

I find this terminology more suitable than its dualistic normal usage because it unsentimentally detaches the cognitive situation from the psychological setting peculiar to humans. Here, cognition conjures up an *objective* world for us — a world that is cut loose from all sorts of irredeemable historical bonds to classic ideas of truth, intersubjective control or solipsistic certainty. Seen from the standpoint of semiotics, objective being is constituted through the creation of an *interpretant* (the putting together, by an agent, of things and objects into signs) while subjective being corresponds to the pre-semiotic substrate (or — to the extent that living beings are concerned — perhaps to the *Innenwelt* in Uexküll's explanatory schema),[6] whose relation to the interpretant doesn't concern it.

The understanding that *things* and *objects* are *not* identical is, for Deely, the distinguishing mark separating animals and humans — and he ascribes the human capacity for making such a distinction as arising from our belonging in a communal linguistic *Lebenswelt*. Animals only know of objects, not of things. The emergence of philosophy in early Greek antiquity exactly consists in the free (and nonmetric) reflection upon the implications of this basic distinction. That the distinction between what Deely calls things and objects now becomes subjected to deep reflection is in itself a gigantic step in human understanding — and yet, nearly another two thousand years had to pass for the further recognition to dawn upon thinkers that such an understanding is only possible because of the properties of the *sign* — and that our universe, just as Peirce said, is literally saturated with signs.

This millennial historic move in the history of philosophy is the theme for Deely's major work *Four Ages of Understanding* (2001). Here, as my point of departure, I am going to focus on just this great move endorsing, as it were,

---

5   Deely (2001), a historian of philosophy, describes in scholarly detail how such a nominalist inversion of these terms has come to have taken hold.

6   "The interior state, both cognitive and affective, on the basis of which the individual organism relates to its physical surroundings or environment in constituting its particular objective world or Umwelt is called an *Innenwelt*. The Innenwelt is a cognitive map on the basis of which the organism orients itself in its surroundings. The Innenwelt therefore is 'subjective' in just the way that all physical features of things are subjective: it *belongs to* and *exists within* some distinct entity within the world of physical things" (Deely, 2001, 6).

the fundamental view that animals, through their Umwelts, are partakers in an *objective* (mind-dependent / individual-experience-dependent) world, whereas they do not, in any significant sense, have entry to the insight that beyond (or hidden behind) the objective world of the Umwelt, there is another world, a (self-subsisting / individual-experience-transcendent) *subjective* world — a world that precisely "is not our own world," to borrow Thomas Nagel's fine formulation (Nagel 1986, 18).

But since language possession is usually regarded as the main key for opening the door to the human *Lebenswelt*, a central task for our theories of human origins will be to explain the appearance in our human ancestors of linguistic competence as such. How could it possibly have happened that the species-specific Umwelt characteristic to a group of apes would gradually, through an evolutionary process of less than a million years, acquire this radically new flexibility, this Lego-brick kind of functional constitution that, in a certain sense, alienates its possessors from the world — while at the very same time makes the world manipulable by its possessors in an extremely powerful (but in the end, perhaps almost unimaginably dangerous) way? These are the issues to which I now turn.

The Puzzle of Discontinuity

Although John Deely clearly places the human *linguistic Lebenswelt* in an evolutionary framework by emphasizing its generic relationship with the Umwelts of animals, the radical nature of the difference that he points to in the semiotic experience of animals and humans nevertheless stands as a severe conceptual obstacle blocking any easy understanding of how the one thing could lead to the other. While biological evolution, according to mainstream theory, implies minute continuity, the evolution of language exhibits a remarkable discontinuity. No candidates are found among the animals for the role of *primitive language users* — and thus the classical question in biology of how improvements of language competence might be correlated to environmental factors cannot therefore be meaningfully posed.

Indeed, the much debated warning calls observed in vervet monkeys have often been posited as sorts of primitive *linguistic* utterances — and this case is illustrative of the kind of misunderstandings about the different levels of sign-processes that mar the whole field of language evolution and animal communication studies. The vervet monkey hypothesis goes like this: In the 1980s, Robert Seyfarth and Dorothy Cheney from Pennsylvania University reported their finding that, in the wild, vervet monkeys had three distinct alarm calls that warned conspecifics against threats from eagles, leopards, and snakes, respectively. The "eagle call" would make the monkeys jump down from the trees, the "leopard call" would, on the contrary, make them climb the trees, whereas the "snake call" would make them raise and look around to identify any eventual threat in the grass or shrubbery. This, of course, was all very rational seen from a survival point of view, and Seyfarth and Cheney (1992) made the apparently obvious suggestion that these warning calls were analogous to *names* for the implied predators — just like people shouting Fire!

The calls of vervet monkeys are different from genuinely *linguistic* utterances in a very crucial sense, however — for even though each of the calls does, in fact, *refer* to concretely occurring threats, the monkey's way of *referring* to these threats is fundamentally different from the way that humans normally use words to *refer* (Deacon 1997, 57). It is of course highly interesting to notice that monkeys do *refer* as precisely to the outer world as is here the case, and it certainly points to an important aspect of their way of life. But if we think upon such reference-based actions as crying, groaning, smiling, or laughter, it is immediately obvious that an *activity* does not become a *word* just because it *refers* to something else. When somebody laughs, for instance, we cannot easily get away from the feeling that there must be some reason for it, and that the laughter refers to that reason (and hopefully that this reference is not something that you have just done to embarrass yourself). Laughter clearly has a reference, if only some inner mental state or memory in the laughing person. Yet in situations where many people are assembled, it may be hard not to laugh when all the others are laughing — even though you may not know what the reason is for this synchronous outbreak. Laughter is infectious, as we say — and this will have some implications for the monkey calling behavior, as we shall see.

Newborn babies at the hospital exhibit a similar infectious behavior when they start crying and screaming by the mere sound of crying and screaming from neighboring babies. And exactly this infectious way of a self-initiated activity spreading like wildfire through a group is characteristic of the alarm call behavior in vervet monkeys where typically all the monkeys in the troop will continue shouting the very same call for some time, once that call has been initiated. And such behavior, in fact, is exactly what is implied by the expression "to ape." Try to consider, however, how utterly different this situation is from normal talk among people. Terrence Deacon (1997, 58), in his masterful book on language evolution, *The Symbolic Species*, observes, "Not only do we seldom parrot what we have just heard from another, such a response is generally oddly annoying. . . . How odd and unnatural it would feel to enter a room where people were echoing each other's speech in the same way that they tend to echo each other's laughter. This may be why certain ritual practices that employ such patterns of language use are at the same time both disturbing and powerful, depending on whether one feels included or excluded."

The point is that you cannot conclude from the mere fact that a vocal behavior is *referential* that it is referential in the same deep, flexible, and multiply meaningful way that a human linguistic word is referential. That many biologists so easily put vocal utterances on an equal footing with linguistic reference shows how much the de-semiotization of biology leaves us with a lack of qualified explanatory tools for making critically important distinctions. To *refer* to the surroundings is of course a fundamental semiotic capacity — it is derived from the Latin, *referre* = to carry back (e.g., signs of the outside world to the interior of oneself) — and this is a semiotic capacity that all living systems possess to some extent (Hoffmeyer 2000b; 2001a). But the essential point in Deacon's analysis is precisely that human-style *language* is based upon a practice that he calls *symbolic reference* — while the alarm calls of vervet monkeys (as well as human crying, smiling, and laughter) are primarily based upon the much simpler and more limited practices of *indexical reference*.[7]

I am going to discuss Deacon's theory in quite a bit of detail below. But before doing so, I want to direct the reader's attention to what language evolution theorists sometimes call the discontinuity puzzle. That puzzle, in short, asks this: Since — as linguists all but unanimously insist — there are no known primitive examples among animal species of the kind of communicative practice that language offers to human beings, how then could this apparently exceptional tool, *language*, have arisen through an evolutionary and nonsupernatural process?

In answering this question, many linguists and communication researchers have been of the opinion that it was necessary to point to a phenomenon that biologists would traditionally call *hopeful monsters* — i.e., the sudden appearance, in a population, of mutants with radically changed phenotypes. That this (in other contexts much-ridiculed) hypothesis has been widely accepted in this case is partly due to the supposed absence of alternative possibilities — but the reason that this is so is that what these alternative possibilities are being called on to explain is itself a somewhat mysterious theoretical construct: influential American linguist Noam Chomsky's theory of a *universal grammar* (and, by extension, a hard-wired Language Acquisition Module in the brain) as a common fundament beneath all human languages. All children *must* have at their disposal, argued Chomsky, an innate universal grammar — for there seems to be no other way to explain how they can possibly learn so fast, so perfectly, and so invariably the complex, rule-bound grammars of their own native languages (Chomsky 1972).

---

7   As I shall discuss in more detail below, Deacon here uses the term *symbolic* in a more restrictive sense than did Peirce, though the admittedly great influence of Peirce's architectonic of sign relations deeply informs Deacon's *nested hierarchy* of iconic, indexical, and symbolic reference, as we shall see.

A linguistic grammarian by training, Chomsky based his theory on a purely *grammatical* analysis of human linguistic competence. He observed that although human languages exhibit a nearly unbelievable variation in their surface structure, they all *must* derive from the same underlying *Deep Structure* (his term) from which each of the specific grammars applying to the individual languages may be inferred through deductive reasoning. Each individual child, in facing the real-time task of learning how to speak, however, finds itself in the opposite position of the analytical linguist. There, she or he encounters only the *surface structure* of the local language — and encounters them in the form of huge quantities of *only* grammatically valid utterances (with no counter *invalid utterances* to compare and use to thereby tease out the patterns of *valid* versus *invalid* grammar) — yet from this *poverty of stimulus* (again, Chomsky's term), the child must from these examples alone deductively infer the deeper logic that alone will make the local grammatical rules understandable.

Chomsky claims that as a deliberately undertaken cognitive reasoning task, this simply is not possible, practically speaking. The amount of data that the child receives in the form of sentences, is not only too small but in many cases these sentences are not themselves absolutely correctly performed (due to "performance errors" on the part of the speaker), and in addition, the child will often not get sufficient feedback to its own erroneous utterances. Since *language* thus, in practice, is not learnable in the traditional sense that math and reading are, *and* since all children nevertheless do learn successfully to use it during the first two to three years of their lives, the conclusion seems quite compelling — that linguistic competence must be assured by an *innate ability*, a *language-acquisition device* inherent in the neural architecture of the child's brain.

Again, as a linguist (and very busy political activist) Chomsky did not himself deal with the question of how biological evolution would have managed to bring forth the neural substrate for this ability in so short an amount of evolutionary time, nor could Chomsky, who claimed no neuroscience knowledge, specify exactly in what this innate language module should consist. But in the absence of any theory (or even neurobiological evidence) concerning the nature of the proposed neural equipment involved, it is not possible even to begin to look for the evolutionary origin of this supposedly innate device, and the *hopeful-monster hypothesis* is essentially the only remaining way to lend Chomsky's intuition some biological credibility.

The last few decades have seen a mounting revolt against Chomsky's theory among certain circles of linguists, who object that the learning of language is not an abstract mental activity (whether innate or deliberate), but always an interaction between situated, meaning-making participants that takes place in a context of mutually semiotic action. The child does not just learn what words to say, but also — and simultaneously — how, to whom, in what ways, why, and in which situations she should use them. It has also been pointed out that a range of communicative functions (e.g., to label, to request, to mislead) must be present in the psychological system of the child already before the child learns to master a formal language in which to express them (Bruner 1990). Learning to speak proceeds much faster when the child has already pre-linguistically conceived the significance of what is talked about, and in which situations the talk takes place. Not only the dictionary (or semantic) aspect of language use, but also its grammar (syntactic appropriateness) is always acquired in a genuinely meaningful, mutually semiotic context. Jerome Bruner (1990, 72) points out, "Not surprisingly, then, I think the case for how we 'enter language' must rest upon a selective set of prelinguistic '*readiness for meaning*.' That is to say, there are certain classes of meaning to which human beings are innately tuned and for which they actively search. Prior to language, these exist in primitive form as proto-linguistic *representations* of the world whose full realization depends upon the cultural tool of language" (italics added).

According to these critics, Chomsky's theory depends on a set of conclusions that only appear reasonable because he has already *separated* language as a cognitive tool (to be learned, privately, through a kind of inductive analysis) from language as a source for the constitution of communal meaning. If, instead, one sees the child's learning of a language as embedded in a corporeal and social developmental process, supported — again both corporeally and socially — by the social environment, then one's perception of the *challenge* of learning to use language becomes essentially different. Rather than being a question of a pre-formation-based "tuning" of the young brain to specific formal parameters in order to master language use (and grammar in particular), the question instead becomes one of accounting for how *linguistically* mediated sign-processing can emerge — ontogenetically and phylogenetically — from the ability for corporeal sign processing that is at the heart of all mediation of *meaning*.

It may be necessary here to warn the reader against a possible misunderstanding of what I am claiming here. One should not from these last reflections infer that *body language* is in any way a *precursor* of speech, for as Gregory Bateson persuasively pointed out many years ago, evolution never develops new tools for mastering functions that are already well taken care of by other tools. More likely, the spoken word is a completely new invention which has been spread on top of the semiotic relations of body language like a layer of whipped cream. Human beings are every bit as capable as animals when it comes to communicating messages effectively with their bodies; just think of dance or mime — not to mention telegenic politicians. So body language is obviously not a communicatory appendix on the brink of extinction. But such extravagance — equipping an animal with two dedicated means of performing the exact same function — is, from a biological point of view, unheard of. The only sensible conclusion, Bateson maintains, must be that the spoken word communicates something quite different from what body language can communicate (Bateson 1968; Sebeok 1987; Hoffmeyer 1996b).

Worded, vocal communication must have yielded an advantage upon our remote ancestors that body language could not possibly have mimed, and I tend to concur with Terrence Deacon's suggestion that this advantage has to do with the unique capacity of language to carry symbolic reference. This does not, however, contradict the idea that language in its remote origins may have been modeled upon the expressive register of the body, as will be discussed later in the light of Maxine Sheets-Johnstone's theory of language origins. It does mean, however, that the increase of semiotic freedom that language use bestows on human animals could never have been achieved by the more indexical and iconic capacities of body language alone — as I will now argue in more detail.

It should be understood at the outset that it is one task to attempt to explain the origin (or starting point) of language as the detachment of human speech from its basis in corporeal expression; and quite another task to explain the continued modulation in the course of human evolution of this more and more cognitively specialized tool for expression and reflection. The discontinuity in the communicative ways of primates and humans is a hard challenge to any theory of language origins, given the relatively short amount of evolutionary time during which modern humans have diverged from our closest primate relatives.

In order to meet this challenge, I believe that it is necessary to divide the discontinuity into two separate developmental dynamics. One of these concerns the relation between discrete *expressive sounds* and their *meanings*, while the other concerns the relation between entire *systems of expressive sounds* and meaning. It is primarily the last of these two dynamics that accounts for the radical discontinuity that we encounter when comparing human language and primate (or other animal) communication in general. But this dynamic itself would never have been possible had it not been for the prior freeing of sounds from their utilitarian enslavement in the service of corporeal functionality. And yet, as we shall see, even the fully developed linguistic universe of expressive sounds remains internally connected to those pre-linguistic expressive forms.

To understand how this is so, we must first review Terrence Deacon's theory of the evolution of language in the primate world as the emergence of a radically new form of reference wherein individual sound signs or words are always webbed into a complex system of other words or sound signs. Deacon's term for this very unique form of reference — rarely, if ever, seen in the rest of the animal world — is *symbolic reference*.[8] Next we shall pursue the origin of this pecu-

---

8   As a comment on the title of Deacon's book, *The Symbolic Species*, it might be noted that already the German philosopher Ernst Cassirer characterized the human being as *the symbolic*

liar system of meaning-carrying sound-signs backwards in evolution to its roots in a diversity of corporeal forms of expressions and impressions, as suggested by Maxine Sheets-Johnstone.

The Awakening of the Cerebral Cortex

Deacon's theory is based upon an original double take, whereby a neurobiologically substantiated rejection (or rather transcendence) of the hypothesis of the cerebral localizability of a "language instinct" (e.g., Chomsky, Pinker) is combined with a Peircean understanding of the *nested* categorical hierarchy of *reference relations* into (ascendingly) iconic, indexical, and symbolic relations.[9] While such localizability may often be observed in connection with indexical relations (e.g., Pavlov's dog), Deacon shows that a rigid anatomical binding would in principle act as a preventative for the establishment of an ability to perform symbolic reference.

Deacon goes to great lengths to show that the human propensity for symbolic reference is not (as is too often supposed) a simple consequence of a general mental agility ascribable to the characteristic "brain size to body size ratio" of our species. Instead the neural substrate responsible for this propensity is intimately connected to the fact that the cerebral cortex in human beings leaves a surplus of synaptic places free (i.e., nondedicated and underdetermined) to be invested in the accomplishment of novel coordinative and associative tasks — as opposed to being neurobiologically reserved for the unimaginative control of the senso-motoric apparatus, as is the case with the vast majority of our animal ancestors (including, to a lesser but still considerable extent, the apes).

---

animal (*animal symbolicum*) contrary to the — at that time — more general expression of *the rational animal* (*animal rationale*), which was explicitly emphasized by Charles Morris (see Petrilli 1999). The idea of symbolism as a key to human mental life was apparently quite widespread in the early twentieth century. Susanne Langer (1942, 21) in *Philosophy in a New Key*, points out, "Quotations could be multiplied almost indefinitely, from an imposing list of sources, from John Dewey and Bertrand Russell, from Brunschwicg and Piaget and Head, from Köhler, Koffka, Carnap, Delacroix, Ribot, Cassirer, Whitehead — from philosophers, psychologists, neurologists, and anthropologists — to substantiate the claim that symbolism is the recognized key to that mental life which is characteristically human and above the level of sheer animality. Symbol and meaning make man's world, far more than sensation." Deacon's book, however, put this insight to use in an elegant evolutionary theory on the origins of human language.

9   *Ascendingly* here denotes *logical necessity*, not value or importance. Specifically, it notes that in order to establish a symbolic relation, preexisting *indexical* relations must first be in place, and in order to establish an indexical relation, preexisting *iconic* relations must first be in place. This is the essence of what Deacon calls Peirce's *nested hierarchy* of sign relations, as will be discussed in more detail presently.

Classical neuroanatomic theory assumed that the development of each brain structure behaved as an independent unit and that different parts of the brain therefore could be influenced by evolution independently of the others. But this has proven not to be the case. Like so much of the rest of our animal bodies, the growth of the brain proceeds, as we have already discussed, very much via an overproduction of cells — followed by the elimination of nonfunctional cells. It is, as Deacon writes, simpler to construct a door by first building the wall and then removing that part of it where the door should be placed. From a human-engineering point of view, such a strategy may seem to involve a needless waste of material — but from the natural engineering, or evolutionary, standpoint, it is a highly efficient way of utilizing information: "It circumvents the difficulties of planning ahead and allows development to proceed with a minimum of design or regulatory mechanisms" (Deacon 1997, 196).[10]

Expressed in the terminology of the present book, we might say that the final development of the brain is guided by an *epigenetic semiosis* — and that the traditional idea of a preformationistic *causal control from the genetic foundation* therefore has to be left behind. For it is striking in this connection that although the human brain possesses perhaps a thousand times more neurons than there are neurons available in the brains of minor vertebrates — and many millions more synaptic connections, indeed it has been calculated that one cubic millimeter of grey matter houses a total of four kilometers of axons (Nunn 2007) — this neural surplus can not be retrieved as any systematic increase in the size of the human genome. *No* gene literally codes for the structure we call the human brain.

For while it's true that genetic changes lie at the basis of the human brain's deviation from the general primate brain plan, these changes do not concern the precise elaboration of specific brain structures but, rather, merely underlie changes in the most overall and basic architectural patterns of the brain. A comparison of human embryonic brain development to the corresponding development in primates shows that the relative increase in brain size in humans is derived from certain dorsal regions of the neural tube (the extended tube-like structure that runs down along the back from head to tail in the undifferentiated embryonic body), that cover a mainly continuous layer from the future cerebellum to the future telencephalon (cerebrum). The increased amount of cellular divisions in these regions, due to their increased general growth, is caused by an enhanced activity of just two *homeotic genes*, Otx and Emx.

These homeotic genes were originally observed in fruit flies (*Drosophila melanogaster*), via mutants that exhibited remarkably altered phenotypes in their

---

10   Note the remarkable parallel here to the results obtained in the field of situated robotics as discussed in Chapter 7.

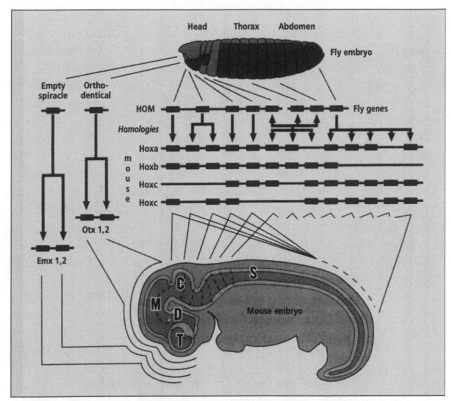

Figure 8.1 Three families of homeotic genes in *Drosophila* (HOM, Orthodenticle, and empty space spiracle families) and the corresponding families in mice (Hox, Otx and Emx). These gene families are also found in the human genome. Each small dark rectangle corresponds to a particular homeotic gene on a DNA thread. Vertical arrows indicate probable homologies. Note the isomorphy between the linear arrangement of genes and the sequence of body segments whose growth rates these genes regulate (from Deacon 1997, 178 with permission).

body structures. It was discovered that some of these mutant flies carried damaged genes possessing a distinct DNA-sequence of approximately 180 base pairs, and this sequence was given the name of a *homeobox*.[11] Using the homeobox as a molecular probe, it was found that a whole series of these homeotic genes (called Hox genes, for short) were actually present in the normal, and not just the mutant, population of fruit flies — and, what is perhaps even more remarkable, that the very same genes (or slightly modified versions of them) were also present in a range of other organisms, including humans. Subsequent research

---

11   In essence, homeotic genes specify transcription factors (see Chapter 5) and the 180 base sequences of the homeobox corresponds to a distinct sequence of 60 amino acids involved in the specific binding of transcription factors to the DNA molecule.

has shown that the task of the Hox genes invariably shows itself to be the control of the growth of distinct body segments (Figure 8.1). Alterations of basic body structure *between species* are therefore largely caused by changes in the relative activity of the homeotic genes — and this applies to changes in brain architecture as well.

Explanations of the incredible abilities of the human brain relative to the brains of other species must therefore proceed from an examination of the increased activity of cell divisions in the embryonic regions that are destined to become the human cerebellum and prefrontal cortex. But in order to understand the later dramatic mental effects that these changes will eventually allow, claims Deacon, one must first also consider the phenomenon that he has called *neuronal displacement*. Behind this phenomenon we find a mechanism we have earlier discussed — i.e., that of *synaptic competition*, which is the competition among growing axons to *find* vacant places for the formation of new synapses.[12] Neurons that succeed in forming functional synapses will survive, while neurons that do not are eliminated.

Yet the term *find* in this connection may need closer examination. An important difference between nerve tissue and other tissues in the body is that nerve cells are specialized for long-distance communication. Whereas, for instance, the formation of the anatomy of the hand is mostly ruled by local cell-to-cell communicational processes, the construction of the brain and the nervous system is to a large extent influenced by signals arriving from far away locations — and especially from other parts of the embryonic (and to some extent, neonatal) brain. In Chapter 5, we saw how embryonic nerve cells develop pathways in an exploratory fashion by extending their axons from an area in the tip, called a growth cone, that literally feels its way forward along the cell surfaces it passes by. Small protrusions from the growth cone called *filopodia* extend themselves or contract depending on the kind of signals they receive from the surfaces of other cells or from the extracellular matrix (see Figure 5.14).

The same basic mechanisms are also involved in the cerebral guiding of axons to suitable destinations in the brain. Differences in the surface chemistry of the cells they pass by, molecules that attract or reject, and differences in the mechanical properties of different tissues all influence this canalizing growth dynamic. Most importantly, specific growth factors, secreted from *destinations* in need of establishing neural connections, attract the exploratory behavior of the growth cones of neurons.

The fact that even the remotest cells in the brain may exert a genuinely causal influence upon the behavior of another cell far away implies that a *nonlocal*

---

12   See the discussion in Chapter 5 on the development of functional blindness in cats that is observed in kittens that are blinded in one eye from birth.

developmental logic is overriding, interfering with, or at least re-shaping the local differentiation process. And due to this complex interaction between local and global influences, brain tissues come to exhibit a more fine-grained variation than tissues of other organ systems. It is for this reason, too, that they contain considerably more potential functional differentiation than can be seen elsewhere in the human body. The brain essentially builds itself, so to say, online. And in fact, this genetic indeterminacy is exactly the unique strength of human brain development:

> Cells in different areas of the brain are not their own masters, and have not been given their connection orders beforehand. They have some crude directional information about the general class of structures that make appropriate targets, but apparently little information about exactly where they should end up in a target structure or group of target structures. In a very literal sense, then, each developing brain region adapts to the body it finds itself in. There is a sort of *ecology of interactions* determined by the other brain regions to which it is linked that selects for appropriate brain organization. This process provides the answer to the problem of correlated adaptations in different parts of such a complex system as brain and body. There need be no "preestablished harmony" of brain mutations to match body mutations, because the developing brain can develop a corresponding organization "on line," during development (Deacon 1997, 205; italics added).

Synaptic competition means that the share of neuronal connections to the brain that a given embryonic gland or organ will obtain, must, for geometric reasons alone, depend upon the size of the gland or organ — and this applies to peripheral tissues, as well as to the brain's own tissues. Deacon (1997, 207) describes the process:

> Among competing structures, the structure that sends the greatest number of axons to a particular target will tend to drive the activity patterns of cells in that target more effectively, and this will give connections that are from a larger source population a "voting" advantage in determining which connections will remain. This has very important implications for understanding the patterns and processes of brain evolution, because it means that modifications of the relative proportions of peripheral and central nervous system structures can significantly alter connection patterns. So, although genetic tinkering may not go on in any significant degree at the connection-by-connection level, genetic biasing at the level of whole populations of cells can result in reliable shifts in connection patterns.

Anthropomorphizing a bit (but perhaps far less than at first sight may seem to be the case), we could say that the efforts of individual tissues to get a neuronal foothold in suitable brain centers acts as a kind of *recruiting activity* — in the sense that relative increases in the neuronal populations of such tissues will tend to result in a yet more efficient recruiting of both afferent and efferent connections. This effect is exactly what Deacon implies is the result of the process of *displacement*. However, that such relatively simple mechanisms have resulted in the dramatic effects that we find in human cerebral development is due also to another surprising circumstance that we must briefly discuss — and this is *the all too small body of the human being*.

The point here is that in a certain sense human beings do not so much have an extraordinarily big brain. Rather, it is the corresponding growth rate of the body that has been disproportionately reduced. This means that an adult human would have a weight of about five hundred kilograms if he or she were like other primates in the brain size to body mass ratio. The consequences of this are notable, for it means that a whole range of tissues have been too weakly represented in the synaptic competition and have not been able to recruit enough connections to their respective brain centers. Those parts of the brain that would normally operate these body parts have therefore diminished their shares of the joint pool of synaptic places in the brain.

And yet this does not necessarily imply any loss of efficiency of the cerebral control of peripheral tissues, muscles, or sense organs. If, for instance, the retina had had a surface proportional to that heavier overall body weight, this would have required that the visual cortex, in order to keep its control function intact, should indeed have conserved a bigger share of synapses. But as it is, the relatively diminished visual cortex has ample supplies of synaptic targets for the diminished retina.

Having a relatively oversized brain does, however, have yet another very decisive consequence. For the displacement effect implies that nerve endings from the prefrontal cortex — which are increased by 300 percent in human beings as compared to our closest primate relatives — by the sheer force of number, will have a competitive advantage that assures them access to virtually everywhere else in the brain. The result is that a *network of causal and communicative relations* is established via the prefrontal cortex that reaches out to all areas of the brain and body. And, as we shall see, this network is of crucial importance for the specific human talent for constructing *symbolic reference* — and therefore, following Deacon, for the linguistic ability and its accompanying cognitive power.

From Iconic to Indexical Reference

Language "seems to be as counterintuitive for us to understand as it is simple for us to use," writes Deacon (1997, 51). And, indeed, for centuries philosophers, psychologists, and linguists have quarreled about the nature of language — joined, in modern times, by the company of new kinds of "experts" such as biologists trying to get apes to talk, and computer scientists who — not surprisingly — suggested the possibility of intelligent machines that will "converse" with us. And yet, a general weakness common to all of these efforts to grasp the nature of language has been, and still is, the failure to appreciate that human language is (in the words of Deacon) an anomaly. For how else should we characterize the finding that there exists no counterpart to genuine *linguistic* reference in the rest of the biological world?

One main reason for this lack of reflection over — or curiosity about the uniqueness of the practice of language use in the natural world may be that it is not acknowledged as such by anyone who takes evolution or biology seriously.[13] Ambiguous definitions of the differences between *words, sentences*, and *language* on the one hand, and *reference, meaning*, and *understanding* on the other, has allowed too much room for metaphoric and misleading reasoning. Among animals, there often occur modes of expression that lend themselves to be conceived by humans as "defective counterparts" or "proto-versions" of words. When, for instance, a dog obeys one's *linguistic* command for fetching or sitting, does that not show that the dog understands the words *fetch* or *sit* just as the person uttering those commands does? Indeed, and as we saw in the case of vervet monkeys, it has felt natural to many scientists to describe such sound-action association pairing as language — albeit, one without, say, the fancy syntactic combinations that characterize human grammar. In other words, if we could just add grammar to this sound-action association, we would have language in these animals!

The deep, unquestioned *anthropocentric* bias of "language science" in general (and of Saussure-inspired semiotics in particular) is probably nowhere better disclosed than through this widespread, tacit (if not *overt*) confusion of *sign*-hood with *word*-hood. And this leads to a critically important misunderstanding that prevents progress in grasping what is the *essential* characteristic of the object of study (i.e., language). At the heart of this misunderstanding, according to Deacon (1997, 52–53), is the fact that the majority of language researchers do not know how to discuss communicative practices *other than* in terms of human language:

> We look for the analogues of words and phrases in animal calls, we inquire about whether gestures have meanings, we consider the presence of combinations and sequencing of calls and gestures as indicating primitive syntax. On the surface this might seem to be just an extension of the comparative method: looking for the evolutionary antecedents of these linguistic features. But *there is a serious problem with using language as the model for analyzing other species' communication in hindsight. It*

---

13   This leaves only those hardcore dualists in the humanities who believe that calling something *part of culture* means that one now need not inquire into the ways in which that thing is also part of nature. I don't spend much time in this book addressing such a retrograde Cartesianism.

leads us to treat every other form of communication as exceptions to a rule based on the one most exceptional and divergent case. No analytic method could be more perverse. Social communication has been around for as long as animals have interacted and reproduced sexually. Vocal communication has been around at least as long as frogs have croaked out their mating calls in the night air. Linguistic communication was an afterthought, so to speak, a very recent and very idiosyncratic deviation from an ancient and well-established mode of communication. *It cannot possibly provide an appropriate model against which to asses other forms of communication* (italics added).

And Deacon does not shrink from comparing this revisionist fixation with human language in the study of animal communication to a fictive situation whereby birds would be classified by biologists according to how closely their wings would resemble penguins' wings — the penguin wing being taken as the exemplar case of what *constitutes* a wing. But language, Deacon argues, is more like penguins' wings than, say, sparrows' wings — i.e., it is a species-specific exception to a more general evolutionary rule, and one that should be studied as such, accordingly, if we are to properly understand either penguin wings or wings in general.

This fixation with the anomaly that is *human* communication leads us to analyze animal communication as a kind of deformed or deficient language — as "some version of language *minus* something" (ibid.). The absurdity of this procedure may be seen most clearly, perhaps, if we apply it to ourselves — for as a consequence we would have to call our own smiles, grimaces, sobs, kisses, and hugs "words without grammar." And yet, the polemic character of Deacon's reasoning in this case appears justified when considering that this narrow-minded language paradigm is exactly the implicit fundament for currently prevailing theories on the evolution of language (Bickerton 1990; Seyfarth and Cheney 1992).

Contrary to these traditions, Deacon has attempted to dig beneath the superficial structure of word reference as a kind of sound-to-action or sound-to-mental-image *association pairing* in an attempt to find a general fundament for explaining the way that the human use of words may be derived both analytically and evolutionarily as a special case of a the more general ability of animals to "know" the world through signs and reference. And this is where the Peircean understanding of sign relations comes to be of help. For the key to the difference that Deacon searches for is found in the Peircean concept of the *interpretant*, which, in the present case, is identical to the cognitive process taking place in the head of the organism that interprets the sound expression. This entails that the dog's interpretation and the human's interpretation of the vocal phenomenon "sit" are *not* the same, as we shall see.

It is exactly here that Peirce's well-known sign trichotomy reflecting the three fundamentally different ways that signs may refer to their object — i.e., the major categorical trichotomy of *icon, index,* and *symbol* — shows itself to be the key that unlocks the understanding as to what joins human language to the evolutionary lineage of other animal communication systems, and what makes it unique. For only humans are able, in natural settings, to communicate using the high-order tools of *symbolic reference*, says Deacon. True, in a few radically atypical and unnaturally human-created learning situations, some few laboratory-raised chimpanzees may eventually become *trained* (by humans) in some limited use of symbolic communication — but in their natural state and habitat, these creatures are, as Deacon expresses it, basically *pre-maladapted* to this kind of communication.

Experience tells me, however, that I had better be careful in the way I choose to specify this ability that apes are invariantly lacking, and that every human child possesses. That chimpanzees are *pre-maladapted* for symbolic reference means no more (and no less) than that they, in their natural state, are not very apt to master this particular kind of reference — just as humans, in their natural state, are not very good at sleeping in an erect position, or in distinguishing foe from friend by smell alone. Thus, the expression *pre-maladapted* implies that the difference between humans and apes is not necessarily based on some sort of general intelligence (much less a specific "gene for" explanation) — but rather on a much more specific *talent* that humans have developed and chimpanzees haven't.

The trichotomy — of likeness (*icon*), physical closeness or correlation (*index*), and conventional law (*symbol*) — has roots long back in the history of thinking, but it was Peirce's genius to see that this trichotomy corresponds to a logical trichotomy of communicational processes.[14] The *icon* represents the most basic and immediate type of reference (in effect, brute sensation); the *indices* (physical associations) and *symbols* (higher-order relations *between* associations) appear through the formation of more logically complex interpretants — with each stage in the process furnishing the necessary *relata* of the next.

Deacon now takes this trichotomy a step further by converting Peirce's *logical partitioning* of these levels of sign processing into an *evolutionary (and temporal) partitioning*, where the capacity for iconic reference is seen as the basis for the development of a capacity for indexical reference — and this again becomes the basis for the later development in human beings of the ability to think and communicate via symbolic reference.

Many semioticians will perhaps think that Deacon, in suggesting this evolutionary temporalization of one particular set of Peircean sign types, is treating

14   The trichotomy of *icon, index,* and *symbol* divides signs according to the way they relate to their objects. In addition to this trichotomy, Peirce also designated a trichotomy of signs reflecting the nature of the quality of the sign in itself and a trichotomy of signs based on the nature of their relations to their interpretants.

Peircean semiotics a bit too heavy-handedly, and we shall shortly consider some criticism of Deacon's way of using the term *symbol*. There is, however, no doubt that Deacon's transfer of these concepts to an evolutionary frame of reference opens up some very interesting and promising agendas for a biological under-standing of human origins. Rather than using such criticism in sterile rejection of Deacon's ideas, we should use them to tighten up and refine the terminology.

For those not yet conversant with these ideas, the *icon* is in one sense the eas-iest — and in another the most difficult — of the three sign types to add to one's conceptual toolbox, for *likeness* per se, of course, is no unambiguous property. When attempting to find one's way about in an unaccustomed milieu (such as, for example, when attempting to use subway lines in a foreign country), one quickly discovers that those *icons* that would guide one's steps immediately in the correct direction (if only one could understand them or pick them out in the first place) are far from obvious to the eyes of a foreigner — though they are almost transparently obvious to the eyes of the customary user. The problem goes much deeper than this, however, for as soon as one does learn to see and understand the meaning of such subway icons as colors, numbers, arrows, and unfamiliar graphical representations, that sign has already more or less ceased being merely an icon, and has instead become an index, that *points to* some meaningful information — *the right direction*.[15]

Deacon gives a telling example in discussing the iconic character of *wing cam-ouflage* in certain species of moths that hide on tree trunks. The whole idea of camouflage, of course, is to avoid being discovered — or, in this case, to fool a bird into thinking that you are just another piece of bark. And if the bird, while passing by, does *not* register anything worthwhile eating while the bark-colored moth is blending in with the real bark of the tree, then the color pattern of the wing is in a way not a sign at all, for in this case no *second* nonbark icon — or differentiating *interpretant* — is formed in the brain of the bird. It just perceives "bark, bark, bark" and *not* "bark, moth, bark." In Deacon's words, the inter-pretive process involved in iconicity is "something that we don't do. It is, so to speak, the act of *not* making a distinction" (Deacon 1997, 74).

This way of formulating the matter may seem a bit extreme, but the example does show that iconic reference does not reside in the primary sign itself (the color pattern of the moth's wings) but in the *relation* between moth wing and its perception (or lack thereof) by the bird. Its iconicity (likeness relation) derives from the *absence* of a differentiation in the recognition process — or, positively formulated, in a mis-recognition. The color pattern of the moth's wings *refers*

---

15   It may even cease becoming either an icon or an index, but instead become a *symbol*, if the knowledge of the significance of the icon is based on a convention rather than on a physical association.

iconically to the bark (for the bird) precisely by its not being discovered as non-bark! Such a failure or (near failure) in distinguishing a *difference* constitutes the *ground* under every icon, even though we are rarely conscious of it.[16] When, for instance, a small gentleman is depicted on the faceplate of an out-of-the-way door, it is reasonable to assume that behind this door, there will be a men's room. The figure in this case is an icon of a man — but at the basis of this iconicity lies an element of indistinguishability that is nicely expressed by the term *re-cognition.*

Another way of expressing this analytically troublesome connection is to say that the icon, like so many other analytical concepts, is a border case. Many signs are actually *degenerated*, which means that although a sign, according to Peirce, always consists in a triadic relation, the *interpretant* may often play a very limited role in the relation (as is the case with indices) — and that when also the object relation of the sign is close to disappearing, we get an icon. The icon thus is essentially a kind of Firstness — a pure quality — and, if it is an icon at all, it is because the degeneration is not total, so that an element of Secondness and Thirdness persist to make it a sign.[17] That a sign is *iconic*, then, just means that *quality* as such is prevailing in the sign process, while Secondness (association) and Thirdness (convention) only play minimal roles.

In short, the process by which something gets recognized as something different is the root form of all *semiosis*. We saw in the preceding chapter, that many ATP-degrading enzymes in bacteria are not capable of distinguishing between cAMP and ATP and will therefore eventually bind to cAMP instead of ATP. The effect of this is that all the usual ATP-based energy supplies are blocked by the *false* substrate — with the result that the energy metabolism of the bacteria is effectively stopped. Seen as a molecule, cAMP in this case functions as an *icon* for the ATP molecule, and this *iconic relation* successively implies that the *concentration* of cAMP in the cell becomes an *index* for a deficient energy metabolism. The evolutionary process then finally utilized this relation for the further development of cAMP as a cellular *symbol* (as suggested by Tomkins 1975; see Chapter 7).

Seen in this framework, the invention of *symbolic reference* was a theme in evolution a long time before humans appeared. The potential of symbolic (conventional, law-bound) reference has, in fact, been used by life processes ever since the replicative sign systems of RNA and DNA first appeared. For if — to stay with the metaphorics suggested in Chapter 5 — a gene plays the same role

16   Floyd Merrell (1992, 271) uses the example of a smoke ring in a room filled with smoke, and observes that this ring "serves as a primitive paradigm (or *icon*) for all systems containing the rudiments of self-reference which has set them on the recursive path to selfhood."

17   Peirce suggested the term *hypoicon* as a designation for icons *understood* as real signs (*CP* 2:276).

in the life of the cell (or the organism) that a menu choice plays in a computer-based inventory-control system, then the semiotic process in which the gene (or menu choice) takes part is obviously not conditioned by *likeness* (icon) or by any *immediate physical causality* (index), but rather by a *conventional*, evolutionarily emergent *connection* between the specific sequence of the gene and the cellular (or organismic) situation that releases its expression (interpretation).

In this case, we are talking about *endosemiotic symbolic reference*, whereas the theme analyzed by Deacon (and in much of this present chapter) is the *exosemiotic processing of symbols* — i.e., the use of symbolic reference by communicating *individuals*. And to create such a system is of course an evolutionary challenge of a quite different kind (although not necessarily more overwhelming). Too, although one may, as we shall shortly see, quarrel with the claim that human linguistic competence is unique in the world when considered as a system for the exosemiotic use of symbolic reference, Deacon is right, in my view, in claiming that normal communication among animals is based upon iconic and indexical reference, and that the development of human linguistic communication therefore constitutes a logical jump to the more abstract cognitive world of the symbol.

One important point (among many) in Deacon's analysis is that the transformation from indexical to symbolic reference contains elements that are very much like the steps taking one from iconic to indexical reference. And since the latter of these transformations is the simpler of the two it is a good idea to start the analysis there.

An *indexical* relation presupposes a prior establishment of a multiplicity of iconic relations — and to show how this works, Deacon uses the relation between *smoke* and *fire* as an example. That animals are frightened by smoke is probably in most cases due to a genetically determined sensibility. But in human beings, this reaction is strongly based on — or at least heavily reinforced by — experience. If we ponder how the indexical connection in this case is brought about, we find that it may be dissolved into two correlated sets of iconic experiences.

Early in life, presumably, we have all been puzzled by the smell or sight of smoke, and little by little, as we encounter still more cases of this experience, we started recognizing it as a token of a *kind* — in other words, we began establishing an *iconic* reference between *each* new occurrence of this experience and *all* preceding occurrences of it. Now, different instances of smoke are not identical — and certain versions of the phenomenon will gradually acquire the character of our prototype of *smoke*. To the extent that we do not differentiate between these instances (e.g., "that is *smoke*" and "there is smoke *again*"), we have established an *iconic relation* for what constitutes this one phenomenon in the world, *smoke*, from all else that is not smoke.

In most cases the child will also, either simultaneously or at least within a very few moments, discover that *something is burning* and the distinct occurrences of *fire* will — just like the occurrences of smoke — come to *iconically* refer to each other. Here, too, certain experiences of fire will come to stand as proto-types of the general phenomenon of *fire*. The only thing now missing for the for-mation of a full-blown *indexical* relation *between* the icons of *smoke* and icons of *fire* is the formation in the child's cognitive system of a *third* icon — namely the *correlation itself* as an icon of *the always experienced correlation* between the presence of smoke and the presence of fire. And as this gradually happens, pro-totype smoke *becomes* an *index* for prototype fire: "What I am suggesting, then, is that the responses we develop as a result of day-to-day associative learning are the basis for all indexical interpretations, and that this is the result of a spe-cial relationship that develops among iconic interpretive processes. It's hierar-chic. Prior iconic relationships are necessary for indexical reference, but prior indexical relationships are not in the same way necessary for iconic reference" (Deacon 1997, 78).

This understanding explains why words *in many cases* — but certainly not in the majority of cases, as we shall see — may also work as indices. For words, like other signs, may become iconically associated with previous uses of the same word, just as the object to which the word refers may be iconically connected to earlier versions of the same object. Previously experienced correlations between these two sets of icons then ensures that the object mentally resuscitates upon the mere hearing of the sound of the word. Exactly this kind of conditioning was the basic methodology behind the training of pigeons and rats in Skinner boxes, and it is a fundamental principle in many forms of animal training.

But experiments with animals in Skinner boxes also disclose a property of indexical signs that makes them very different from normal nouns in language. An index must be *reliably and repeatedly* correlated to its object *in space and time*, otherwise the *association* between its icons will quickly become weakened and eventually lost. Once a rat no longer receives any reward (e.g., food) when it presses a certain button upon perceiving a signal, it will quickly stop pushing that button, and the indexical significance of the signal will thus be lost (to use behavioral psychology's own words, "the association will be extinguished").

Correspondingly, we see in human beings that an often repeated but contin-ually unfulfilled threat gradually ceases to be taken seriously. Literary examples are numerous, but the story of the boy shouting, "The wolf is coming!" is proto-typical. It is important to note, however, that even within the logic of the story, the warning does not lose its *symbolic* signification — none of the villagers, of course, ever doubt what the outcry is meant to *mean* — but the indexical signi-fication is eventually lost due to the high number of times that the sentence was repeatedly used out of correlation with the event it reported.

In normal talk, the *significance* of a word is quite independent of repeated and reliable *physical correlations* with the object or event that the word refers to. Angels, unicorns, ghosts, Santa Claus, pink elephants, and the Fountain of Youth are all sensible words in English but very few of us, if any, have ever come upon their *referents*. Yet nor do we have difficulties in understanding the manifold signification of words whose referents we have never had any direct experience of. Only rarely, for instance, have I been asked to a *torskegilde* (a Danish word for tax audit that literally means "cod feast"!) — but I have no doubt that I would greatly prefer to avoid it.

Words are not normally, then, indices — and this is most evident in their all but unbreakable interdependence. A weathercock is expected to indicate the direction of the wind, and a compass should point north quite independently of the wind's direction. The one index — at least considered *as* an index — has nothing to do with the other index. And if the compass should, for some reason, be placed upon the weathercock, it would not stop pointing north. But words have *everything* to do with each other, for words *acquire* their signification as much from their connections with other words as they do from their connection to single objects or events — probably more.

For example, consider the sentence: The ham sandwich at table five is becoming impatient. Here, we must give up any attempt at interpreting the words sensibly if we take it that *ham sandwich* refers indexically to that popular arrangement of food.[18] But the sentence will make perfect sense if we assume, for instance, that the term *ham sandwich* refers metonymically to the guest at a restaurant who has *ordered* such a dish.[19] A hint of why this might be the case is found in the words *table five* — a rather exceptional designation for what is, after all, just another table — unless by chance we are working in a restaurant. One might then imagine that the utterance was meant as a reminder from the waitress to the cook to hurry up. Most of us spontaneously make up interpretations like the one offered here, by placing the utterance in an imagined *sense-making* context that cannot be deduced by merely considering the individual words — but only from considering their internal play with one another, *and* the whole system of worded relations wherein such play makes sense. Such systems, when considered as a whole, we call *languages*.

The *symbolic reference* of language thus depends on the fact that the real communicative and knowledge-bestowing power of words and symbols derives primarily from their ability to refer to other words and symbols — and only

18   I am vaguely aware of having borrowed this example from John Searle, though I cannot locate the exact source for citation.

19   *Metonymy* being the everyday linguistic practice of referring to something by naming a thing that it is closely linked with, rather than the actual referent itself, as in "*Russia* asserts . . . " and "*The White House* denies . . . ." As in the ham sandwich example, one has to transcend the temptation to interpret these words *indexically*, and must interpret them instead *symbolically*, if they are to say anything true about the world.

indirectly, through their linguistic situatedness, to the reality of *things* implicated in the linguistic utterance. The symbolic layer in communication is thus
subsuming the original indexical layer — over-layering it so to say — nourishing
itself upon it while taking it in new directions of its own.

The symbolic power of words is, then, as Deacon puts it, "*distributed* in the
relationships *between* words." Deacon (1997, 83) writes, "Symbolic reference
derives from *combinatorial* possibilities and impossibilities, and we therefore
depend on combinations both to discover it (during learning) and to make use
of it (during communication). Thus the imagined version of nonhuman animal
language that is made up of isolated words, but lacking regularities that govern
possible combinations, is ultimately a contradiction in terms." In this understanding of the inner anatomy of linguistic reference, the problem of human
evolution very much changes from one of how our ancestors could possibly
learn to speak to one of how our ancestors managed to *unlearn* their propensity
for making (what we must assume were largely successful and reliable) indexical references. How did our ancestors get rid of the much too narrow binding to
reality exhibited by pure indexicality?

Entrance into the landscape of symbolic understanding clearly enough
depends on the *prior* establishment of a rich inner world of relationships that
are themselves detached (in some meaningful sense) from an iron-bound physical relation to the brute cause-and-effect relations of the external world. The
capacity for imagining a world that does not necessarily exist — in other words,
to think the impossible — thus appears as the key to the distinction between
the *world-in-itself* and the *world-for-us*, or in Deely's terminology, the dividing
line between *things* and *objects*. This distinction, as we saw, began to be seriously
reflected upon in a philosophical sense only starting in Greece around the sixth
century BC — but it was latently present in the *Lebenswelt* that our hominid
ancestors placed themselves in by their growing development of language.

Indeed, Deacon analyzes in depth the many attempts made by human experimenters to teach apes to express themselves by using language-like tokens, and
concludes that these animals are not prevented from doing so either from a
lack of general intelligence nor by the lack of the appropriate anatomical means
for vocalization. Instead, it turns out that it is the apes' strong cognitive and
practical dependence on *indexicality* that is the stumbling block that prevents
them from becoming the kinds of uninhibited language users that virtually all
human children become without further ceremony.[20] Thus, because apes cannot

20   The French cognitive philosopher Dominique Lestel has elegantly expressed this point in
noting that "anybody taking an interest in this question of 'speaking apes' is struck by the fact
that although it is beyond doubt that the nonhuman apes exhibit language abilities (primitive,
but real), *they have nothing to say*" (emphasis in original). "Qui s'intéresse a cette question des
singes parlants est frappé par le fait que si les primats non humaine montres indiscutablement des

unlearn how to understand the world *indexically* they stay pre-maladapted for a life based on *symbolic* reference.

And now, perhaps, Deacon's evolutionary reading of Peirce's hierarchy of sign logic becomes clear. For just as we saw that *indexicality* presupposes the transcendence of the nondynamic, self-catching operation of *iconic* relations, we now see that *symbolic reference* presupposes a transcendence of the *indexical* bonds. A mental inversion must take place that pushes the objects and events of the surroundings to the mental background in order to permit the establishment of a new systemic web of word-to-word (or more accurately, symbol-to-symbol) relationships capable of imprinting meaning — one based on the network of relations between words and other words (symbols and other symbols) rather than on the more fixed, dyadic relations between words and reality — upon the flat indexical backdrop.

To understand the puzzle of language is to understand the emergence of this experiential reconstruction in the brain of the little child. But how our hominid ancestors acquired the capacity for such a revolutionary re-construction of their pre-linguistic Umwelts in the first place is the central question that an evolutionary theory of human language origins must answer.

Language Occupies the Brain

The question posed at the end of the preceding section may be divided into two. We must first explain how in the human being a brain could be developed that would allow this new (and, in the animal world, decidedly unique) talent for *symbolic reference* to grow up. Secondly, we must delineate what reproductive advantages, if any, this talent for symbolic referencing might have bestowed upon the populations of hominids that originally developed this trait that directly contradicts an eons-long background of successful indexical reasoning.

In answering the first of these questions, Deacon suggests what is probably his most fruitful idea in this context. Classical theories about the brain-language relation have placed the cart before the horse, says Deacon. These tried to explain the emergence of language as resulting from the growth of the brain volume of hominids — connecting this physical increase with the assumption of an increased general ability for information processing. Yet whether this idea is accompanied by a theory of the formation in the human brain of a special neuronal *language device* or whether it is accompanied by the theory that language development is simply the natural result of the

---

capacités linguistique (primitives mais réelles) *ils n'ont rien à dire*" (Lestel 1995, 6). Lestel, however, suggests that the indexicality of the actual talk of apes is due to the positivistic paradigm behind their training programs.

general increase in brain complexity, the basic understanding, at any rate, is that the *humanization* of the brain *preceded* the invention of language. Deacon now suggests the opposite.

He sees the nascent ability to make symbolic reference as the single most important factor characterizing the selective pressure that pushed human evolution forward by rewarding individuals possessing a brain architecture approaching that of contemporary humans. In other words, according to Deacon, it is this nascent linguistic ability that explains the specific growth of the human brain, rather than the growth of the brain explaining the origin of language:

> If neither greater intelligence, facile articulatory abilities, nor prescient grammatical predispositions of children were the keys to cracking this symbolic barrier, then evolution of these supports for language complexity must have been consequences rather than causes or prerequisites of language evolution. More important, these adaptations could not have been the most critical determinants of brain evolution in our species. Approaching the language origins mystery from this perspective is like stepping out of a mirror to find everything to be the reverse of what we assumed. From this perspective language must be viewed as its own prime mover. It is the author of a co-evolved complex of adaptations arrayed around a single core semiotic innovation that was initially extremely difficult to acquire. Subsequent brain evolution was a response to this selection pressure and progressively made this symbolic threshold ever easier to cross (Deacon 1997, 44).

This suggestion takes Deacon to a theoretical position not far from that of the Baldwinism that was discussed in Chapter 6.[21] Essentially, what Deacon suggests is that an inseparably mental *and* social property — the ability to communicate with symbols — has influenced and even shaped the evolutionary process in human beings. On the surface, this view would seem to stand in considerable opposition to firmly held intuitions in the scientific world holding that physical changes (like the evolution of the brain) must similarly have physical causes. But Deacon's theory does not conflict with such intuitions. For one thing, Deacon — as we shall see — anchors the ability for symbolic thinking solidly in the specific organization of the human brain. In addition, his scenario for selection — like any other neo-Darwinian selection scheme — assumes that mutations are the concrete source for inherited changes. The only deviation from prevalent scenarios is that Deacon ascribes a *social* factor — i.e., the fact that our ancestors gradually learned to communicate via symbolic signs, and then came to depend on this process — as a decisive influence upon determining those

---

21   Deacon, of course, is fully aware of this and he explicitly recognizes the relationship. However, as he has later pointed out, Baldwinism per se is only half the story — the unique nature of *symbolic reference* is what really tells the tale of human origins (see footnote 24 to the section "The Biosemiotic Core of Evolution" in Chapter 6; Wiles et al. 2005.

changes in the brain that would win in the competition for reproductive success among our remote ancestors.

It is important for an evaluation of Deacon's conception to recognize that not even contemporary humans are born with the ability of symbolic reference. Small children may well — unlike the pre-maladapted apes — be biologically prepared for the task, but it remains nevertheless a hard exercise, one that it takes the child at least two years to master to a significant extent. And compared to the situation that our remote ancestors who originally tried to acquire the art must have had to cope with, contemporary children have the advantage of growing up in a semiotic *milieu* where language, virtually from the very beginning of life, plays a decisive role. And yet, the hominid young of those past times would have been not only handicapped by the fact that their brains were more ape-like than human-like, but they would have also had to master the trick of symbolic communication *de novo* — i.e., without any help from the adults. How can they possibly have managed to do this?

Here we have a typical problem for all theories of evolution. We know that the process of language invention has indeed taken place, but we do not know how — and no matter what scenario we posit in order to explain it without invoking the use of supernatural events (e.g., miracles and, to a lesser extent, *hopeful monsters*), we must recognize that such "just-so" stories necessarily feel impoverished when compared to the enormity of the phenomenon that they are attempting to explain. I have inserted this remark only to prevent the reader from concluding that Deacon's scenario is uniquely not a fully satisfactory one — for indeed I believe that such an objection may be launched against any and all naturalistic scenarios for the origin of language.

Deacon suggests that the ability for symbolic communication originally may have arisen as a solution to the need in early hominid groups to establish a *binding social cooperation* that, in the end, would imply the creation of marriage-like institutions. The need for such institutions ultimately derives, according to prevailing theories, from an imbalance between the genders in life in close groups where the male's contribution to parental care stands in logical opposition to the male's principled ignorance of paternity to the offspring. Moreover, one major advantage of having big brains is the ability to learn — which means that big-brained babies are born with a lot to learn or, differently stated, that they are born very unfinished. The implication is that parental care is a much more demanding task for humans than it is for other species — and, very likely in fact, it could not have been successfully accomplished in the hominid lineage without the participation of both sexes.   Marriage-like institutions might have solved this potential conflict by providing a ritual consolidation of paternity. However, Deacon (1997, 401) claims that pure indexical thinking would never have been able to support such institutions, whereas even a limited ability for symbolic understanding would have sufficed: "The need to mark these reciprocally altruistic (and reciprocally selfish) relationships arose as an adaptation to the extreme evolutionary instability of the combination of group hunting/scavenging and male provisioning of mates and offspring. This was the question for which symbolization was the only viable answer. Symbolic culture was a response to a reproductive problem that only symbols could solve: the imperative of representing a social contract."

Deacon's theory of the possession of symbolic reference as a motive force in the evolution of the human brain does, of course, presuppose that eventual gains in the ability for symbolic communication would also increase the evolutionary fitness of the concerned individuals. Rather than entering into a detailed discussion of Deacon's specific attempts to support this concept, however, I shall content myself by noting that the theory does indeed illustrate how social advantages might be gained from the mastering of symbolic reference. But scenarios implying social institutions other than marriage may, of course, just as convincingly be suggested.[22]

To begin with, our remote hominid ancestors were hardly any more capable of symbolic thinking than are present-day chimpanzees, and it is not unlikely that this new capacity may have first appeared simply as a fluke. It may then have been conserved in the group in the same way that laboratory chimpanzees born to parents that with great difficulty (and strongly supported by their trainee) acquired a limited ability for symbol use, have been reported to be able to learn this skill with much less ado (Savage-Rumbaugh and Lewin 1994). Whether it was just for fun or because it was, in fact, advantageous to the population, the habit of symbolic referencing must at some time have persisted long enough to be incorporated into the social network in a way that eventually stabilized it.[23] If so, then hominids that exhibited the most talent in this regard may have also managed to thrive above average in the social game, and their share of the gene pool would thus have tended to grow. The main conception can now be summarized in a simple way: "The human brain should reflect language in its architecture the same way birds reflect the aerodynamics of flight" (ibid., 45).

By thus posing the invention of the ability to manipulate symbols as a *precondition* for further brain evolution, rather than vice versa, Deacon establishes a strong argument against Chomsky's explanatory hypothesis of an innate universal

22   As a loose speculation, let me point to the phenomenon of *children's play* as a possible source of this new ability. Even today we have the expression *playing with words*, and symbolic referencing in the beginning simply may have been fun. Children capable in this respect may perhaps have had acquired a lifelong advantage. It is, after all, in children that the linguistic ability must first have appeared (since the capacity to learn this skill seems to decline with age) and children's play is the root form of most happenings in the social life of adults.

23   See the discussions on *semiotic scaffolding* in Chapters 4 and 5.

grammar. Now, what was thought to be the insoluble dilemma posed by grammar — its supposed unlearnability — may be solved in another way. For not only did the brain gradually undergo evolution, but so did language, hand-in-hand.

In its beginnings, language must have been primitive and clumsy — but the increasingly mandatory social use of this new semiotic tool must have gradually tuned it to result in the establishment of simpler and more efficient linguistic patterns that everybody could easily learn (indeed, we see this process still at work today, in our shorthand words and phrases for otherwise cumbersome linguistic constructions — *fax* for a facsimile sending and receiving machine, OPEC for Organization of Petroleum Exporting Countries — even the conscious misspelling induced by text-messaging may in a way testify to the universality of this mechanism).

The point here is that while biological evolution of even the smallest morphological change takes numerous generations and is exceedingly slow, language — and especially symbolic language, which is not tied to any physically necessary relations of cause-and-effect — may evolve with lightning speed. In fact, linguists confirm that a language may change to unrecognizability in just a thousand years. Seen in this light, Chomsky's puzzle may be elegantly solved: "Children's minds need not innately embody language structures, if language [evolve to] embody the predispositions of children's minds" (ibid., 109). Deacon's scenario points to the obvious possibility that language gradually changed in such a way that its grammar came to correspond optimally to what children would immediately intuit. Since changing human brains is an extremely slow process, what happened instead was that language adjusted itself to the patterns of children's brains — and that the human brain the next time round adapted to the new linguistic challenges.

That this last process also occurred is evident, of course, and well-studied — although not well understood. Obviously we are dealing here with the development of brain regions that have to do with the production and deciphering of speech sounds, but before anything else the human brain is a deviant in nature because of its oversized prefrontal cortex. This immediately confronts us with a new puzzle, for patients with injuries to the prefrontal cortex are only rarely hampered in their ability to produce or understand speech or to correctly analyze grammar. Destruction of basic language abilities are connected instead to two areas in the lower brain that seem to be involved in hearing and speech movements (Warnicke's and Broca's areas, respectively). Also, up until relatively recently, it had been customary to treat an epileptic by performing a pre-frontal lobotomy — yet this treatment did not generally result in any immediate language-processing damage, nor lowering of the IQ.[24]

24   However, the implications of this intervention do, as we now know, show up in a number of special learning contexts and there they may be severe.

In general, then, the functions of the prefrontal cortex are not well understood and have resisted attempts at unambiguous function-to-structure mapping. Unlike in the topography of the sensoric or motoric brain centers, there are apparently not, in the prefrontal cortex, any direct connections between distinct areas and distinct functions. Rather, and as mentioned earlier, the relative increase in the volume of the prefrontal cortex in humans is accompanied by an increased recruitment of synaptic connections to all the other parts of the brain. Information processing in the prefrontal cortex must therefore be assumed to influence or even dominate nearly all facets of brain function. As Deacon puts it, the prefrontal cortex makes its influence felt everywhere in the brain, not so much because it has a greater capacity as because it simply "has more votes" (ibid., 257).

A comparative analysis of the many disabilities that are seen as the effect of prefrontal damage in humans and monkeys alike, however, does gives us a hint as to the kind of skills and abilities that this part of the brain is responsible for. Deacon (1997, 264) sums up the case:

> In general, tasks that require convergence on only a single solution are minimally impacted by prefrontal damage, but those that require generating or sampling a variety of alternatives are impaired. This capacity has been called *divergent thinking* by J. Guilford (1967), and it may explain why prefrontal damage does not appear to have a major effect on many aspects of paper and pencil IQ tests. *Prefrontal damaged patients exhibit a tendency to be controlled by immediate correlative relationships between stimuli and reinforcers, and this disturbs their ability to entertain higher-order associative relationships.* In summary, these human counterparts to different frontal lobe defects in monkeys also involve difficulties in using information negatively. Impairment of the neural computations provided by these cortical areas makes it difficult to subordinate one set of associations to another, especially when the subordinate associations are more immediate and salient (emphasis added).

Returning to the discussion of the cognitive leap from *indexical* to *symbolic* thinking, we may now see that this jump relied on precisely those abilities that seem to be impaired in patients with damages in the prefrontal cortex. This corresponds exactly with what I was referring to earlier when I wrote that "a mental inversion must take place that pushes the objects and events of the surroundings to the mental background, in order to permit the establishment of a new systemic web of word-to-word (or more accurately, symbol-to-symbol) relationships capable of imprinting its meaning — one based on the network of relations between words and other words (symbols and other symbols) rather than on the more fixed, dyadic relations between words and reality — upon the flat indexical backdrop." The prefrontal areas are absolutely necessary to produce those associative connections that presuppose the subsuming of one learning process by another — as must be required if the indexical bindings between signs and objects are to be loosened enough for the signs to be networked instead into systems of other signs.

Neither general intelligence nor the presence of a specific language center is thus to be found at the root of the peculiar human adaptation for language acquisition. Instead, prefrontal involvement in even the remotest corners of intensely intercommunicating brain activity does the job. This does not mean, however, that the construction of symbolic relations therefore can be localized in the prefrontal cortex in the same way that indexical relations are rooted in specific senso-motoric pathways that may be connected with one another in the brain.

Rather, the symbolic associations generating the changing contextual significations of words are probably highly dependent on mnemonic support from sensorically based "pictures." This conception is confirmed by the frequent occurrence of semantic disturbances observed in patients suffering damages of the posterior cortex, and is borne out, too, by our everyday introspective intuitions of having pictorial representations while reading or listening to stories. Such pictures are not as such *symbols*, but should rather be considered as neurological buoys. Deacon (1997, 267) says, "Like buoys indicating an otherwise best course, they mark a specific associative path, by following which, we reconstruct the implicit symbolic reference. . . . The critical role of the prefrontal cortex is primarily in the *construction* of the distributed mnemonic architecture that supports symbolic reference, not in the storage and retrieval of symbols." Symbols, then, cannot be localized, for they are essentially *relational*; they appear and disappear like the relations between buoys in the neural continuum.

The shaping of the human brain therefore corresponds nicely with a scenario that has the ability for *symbolic reference* as its decisive mark of separation, by which the branching evolutionary hominid lineage sorted itself out into those that did and did not enter into the niche of the linguistic world — and in which the contemporary representatives of one of these lineages, *H. sapiens*, now scampers about with such exquisite symbol-using agility. For when the habit of playing with symbols, no matter how primitive, first took hold in a group of hominids, the development of this peculiar ability became an essential factor for the continued evolution of that lineage. Like any other social resource, then, language ran through a process of adaptation that optimized its usability and intelligibility. And gradually, as language use spread, it must increasingly have become the key to social success. The resulting *culture of symbolic reference* then established the evolutionary pressure on children's brains that, over time, resulted in the reconstruction of the associative patterns of brain in the image of the associative patterns of symbolic language.

As intimated previously, however, one objection to Deacon's theory concerns his use of Peirce's *symbol* category in a considerably narrower sense than has been usual among Peircean semioticians (Stjernfelt 2000). Peirce developed a complex taxonomy of sign types which Deacon has chosen to simplify for the purpose of its use within the biological — and particularly, the evolutionary — context. This makes Deacon's *symbol* concept at the one time both broader *and* more specific than the Peircean concept. For example, the kind of conditioned Pavlovian reflexes that in Deacon's analysis are described as examples of *indexical* relations, would have been considered instances of *symbolic* (law-like, conventional) relations if the Peircean terminology were followed, for they instantiate a kind of *future directedness*, an *esse in futuro* — i.e., a *habit*-like existence. Stjernfelt (2000) writes, "Accordingly, Peirce's symbol concept includes a wide range of subtypes of very different complexity degrees, ranging from simple terms over propositions to arguments — each of these in turn including a whole fauna of further subtypes. Thus, symbol use is neither as simple as Deacon presupposes (with respect to symbol subtypes) nor as complex as he presupposes (with respect to the simpler types)."

Too, behaviors that, at least according to Peirce's criteria, deserve to be called *symbolic* are indeed widespread among animals, as discussed at quite some length by Thomas Sebeok, among others (see Sebeok 1979; see also the analysis of Sheets-Johnstone discussed below). Stjernfelt claims that Deacon's criterion of *symbolic reference* is too weak to distinguish the cognitive capacity of humans from that of monkeys, and suggests that Peirce's analysis of the human abstractive forms offers a better starting point for the establishment of such a criterion. Particularly, according to Stjernfelt, the human capacity for making *hypostatic abstractions*[25] should be considered the core ability that distinguishes the human mindset from that of chimpanzees. In support of his view, Stjernfelt reminds us of Peirce's explicit emphasis on the capacity of language as a medium for self-control:

> For thinking is a kind of conduct, and is itself controllable, as everybody knows. Now the intellectual control of thinking takes place by thinking about thoughts [as in hypostatic abstractions]. All thinking is by signs; and brutes use signs. But they perhaps rarely think of them as signs. To do so is manifestly a second step in the use of language. Brutes use language, and seem to exercise some little control over it. But they certainly do not carry this control to anything like the same grade that we do. They do not criticize their thoughts logically (*CP* 5:534).

Stjernfelt (2007, 254) notes,

> Man as well as animals are consequently rational beings, probably even necessarily so. . . . But what enables man to build up his symbol systems and its resulting more

---

25   As when the experience of hot things gets abstracted into the concept of *heat*, which then may be the subject of further theorizing such as for instance the kinetic theory of heat.

acute and accelerated rationality is precision and abstraction working together, making it possible to isolate and make explicit single phases in the ongoing chain of arguments in order to control them, scrutinize them, experiment upon them, combine them, and make them better.

This capacity for making hypostatic abstractions, however, also carries with it an increased risk for committing serious errors, a capacity for being fooled or for lying. Of course, all higher animals possess such abilities, writes Stjernfelt, but "abstraction adds the possibility for the construction of enormous subdomains of discourse: myth, religion, literature, science whose vast capacity for general truth mirrors an equally large capacity for general fallacies" (ibid.). And Stjernfelt adds that this all points to the reasonable assumption that a very strong selection pressure must have prevailed against this increased risk for fallacies "especially against the formal logical fallacies without any empirical content" (ibid.).

While these objections seem well-founded, I do not think that they decisively undermine Deacon's fascinating suggestion of human brain evolution as in some sense shaped by the needs of linguistic communication rather than vice versa. But they do accentuate the need for an even *more* nuanced elaboration of the idea of human language evolution as the finalizing step in the temporalization of the Peircean sign trichotomy. If the term *symbol* is used in its more usual sense, as in Peirce, and not in the very elaborate sense given to it by Deacon, then, obviously, symbolic reference is not what the enlarged prefrontal cortex enables us for. Rather, in this case, hypostatic abstraction may be seen as the talent that presupposes prefrontal cortical interference in all and every part of the brain. When Deacon (1997, 267) says, as noted above, "The critical role of the prefrontal cortex is primarily in the *construction* of the distributed mnemonic architecture that supports symbolic reference, not in the storage and retrieval of symbols," then this very term *mnemonic architecture* itself implies a capacity for figurative reasoning that might serve as a tool for supporting hypostatic abstractions.

Proposals of radically new theories probably always suffer from the Procrustean-like weaknesses that lead us into the even more fruitful phases of clarification and specification. But in sum, I believe that Deacon's central thesis — i.e., that linguistic ability arose through the transcendence of a more primitive, indexical level of communication, one where individual indexical bindings of signs to their objects prevented a more autonomous system of referential relations from breaking loose — seems both compelling and well-founded, all the same. Whether one should best classify linguistic signs as *symbolic markers* or rather envisage them as *analytical units in the abstractive schemes of human cognition* — or perhaps even some innovative synthesis of the two — is a question that shall not be further pursued here. This work must be left for future biosemioticians to do.

Language and Biosemiosis

Terrence Deacon's theory is presented in the context of evolutionary anthropology and neurobiology — but both thematically and theoretically, it belongs naturally in a biosemiotic complex of theories. For a cornerstone in the theory is that language is not a unique invention on the planet, but should be seen instead as a sophisticated new *semiotic resource* that grew from the web of semiotic activities already well established in the animal kingdom, long before human beings appeared. Too, rather than seeing language as an either-or phenomenon, with the *discontinuity puzzle* as its inevitable consequence, Deacon's evolutionary temporalization of Peirce's *icon-index-symbol* trichotomy allows us to understand language as a special case of a more general biosemiosis. And seen in this way, i.e., as a particular species' instantiation of more general biosemiosis, language emerges as a "more-or-less" phenomenon in continuity with the rest of the animal world.

This Deaconian solution of the discontinuity puzzle on the exosemiotic level may then be seen as a parallel to a complementary solution of the discontinuity puzzle on the endosemiotic level. This is the duality that Pattee termed *the epistemic cut* — i.e., the complementarity between the symbolic and the dynamic *modes* of the cell. Let me remind the reader of my comment on this puzzle in Chapter 4:

> The tendency of nature to generate regularities — or, as it is more commonly described today, to self-organize — may be understood as the very first exposition of the principle that will develop through cosmic evolution to become semiosis, the ability of living systems on planet Earth (and possibly many other places in the universe) to form *interpretants* (or self-maintaining and self-perpetuating habits, if you like).

> This change of basic viewpoint will allow us to reach a solution to Pattee's paradox of the epistemic cut, because now we can assume that not only the symbolic functional modes (related to DNA function), but *also* the dynamic functional modes (related to the functional cytoplasm) are both, in the end, semiotic functional modes. For the difference between the two modes is, at bottom, a difference in the kind of semiotic dynamics involved. Thus, the sign processes characteristic of the dynamic functional mode — i.e., the protein world so to say — are indexical and iconic (i.e., analog-coded) rather than symbolic or digital as are the sign processes connected to DNA function. The analog-coded signs correspond to the jumble of topologically organized indexical and iconic sign processes in the cells that are responsible for the interpretation of the digital genetic instructions as well as for the execution of them (p. 95).

Drawing this parallel between the origin of exosemiotic symbolic functioning (language) with the origin of endosemiotic symbolic functioning (Pattee's *symbolic mode* of life) poses the problem of life in a new way, in its suggestion that a temporalization of the Peircean sign trichotomy in the exosemiotic sphere may also enable us to gain new insights in the endosemiotic sphere. Might, for example, the establishment of the very first genetic reference relations some four billion years ago — based on the conventional (not physically necessary) recodings of peptide sequences to nucleic acid sequences — be considered a shift from a purely *indexical* world of sign relations to a *symbolic* world?

We know for certain that the single genes of contemporary prokaryotes are not independent of each other, but rather, constitute a system of "menu choices" whose activation always occurs in a cellular context that reflects what other "menu choices" are presently activated. The *signification* of the single gene may then well be *indexically* related to the synthesis of a given protein, but at the same time, its full *significance* depends on a range of continually shifting gene-to-gene relations.

We saw for example in chapter 5 (section on "Organic Codes") that while the gene *ApoB* in liver cells specifies a version of apolipoprotein B of 4,536 amino acids, the very same gene, if situated in the intestinal glands, will specify apolipoprotein molecules of only 2,152 amino aicds. This reflects — as we saw — the necessity of the liver apolipoprotein B (but not the intestinal lipoprotein) to not only assist in transporting lipids but also to activate further enzymes involved in lipid biosynthesis. The significance of the *ApoB* gene thus cannot be confined to an indexical interpretation (which in both cases would refer to the same primary mRNA transcript of 14,100 base pairs), but must include the full system of operative genetic settings in the particular cell, where the gene is expressed. The gene here is a symbol rather than an index.

The origin of life, too, must have been closely connected to the formation of a cellular system that would allow the loosening of an all too rigid binding between the gene and the metabolic process that the gene contributes to — thereby facilitating the formation of *contextual relations* between the individual genes. The operon structure and the establishment of regulatory activity at different levels exemplify the concrete realization of such contextual mechanisms for directing genetic signification. The molecular geneticist James Shapiro (1999, 26) even suggests that the pattern of repetitive DNA elements determines the systems architecture of a species: "From the systems architecture perspective, what makes each species unique is not the nature of its proteins . . . but rather a distinct 'specific' organization of the repetitive DNA elements that must be recognized by nuclear replication, segregation, and transcription functions. In other words, resetting the genome system architecture through reorganization of the repetitive DNA content is a fundamental aspect of evolutionary change." Depending on one's conceptual perspective, such a scenario might perhaps be construed as a *naturalization of language* or as a *linguification of the processes of life*.[26] No matter what one chooses to call it, however, there is no denying that it certainly constitutes an emergent jump in the development of semiotic freedom that is the history of life on Earth.

Corporeal-Kinetic Forms

Touching and being touched probably counts as the most basic dimension of human experience, as well as of all other animals.[27] "Without touch, it is impossible for an animal to be," said Aristotle. (The earliest of the great thinkers, for good reasons, formulated the most fundamental truths.) The idea that movement and touch underlie the formation of language constitutes the central thesis in Maxine Sheets-Johnstone's *sensoric-kinetic model* for the origin of language. "No language can be spoken for which the body is not prepared," she says (1990, 135) as she sets out to show how this corporeal preparation for the realization of language is decisive for an understanding of its nature and its origin. To touch and to move are two sides of the same coin, she argues, and nothing in human life makes sense in the absence of those two activities — not even language.

Sheets-Johnstone rejects the now fashionable idea in cognitive science of *embodied image schemata* (e.g., Lakoff 1987; Lakoff and Johnson 1999). This idea she sees as a kind of grammatical, cosmetic covering-up of an underlying Cartesian distinction between the self or the subject, on the one hand, and the body, on the other. When calling something *embodied* one has already presupposed some essential belonging to a de-corporealized hypothetic domain of reality that could, in principle, be somehow separated from the body. But how can anything be embodied if it *is* itself a body?

Sheets-Johnstone (1999, 151) instead prefers to talk about corporeal archetypes or *corporeal-kinetic forms* — as, for example, the fundamental corporeal-kinetic form *in*, *inside*, *to be inside* or, in general, *insideness*. The preposition *in* has been shown to be an early "locative state and locative act type of being" that pops up in the language of a child.[28] She explains,

---

26   Again, Susanne Langer (1942, 25), more than sixty years ago, saw this: "Man is doing in his elaborate way just what the mouse in his simplicity is doing, and what the unconscious or semiconscious jellyfish is performing after its own chemical fashion. The ideal of '*Nihil est in homine quod non prius in amoeba erat*' is supported by living example. The speech line between man and beast is minimized by the recognition that speech is primarily an instrument for social control, just like the cries of animals, but has acquired a representative function, allowing a much greater degree of cooperation among individuals, and the focusing of personal attention on absent objects."

27   See also the discussion of Guillan-Barrés syndrome in Chapter 2.

28   Commenting on this claim, Don Favareau (in a personal communication) suggested that "the deictic *there* and its equivalent pointing/gesturing-towards comes earlier, but the point still holds if amended with *among*."

Being in or inside something, and placing in or inside something, constitute prime and ongoing experiences in infant and early childhood life and are the basis of the corporeal concept of *insideness*. Archetypal forms and relations are thus not *preconceptual* entities, as embodied image schemata are consistently described (Johnson 1987), but are the substantive conceptual backbone of language when it appears. Neither are archetypal forms and relations vague mental means whereby we forge concepts, something on the order of Platonic conjurations having no direct cognitive reality. Archetypal corporeal-kinetic forms and relations are conceptual by their very nature.

The decisive point is that insideness is a corporeally constituted and absolutely basic experiential category in the life of the child, one that cannot be contained in or covered by concepts such as *image schemata* of *containment*, *in-out orientation*, or *boundary*.

Other examples of archetypal corporeal-kinetic forms are *thickness*, *thinness animate being*, *power*, *verticality*, and *force*.[29] All such concepts are derived from basic human experiences that precede the child's use of them in language, and Sheets-Johnstone therefore finds it unfortunate that the nonlinguistic body is often called "*pre*-linguistic" in as much as the acquisition of language in her conception is, instead, "*post*-kinetic" and should be considered as such. The fundamental concepts of language are anticipated in the experiential dynamics of corporeal movements. Therefore, language can only retrospectively conceptualize this primary dynamic.

And evolutionarily speaking, long before the emergence of the sophisticated linguistic mind, forerunners of this *linguistic mind* were already at work in the world, in the form of even more obviously corporeal forms of communication.[30] In Sheets-Johnstone's understanding, such forms of communication were — and are — *symbolic*. One obvious example is the type of animal behavior that ethologists have called *ritualization* — i.e., stereotypic behavioral patterns with communicative content that are addressed to conspecific individuals (see Chapter 6). A baboon displaying its teeth in an aggressive threat is, in a fundamental sense, doing exactly what the words say — it is *threatening* the recipient of the display. This display symbolizes what *might happen* in just the same way

---

29   Thickness and thinness are important concepts for the production of tools and derive ultimately from the experience of one's own teeth, claims Sheets-Johnstone (1999, 176): "Stone tools replaced teeth." The concept of being alive is the basis of the origin of the concept of death. *Force* is a concept that is articulated in the context of a diversity of situations that involve bodily reciprocity, etc.

30   I insert the modifier, *even more obviously*, to draw attention to the (too often under-noticed) fact that all of our human linguistic communication today — from mouth and tongue to ear and eye and brain — is fully *corporeal* through and through as well. Cartesianism still penetrates enough of our thinking about mind and language that we still often need to be reminded of such obvious facts.

that the clenched fists of people do when used in a threatening way — and by so doing, the baboon engages itself in a communicative behavior implying that an explicitly *demonstrated* dangerousness might be *realized* in a closely approaching but still future act. Here it should be noticed, Sheets-Johnstone reminds us, that the fundamental iconicity in the open-mouth-threat of the baboon does not obliterate the ritualized *distancing*: "Iconicity is not the opposite of symbolicity, but its fundament" (ibid., 160).

In general, then, corporeal communication in animals occurs precisely as a *symbolization* of the spatial and tactile dynamics that the animal itself experiences — experiences that its biologically similar conspecifics also can be counted on to share. Sheets-Johnstone in this connection refers to the concept of a *com-sign*. As suggested by the primatologist Stuart Altmann (1967, 336), these are signs that "are part of the behavioral repertoire of all members of the society." Common to these signs is the characteristic that the thing represented in the communicative act — be it of a visual, auditory, or tactile-kinetic kind — has a direct connection to the preceding corporeal experiences of the receiving animal. The invariance of bodily experiences (which is itself a consequence of the invariance of a species' body plan and its repertoire of possible activities) in this way assures the understanding of the com-sign by the entire group. These tactile-kinetic invariances predispose organisms for *iconicity*, writes Sheets-Johnstone (1999, 126), "since the most easily formulated, consistently utilizable, and readily understood signals are those that are similar to bodily behaviors and experiences shared by all the members of a species."

The Linguistic Organ

In Sheets-Johnstone's understanding, the first approaches made by humans towards the performance of linguistic skills — the human *ursprache*, as it were — must have departed from this kind of communication. And the focal point in this context is the tongue as considered as a linguistic organ, for the tongue was — and is — the center for sensoric-kinetic linguistic competence. The tongue that, according to Piaget, was not at first used for the experiencing of taste at all, but exclusively for touch — an activity that, as we know, is highly engaging for the baby — is not only the place where the infant's first explorative gestures towards the world is located, but also the center from which a qualitative world first comes alive (Piaget 1968, 51, referenced in Sheets-Johnstone 1990, 159). The whole area of the mouth, and the lips especially, are involved in these first explorative experiences, writes Sheets-Johnstone, but the extraordinary mobility of the tongue gives it a special role: "While the tongue can glide over the lips, for example, the lips cannot glide over the tongue" (ibid.).

The tongue also plays a central role in small children's experiments with sounds in the form of babbling during the first year of life. Babbling is often explained as the child's way of acquiring the skill of expressing itself in the universe of sounds characteristic to the local language. But Sheets-Johnstone emphasizes that the most significant quality of the child's babbling is that it establishes an intimate continuity in the passage from one kind of sensoric dominance to another: "A purely tactile world awakens a world of sounds, and in consequence, a new range of powers is discovered" (ibid., 160). This last point is essential, for the child must not only learn to produce the distinctive sounds of language in order for speech to be a communicative tool, but it must also consciously experience the power of itself being a sound maker. It is, as previously mentioned, one of the deep insights that phenomenology has given us that human experience and consciousness is rooted in a corporeal arsenal of "I can"s.[31] "Consciousness is in the first place not a matter of 'I think that' but of 'I can's," wrote Merleau-Ponty (2002 (1945), 137; Hoffmeyer 1996a).

A third major element in the passage into a world of expressive sounds is the integration of the visual world into the already established domain of touch and sound: "The tactile/aural tongue becomes a tactile/aural/visual organ. Its enunciations are things — or relationships — *seen*. The most distant sensory world is thus brought within the realm of *touch*" (Sheets-Johnstone 1990, 161; italics added). "[And further,] a verbally capable creature can tame the visual world — however out of reach it might be — and make it her/his own. The appropriation is not metaphorical. What is distant and not ready to hand must be corporeally appropriated in some way if it is to be part of a shared world — indeed, if it is to have a fixed place in a creature's world to begin with" (ibid., 162).

An interesting confirmation of Sheets-Johnstone's theory of corporeal-kinetic forms as the starting point for the earliest steps in language evolution comes from studies of speech perception. These show that the listening person responds as if he has interpreted the acoustic signals through the very same articulatory gestures he should himself have used to pronounce the words. The perception of human speech is generally structured in terms of the articulatory gestures that underlie the acoustic signal (Lieberman 1975). An entirely different kind of confirmation comes from studies of mother-child communication.

In traditional communication theory, communication is conceived as consisting of a series of delimited messages that flows through a *channel*. For instance, when the child hands an object to the mother, this is seen as a sequence in which first the child signals its wish to give the object by holding out the arm, and then the mother signals her willingness to receive. The American psychologist Alan Fogel, however, got the idea of analyzing this kind of interaction through the

31   See the section, "Intelligence and Semiosis" in Chapter 6.

slow-motion replaying of videotaped recordings of such naturally occurring inter-action, where he discovered that, in practice, the temporal sequence of autono-mous agency posited by the naïve model collapsed to the extent that it was indeed not possible to distinguish between who was the sender and who the receiver. The movements of mother and child are practically simultaneous and the process of delivering the object appears as a smooth reciprocally coordinated act, says Fogel (1993). There is no exchange of fully-finished, mentally constructed and then physically enacted messages going on here — rather, what we see happening is a social loop being co-constructed by the participants for the pleasure of them both. *Communication*, in the most proper sense of this word, is *happening* here.

Although Fogel's work does report directly on *speech* communication, it is nevertheless remarkable how well the observed reciprocity in the senso-kinetic relation between mother and infant supports Lieberman's observation of the reciprocity between listener and speaker in the senso-kinetic foundation of language acts. Similarly, Favareau (2008) has argued that data from the rela-tively recent fields of Interaction and Conversation Analysis (e.g., Schegloff 2007, Goodwin 2003) overwhelmingly confirm that the fine-grained, moment-to-moment choreography of human beings in everyday conversation with one another derives as much, if not more, of its meaning-making efficacy from the minute — but ever-vigilantly perceived — tracking of one another's eye gaze, head turns, breathing rhythms, body positions, and muscle movement, than could ever be possible by a detached mental interpretation of sounds and words alone.[32] This, perhaps, is what Gregory Bateson (1972, 11–13) was getting at when he wrote, "The notion that language is made of words is nonsense. . . . It's all based on the idea that 'mere' words exist — and there are none." Thus our everyday experiences in interacting with one another linguistically do also, I suppose, largely support the feeling of real communication as something like a smile that breaks through without our knowing.

The Tone in the Talk

The notion that spoken language, in its earliest instantiation, was based on recod-ings of basic (and therefore invariant) senso-kinetic schemes of experience, leads to the further hypothesis that speech in its *acoustic* origins must have also been *emo-tionally* connected to these schemes. This is a rather radical dimension of Sheets-Johnstone's theory, for the *fundamental arbitrariness* of the linguistic sign has been a cornerstone in nearly everything written about language. That the arbitrariness of the words — i.e., the realization that the word spelled *h-o-r-s-e* itself bears abso-lutely no likeness to *any* aspect of what a horse is as a real thing in the world, and

32   A biosemiotic analysis and review of this work and its data appears in Favareau (2008).

that a virtual infinity of entirely different words (such as *cheval, uma, hest, kuda, hobune, caballo*, and *Pferd*) may just as legitimately mean horse — applies to the linguistic surface of contemporary languages is not contested. But in the history of language origins, this arbitrariness must itself have emerged (or in some way have been segregated from) the nonarbitrary use of sound expressions. Sheets-Johnstone claims that the alternative to this view would be some kind of hopeful monster hypothesis — a "Gee, I ought to use arbitrary sounds and start talking!" hypothesis that, from an evolutionary point of view, is not only quite unsubstantiated, but virtually incomprehensible (ibid., 140).

It is a telling fact, however, that the babbling that is unique to babies in the human species does not at all occur in young of other primates. And it is tempting to infer that this playing with sounds could have been a point of departure for the discovery of the potential benefits to be obtained from the free (i.e., unmotivated) use of sound expressions. For the vocalizations of babbling are not just empty forms — as is evident from their mutual variability concerning qualitative properties such as pitch, timbre, intensity, rhythm, and duration. Rather, sounds may be subdued, whispered, moaned, or mumbled; they can be stressed or unstressed, full, thin, painful, explosive, soft, brief, etc. At some point in time, our remote hominid ancestors must have discovered that one might communicate with the help of such *more or less* arbitrary — but still emotionally loaded — sound expressions.

At first, the relationship to the nonarbitrary iconic-indexical sounds that all primates use in communication may have had an infectious influence on the significative possibilities connected to arbitrary sounds. But in the end, these hominids must have realized, as Sheets-Johnstone puts it, "that a 'sheer sound' could stand for something in the same way that an *already* significatory sound stood for something" (ibid., 143). And this discovery must, by necessity, have taken place within a sound-meaning paradigm that had no alternative models to build upon, other than the *non*arbitrary universe of sound-meaning associations that preceded it in evolutionary history.

Against the background of these observations, it may appear less surprising that the American linguist and anthropologist Mare LeCron Foster has hypothesized that the *symbolic structure* of the human *ursprache* was firmly attached to what Foster calls elementary *articulatory gestures* — *movements* such as clicking the lips to produce the sound *mm* — and that such *phenemes* (corresponding to Sheets-Johnstone's *senso-kinetic forms*) will have in themselves a basic *significative* content, analogously connected to those movements of the mouth, tongue, and vocal tract that call forth the particular pheneme.[33]

---

33   For example, Foster (1978) claims that all reconstructed root forms of the sound *mm* refer to *bilateral relationships* (e.g., the fingers and hands in taking or grasping; two opposed surfaces in

Syntax as Movement

The word *syntax* comes from Greek *syn* + *tassein,* which in contemporary English means "to put in order." Implicit in this word lies the idea of a temporal sequence — one where the thing to be ordered logically precedes the ordering of it. It is, however, highly unlikely that individual words, in this way, *preceded* the syntax that determines their sequence in linguistic utterances. The separation of a syntactic level from a semantic level may be practical for analytical purposes, but as far as their origin and contemporary use is concerned, the syntactic and semantic levels are internally connected. Seen from Sheets-Johnstone's theoretical position, syntactic order must *from the beginning* have been wrapped into corporeal movement: "There is an order in which things occur which is contingent upon movement" (ibid., 168). Grammar (the sequences by which we put our meaning-making tokens in order) — as much as semantics (the meanings that we are making with these tokens) — is inextricably rooted in the body's corporeal-kinetic forms.

The Problem of Freedom

There are, of course, categorical dividing lines between humans and animals other than the use of language. But from the moment that language grew its hyphae into the vaults and architecture of the human brain, it became impossible for us to somehow "think it away." For what kind of creature would a *non*linguistic (or *non*symbolic) human being amount to? The question is absurd — not just because language may be seen as at the very source of *human* origins, but also because it is impossible to abstractively look "behind" language competence in an analysis of human nature.

Morality, self-reflective consciousness, and *free will* are some essential traits of humanness that have been used to argue for the absolute unique position of humans in the natural world. And, in fact, it is this central supposition of humans as free and animals as unfree that lies at the root of the discourse on human rights that arose in the wake of Enlightenment philosophy. Let us here quote the famous passage from Rousseau (1910 (1755)).

> I can discover nothing in any mere animal but an ingenious machine to which
> nature has given senses to wind itself up, and guard, to a certain degree, against
> everything that might destroy or disorder it. I perceive the very same things in the
> human machine — with this difference: that *nature alone* operates in all the oper-
> ations of the beast, whereas man, as a *free agent*, has a share in his. One chooses by

---

tapering, pressing together, holding together, and crushing; chewing food) that are analogous to the act of bringing the lips together in forming the sound *mm* (cit. in Sheets-Johnstone 1999, 165).

instinct; the other by an act of liberty; for which reason, the beast cannot deviate
from the rules that have been prescribed to it, even in cases where such deviation
might be useful, and man often deviates from the rules laid down for him to his
prejudice. Thus a pigeon would starve near a dish of the best flesh-meat, and a cat
on a heap of fruit or corn, though both might very well support life with the food
which they disdain, did they but bethink themselves to make a trial of it: it is in
this manner dissolute men run into excesses, which bring on fevers and death itself;
because the mind depraves the senses, and when nature ceases to speak, the will still
continues to dictate (see also Ferry 1992, 39).[34]

The final sentence in this quote is telling. We humans are *so* free that we may
die from it. For Rousseau, the possibility of evil is inherent in the whole phenomenon of human freedom. As the proverb has it: Man sees what is good but
chooses what is evil.[35] And, paradoxically, this is exactly what justifies the unique
concept of human rights in the first place. For, seen historically, human rights
appeared as a defense against the inhumanity of the somewhat "more than mere
human" rulers (e.g., the English colonial power in the case of the United States,
the monarchy and nobility in the case of France). Human rights were meant to
function as the basis for a constitutional state — one that would be capable of
protecting human individuals against the evils of other individuals, not the least
of whom included the potentially evil-doing rulers.

Biosemiotics rejects Rousseau's narrow conception of the animal as an *inge-
nious machine*. But by so doing, does biosemiotics bring itself dangerously close
to a denial of the value set manifested in Enlightenment thinking and realized
in the most fundamental social institutions of the Western world? Other critical positions in the discourse regarding the interface of nature and culture in
Western societies have indeed embraced the anti-Enlightenment position that
human rights expresses an inadmissible *species-ism* — i.e., an oppression by the

34   "Je ne vois dans tout animal qu'une machine ingénieuse, à qui la nature a donné des sens pour
se remonter elle-même, et pour se garantir, jusqu'à un certain point, de tout ce qui tend à la détru-
ire, ou à la déranger. J'aperçois précisément les mêmes choses dans la machine humaine, avec cette
différence que la nature seule fait tout dans les opérations de la bête, au lieu que l'homme concourt
aux siennes, en qualité d'agent libre. L'un choisit ou rejette par instinct, et l'autre par un acte de lib-
erté; ce qui fait que la bête ne peut s'écarter de la règle qui lui est prescrite, même quand il lui serait
avantageux de le faire, et que l'homme s'en écarte souvent à son préjudice. C'est ainsi qu'un pigeon
mourrait de faim près d'un bassin rempli des meilleures viandes, et un chat sur des tas de fruits, ou
de grain, quoique l'un et l'autre pût très bien se nourrir de l'aliment qu'il dédaigne, s'il s'était avisé
d'en essayer. C'est ainsi que les hommes dissolus se livrent à des excès, qui leur causent la fièvre et
la mort; parce que l'esprit déprave les sens, et que la volonté parle encore, quand la nature se tait»
(Rousseau 1910 (1755)).

35   The Danish philosopher Villy Sørensen once said, "A human person can be inhuman, but a
wolf cannot be unwolfish." This elegant formulation should not be romanticized, however, for the
internal relations between animals is no idyll! But the human talent for what seems to be *gratu-
itous* evil is, I think, somewhat striking.

human species of all other species on Earth (Regan 1983). It is therefore important to state in exactly which ways biosemiotics differs from such ideas.

Biosemiotics, admittedly, does not accept any essentialist, absolute, or a priori distinction between the human being and the rest of the creatures of this world. But it also concurs with the previously mentioned characterization of human beings by John Deely as those singular animals that — at least potentially — may distinguish their own *objects of consciousness* from the world of *things as they are in themselves independent of such consciousness*. Biosemiotics is therefore prevented from endorsing the reductive position that sees the human being as nothing but one more animal absolutely indistinguishable from all the other animals in any important sense. For by being linguistic creatures — and, perhaps what is even more primary, *by setting up symbol-based communal cultures* — humans come to acquire a semiotic freedom that is unparalleled anywhere else in the animal kingdom.

What Sheets-Johnstone's and Deacon's analyses show (and especially so when they are seen in combination) is that human freedom does not divert itself from that of the animals by being merely semiotic. All creatures are semiotic creatures in some sense, as I have argued extensively throughout this book. Rather, the distinctive property is not *that* we are semiotic creatures but *the way* that we are so — i.e., in the possession by the human species, held jointly by culture and by biology, of a linguistic resource for thinking and communicating that reaches far beyond the semiotic possibilities available to even the most intelligent of other animals.

Human rights, therefore, cannot be justified through the positing of an absolute distinction between human and animal — but by the language- and history-creating nature of human beings alone. The discourse of human rights — and by extension, of the constitutional state — is therefore, in a deep sense, the consequence of a dawning self-recognition of the human being *as* a historical being.[36]

Neither, then, does biosemiotics see the division of knowledge production into *scientific* and *humanistic* domains as one based on either essential or absolute distinctions — such as is commonly done when the first regime is defined as concerned with those parts of reality that are governed by unbreakable *laws*, while the second regime is confined to some small pockets in the universe where *freedom* has attained (or preserved?) a certain latitude vis-à-vis these natural laws.

36 Human rights, then, amount to just this: the articulated realization of a *historical experience* that human civilizations have obtained. Attempts to legitimize human rights by appeals to the (supposed) constitution of human nature — or to some kind of supposed ethical superiority — are not only misdirected, but bound to fail.

The absolutization of this distinction, although rarely explicitly declared, seems implicitly present in most contemporary scientific work — where it often permeates, and is hidden in, the value systems and exemplars (in the Kuhnian sense) that ultimately *legitimize* scientific thinking as it unfolds in peer review evaluations and editorial decisions. Such an absolute dichotomizing of the world of inquiry and knowledge necessarily leads our thoughts into reductionism since, logically speaking, freedom cannot possibly evolve in an un-free world.[37] The result of this is the well-known tendency of science to consume increasingly larger chunks of the explanatory and research funding possibilities, at the expense of the human and social sciences — in the process, reducing these areas of human life to nothing but lawful causality.

But again, this division into two fundamental knowledge categories is — like human rights — a product of a historical experience. For centuries, it was an exceedingly fruitful strategy to consider nature as a machine-like, law-bound system — since this permitted the discovery of an apparently infinite series of new deterministic relationships that could, in turn, support the more and more powerful technological control of natural resources for use in human societies.[38] However, the elevation of this historically useful *working hypothesis* to the status of *ultimate ontological truth* has allowed an unwarranted arrogance to take hold in the scientific society that has dangerously blinded our critical facilities to the principle shortcomings — and, indeed, the often obvious inconsistencies — of this hypothesis.[39]

The basic epistemological problem of how to legitimize an internal relation between scientific knowledge — which like all kinds of knowledge is necessarily wrapped in language, even when it appears as couched in a mathematical formalism — and the fundamental world of things that science is supposed to be concerned with, has, in fact, never been satisfactorily solved. Instead, the problem has been set aside — with reference partly, I suppose, to Kantian-type nominalist presuppositions, and partly (maybe especially) to the undeniable fact that scientific knowledge in most cases actually *does work* in our real lives, and is thus

37   A natural law that stipulated a dynamics leaving some consequences *undetermined* would hardly qualify as a true law — and even if it did, such a "law" would still have to be obeyed by (what science often sees as) the passive pawns that inhabit physical reality. So, there is no freedom in this worldview after all.

38   That the order of causality was often inverted, so that technological innovations would lead to the construction of new scientific lawfulness rather than vice versa, doesn't detract from the functionality of the essentially dualistic logic at work here.

39   Such arrogance and hubris is often mirrored in the beliefs of those politicians who so admire the technological strength and deterministic "single answers" of the scientific approach.

in no need of further legitimization.[40] And in fact, I shall be the first to admit that empirical confirmation — not the least of the incredible achievements that science has afforded us — is a strong reason for believing a theory. But it is nevertheless strangely unsatisfactory that, *within this very same framework*, we *cannot* explain how something as elemental and critical to the entire success of the project as *linguistically expressed knowledge about an otherwise mind-independent world* is possible at all.

In Peirce, the possibility of *knowing* is, in the end, justified by the adoption of an evolutionary and realistic philosophy:

> In itself, no curve is "simpler" than another. A system of straight lines has intersections precisely corresponding to those of a system of like parabolas similarly placed, or to those of any one of an infinity of systems of curves. But the straight line *appears to us* [as being] "simple," because, as Euclid says, it lies evenly between its extremities; that is, because viewed endwise it appears as a point. That is, again, because light moves in straight lines. Now, "light moves in straight lines" *because* of the *part* which the straight line plays in the laws of *dynamics*. Thus it is that, our minds having been formed under the influence of phenomena governed by the laws of *mechanics*, certain conceptions entering into those laws become implanted in our minds, so that we readily guess at what the laws are. Without such a natural prompting, having to search blindfold for "a law which would suit the phenomena," our chance of finding it would be as one to infinity. [From this it follows that] the further physical studies depart from phenomena which have directly influenced the growth of the mind, the less we can expect to find the "laws" which govern them [as being] "simple" — that is, [as being] composed of a few conceptions natural to our minds (*CP* 1:10; italics added).

An *evolutionary realism* of this kind presupposes an anchoring of human cognition within a frame of evolutionary theory. And such a frame of understanding is exactly what can be constructed on the basis of a biosemiotic approach to nature as a whole. Language then, as we have seen in this chapter, becomes an *extension* of the principles discoverable in a general biosemiosis — and this is the reason why *veridical linguistic knowledge* is *possible*. The perception by organisms of their surroundings is, from the beginnings of life, already embedded in their survival project, their corporeal intentionality. Thus the fact that language

40   One is, of course, often labeled *anti-science* or *a crank* if one points to things such as environmental pollution, medical side effects, the rise of allergies, desertification, forest death, stress, psychosomatic disease, and the general unease of living in modern society as evidence of the kinds of *nonknowledge* (or *anti-knowledge*) that *also* seem inherent in this determinedly dualistic approach to science and technology. Biosemiotics is attempting to overcome science's preemptive dismissal of its critics by engaging with arguments and evidence that are drawn from, and informed by, the scientific literature itself. For what biosemiotics is working towards is not *anti-science* but *better science* (in the original sense of *science* as a knowledge-bestowing explanation of the natural world as it really is).

permits humans access to an entirely unusual *Lebenswelt,* is in no way opposed to the basic role of language (and semiosis more generally) as instrumental for survival. As such, language is internally adjusted to the necessities that any possible human acts must obey, because such acts take place in a world that is itself *independent* of language use. Language opens the door for humans to invent all kinds of nonexistent and even impossible worlds, but it also endows us with the ability to *distinguish between* all these nonexistent worlds, on the one hand, and that one world, on the other hand, that we — like all the other animals — do not long get away with consistently misjudging.

As such, biosemiotics then offers a new fundament for an evolutionary epistemology. It does not derive an evolutionary epistemology from the neo-Darwinian belief that the cognitive worlds of organisms can in any simple way be derived from *survival-mechanical* or *game-theoretical* scenarios operating on populations of *asemiotic* entities and individuals. Rather, by recognizing *semiotic capacity* as an *inherent* property of *all* living systems, the task set for biosemiotics in explaining the origin and nature of knowing in the biological world becomes much less Sisyphusian. The human form of cognition is not, biosemiotically speaking, solitary in the world (*sensu* Monod) but is an admirable new and sophisticated *elaboration* of cognitive forms that, by the time hominids finally appeared, had been at work in nature for hundreds of millions of years (and, of course, still are).

With this realization, the idea of human cognition as constitutive of an ontologically separate mental sphere, different in kind from all other ordinary natural phenomena, loses its raison d'etre. A biosemiotic evolutionary epistemology deals in *continuity* and therefore encounters no puzzle of discontinuity that must be solved by recourse to genetically improbable hopeful monsters, radically disjunctive evolutionary "just-so" stories, and the like. The explanatory victory, rather, has in the end been obtained simply by accepting that living systems from the very outset have been engaging in genuinely semiotic activities — and *must* have been so engaged if these systems were to successfully negotiate the world, adapt to it, and survive. Thus, a relatively modest move in the basic ontology of cognitive science — i.e., admitting the reality of sign processes in even the simplest of living systems, an observation that is, in addition, well supported by a century of empirical evidence — thus seems to help us solve one of the hitherto most intractable problems in human evolution, as well as in natural philosophy.

Biosemiotic epistemology, then, at last takes seriously the insight that Charles Darwin expressed in *The Descent of Man* in 1871: "The fact that the lower animals are excited by the same emotions as ourselves is so well established, that it will not be necessary to weary the reader with many details. Terror acts in the same manner on them as on us, causing the muscles to tremble, the heart to palpilate, the sphincters to be relaxed, and the hair to stand on end"(Darwin 1981(1781), PAGE?). It is time that contemporary Darwinists start taking their own hero just as seriously as we biosemioticians do as far as the nonanthropomorphic existence of cognition and feelings in animals are concerned. When they do, they discover that Darwin essentially took the biosemiotic conception to be obvious — although he did not, in the nature of the case, know the term, much less possess the means to make this idea useful in his own theoretical work. We today, however, have no such excuse.

9
Perspectives

Experimental Biology

When experimental biologists are confronted with the biosemiotic perspective, they often ask what advantages might come from such an approach that, on the face of it, appears to only make everything more complicated — and therefore, more troublesome. This kind of response is often nourished by a widespread instrumentalist attitude to scientific work, according to which theoretical positions must be judged on their ability to fruitfully guide experimental work.

The concern is a valid one, for experimental work consists in creating intelligent simplifications of the systems under study, and dyadic relations are for obvious reasons more easily adapted for this purpose than triadic. Thus, although the experimentalist again and again will be challenged to explain the biochemical signals he or she finds so clearly implicated in all life processes, trying to investigate the cause and effect relations of such signaling is problematic in that such relations are dependent, ultimately, on circumstances that cannot easily be controlled in the experimental setup. Therefore, mechanistic dyadic models with many variables are still preferred in contemporary science over triadic models that can not easily be mathematically formalized, and that carry ontological undertones that are experienced as alien. In fact, the adoption of a genuine (i.e., irreducibly) triadic model in place of a matrix of dyadic relations will perhaps even be conceived as cutting corners, or as relinquishing the possibility that the standard dyadic approach will eventually result in an understanding that goes to the very bottom of the matter.

Regarded as a working hypothesis, this position must be respected. For when it comes down to it, the whole argument of this book is based on millions of reliable biochemical and biological experimental studies aimed at discovering the dyadic relationships between phenomena. It is only due to the many insights laboriously procured through this (standard dyadic) methodology that enough genuine knowledge is now in place to suggest the hypothesis that such mechanistic and dyadic logic itself, in a more profound analysis, should be seen as subsumed under a more biologically primitive, interactive (triadic) logic. Experimental biology thus, far from being a stumbling block to the biosemiotic approach, is its necessary precondition. In the absence of the efforts of thousands of experimentalists, whose work was — and is — driven by instrumentalist values, the biosemiotic approach would not have been capable of qualifying itself. This said however, it should also be realized that the instrumentalist position is itself seriously insufficient if the aim of the investigation is to understand nature — not simply to continue the line of successful experimental work. It may well be — to repeat a cliché — that the lost key is easiest to look for where the streetlamp is shining, but if the key to nature's principles for creating living systems that can *sense* the world is hiding somewhere else, then this instrumentalist approach will never lead to the desired understanding.

That many experimental biologists do not find it worth the trouble to examine this problem may seem legitimized by the sheer success of modern biochemistry. Confronted with the capacity of biochemistry to unite formerly separate phenomena into one coherent picture by proving their common basis at the biochemical bottom line (not to mention its function as the paradigmatic fundament for a technological command of medicine and manufacture without parallel in history), it is understandable that the practitioners of the discipline do not themselves see any pressing need for questioning the theoretical fundament of their scientific practice.

In the long run, however, they too, may discover that it is not healthy to rest on one's past successes. And in this particular case, one may even add that is has been precisely the extreme success of experimental biology that has now brought us to a stage in scientific development where biochemistry should begin to address, rather than continue to avoid, the recurrent problems inherent — in spite of its success — in the chemical paradigm. For, although the strategy of biochemical reduction may presumably still have many triumphs to celebrate, it has nevertheless been taken so far now that, little by little, it has begun hitting its own natural limits.

That this may be the case was surprisingly revealed in the episode of what surely must be considered biochemistry's most striking accomplishment to date — the mapping of the human genome. For the results of this gargantuan project made it abundantly clear that genes do not constitute that unambiguous key to an understanding of human biology (much less human nature) that many scientists had expected — and that, indeed, several outstanding celebrities in the scientific world had predicted. For it turns out that there are simply far too few unique human genes to support the naïve belief that genes can, in any straightforward or simplistic way, be the singular *causes* of specific human traits such as personality characteristics or mental diseases. As we have seen already — and as the studies of the neural and genetic grounding of behavior in the worm *C. elegans* make clear — the conception of *genes for* specific traits is and always was a gross misunderstanding. Even many geneticists of the old school may now admit that a biosemiotic understanding of the genome as a sophisticated inventory-control system that minutely furnishes components (enzymes) for the semiotic scaffolding of ontogenesis on an as-called-for basis offers a more realistic model of the relations between genes and traits.

Exactly the uncovering of this previously assumed bottom line for the organization of life made it clear that evolution has worked an interactive, triadic logic into the organization of life processes — one that subsumes the otherwise organismic chaos of biochemical activities under its controlling agency. Dyadic modeling (and dyadic thinking) simply does not suffice as an *explanatory* strategy here. Thus, even though biochemistry as an experimental discipline may still, and for a long time, profit from basing its lab-isolated piecework on a dyadic simplification of naturally occurring biological chemistry, the discipline ought to recognize this approach for what it is — a very practical but necessarily artificial and incomplete working method, not a satisfactory theory or fully explanatory model of the actual organization of life processes as they occur in nature.

Biochemistry needs to be put into its rightful place among the natural sciences as the tool for analyzing the bottom-line processes of life, which is what this science is about. And the more we learn about the biosemiotic logic that organizes the processes of life, the more we must expect that a triadic and semiotic understanding will replace the mechanistic and dyadic models of biochemistry. Indeed, it is already now difficult to open a scientific journal (let alone a textbook) in the area of biochemistry without encountering, on every page, semiotic terms such as *signal, messenger, communicates,* and *sign* — yet these very terms hardly make any sense inside of the reductionist framework adhered to by most biochemists (see Yates and Kugler 1984; Yates 1985; Hoffmeyer 1997a). Claus Emmeche's apposite expression for this odd kind of doublethink is *spontaneous semiotics* (Emmeche 1999a) — and it can presumably coexist effortlessly within the hegemonic paradigm of biochemistry because of a tacit consensus that one is merely "speaking in shorthand" and could, in the end, avoid the semiotic implications altogether, if only one expressed oneself carefully enough (Emmeche 2002). The increasingly widespread *need* to express one's scientific findings in semiotic terminology does, however, indicate that real and important insights are indeed hidden in this unavoidable semiotization of the terminology — and that our overall understanding of the organization of living systems would suffer a decisive loss if terminology more appropriate to the current dyadic paradigm were imposed upon authors by the editors of scientific journals.

In fact, after reading the current biochemistry literature, one might claim — just a little ironically perhaps — that biochemistry has already become biosemiotics, it just has not realized it yet. And this is why it is currently incapable of drawing the necessary theoretical implications of its own research data. In the long run, however, this selective eclecticism probably cannot be sustained.

Man's Place in Nature

The instrumentalist objection to biosemiotics may reflect a deeper source of conflicted thinking in the general conception of science of the Western world. Perhaps because many people see it as the task of religion, and not of science, to understand the world as a coherent whole, the conception has gained footing that science is to be defined by its methodology, rather than by its achievements, as a tool for *understanding* (and not just for successfully manipulating) our world. This limitation in the scope of science is probably reinforced by the Anglophone conception of *science* as comprising mainly the *natural sciences*. The Scandinavian term for science, *naturvidenskab* (like the German equivalent *Naturwissenshaft*) implies an activity which concerns both nature and knowledge — or, in fact, knowledge about nature. The term thus is more comprehensive than the English *science*, and it unambiguously indicates the aim of science as coming to understand the workings of nature. The conception of *science* as limited to an approach based on a certain (if rarely well-defined) *scientific methodology* has however now become widespread even in continental Europe and Scandinavia.

But, according to this latter, more restrictive conception of science, a project like that of quantum mechanics already seems seriously close to being asked to leave good company. Quantum mechanics, however, was perhaps freed from this dangerous lure by complementarity theory. By epistemizing the inherent holism of the subject-object relationship, complementarity theory protected physics against the troublesome subjectivization of nature that would have brought it into conflict with the *ideal* of science. The same maneuver was once again put to use by John von Neumann and Howard Pattee when they — as we saw in Chapter 4 — claimed complementarity between the symbolic and the dynamic functional modes in living systems.

By positioning the sign process (*semiosis*) as an irreducible ontological category in our universe, biosemiotics (standing on the shoulders of Peircean cosmology) has definitively put itself beyond this methodologically justified limitation of the scientific understanding of our world. And while it takes an agnostic attitude to the question of religion, biosemiotics does aim toward an understanding of the world of nature as a place that has produced human beings — and that therefore may hold *subjects*. Biosemiotics has then, from its very point of departure, refused to accept the absolutization of the Cartesian subject-object distinction as an ontological reality in the sense of subject-object dualism.[1]

Breaking the hold of the Cartesian mischief somewhat, Charles Darwin indirectly established a cosmology that allowed humans to belong in the world

---

1   I am speaking here in the normally understood sense of this *dualism* (and thus not in the sense given to it by John Deely; see Chapter 8).

without having to invoke any supernatural causes or beings. Since Darwin, as we saw in the preceding chapter, had no trouble accepting the idea that animals possess feelings — and that therefore in a very elementary sense they are *subjects* — his conception of the human situation, vis-à-vis nature, was not very far from the biosemiotic one. However, with the fusion of Darwin's theory with molecular genetics, a burden was placed upon the shoulders of *natural selection* that slowly annihilated the original vision of Darwin. With that, even thoughts and feelings had to be understood as genetically-based components in the behavior of an animal — a kind of behavioral spasm or, at least, involuntary reflex, like the movements of a compass needle. And with this new perspective, gone was the possibility of visualizing animals as *subjects* in the same sense that we see ourselves.

So, while Darwin himself placed the human being safely and understandably within the masterpiece of nature, the neo-Darwinist (not so safely) threw us out again — telling us that our *genes* may well belong to the reality of nature, but our *thoughts* and *feelings* are well outside of the reach of science. Consequently, those previously least deniable aspects of our own biological experience — because they could not be reduced to a molecular explanation — came to be viewed with increasing suspicion, as if they were *Fata Morganas*, or epiphenomena without proper autonomous ontological reality.

Biosemiotics, by denying this construction of living creatures as gene-determined robots — or *survival machines*, in the hard-hitting terminology of Dawkins (1976) — reestablishes the conception of living beings as *subjects* whose activities are governed by internal models of their surroundings.[2] True, we currently have little verifiable knowledge about what precisely these *models of the surrounding worlds* actually consist in — we need a model of how it is that animals can model, in other words — but we must at least recognize the reality of the phenomenon. And in so doing, biosemiotics explicitly rejects the conception of nature that the French molecular biologist, Jacques Monod, in his influential book, *Le Hazard et la Nécessité*, characterized with the following words: "Il faut bien que l'Homme enfin se réveille de son rêve millénaire pour decouvrir sa totale solitude, son étrangeté radicale. Il sait maintenant que, comme un Tsigane, il est marge de l'Univers où il doit vivre. Univers sourd à sa musique, indifférent à ses espoirs comme à ses souffrances ou à ses crimes" (Monod 1970, 187–88).[3]

---

2   See Sebeok (1988, 72): "The recalcitrant term *Umwelt* had best be rendered in English by the word *model*."

3   English translation: "If he accepts this message — accepts all it contains — then man must at last wake out of his millenary dream; and in doing so, wake to his total solitude, his fundamental isolation. Now he at last realizes that, like a gypsy, he lives on the boundary of an alien world. A

Biosemiotics, of course, agrees with Monod that the universe does not care for human hopes or traumas and is deaf to our music. But the simple fact that there *is* life in the universe already indicates that human solitude is less severe than claimed by Monod's heroic philosophy of objectivity. For with the creation of living systems, the universe has disclosed an inherent propensity that — although it surely does not include the possession of anything remotely resembling human emotional life of the kind that Monod mocks — nevertheless seems much more related to that dimension of existence we call mental, spiritual, or (with a better word perhaps) *minded*, than it is to the more mechanical aspects of existence that the universe otherwise seems to specialize in.

Peirce, as we saw, claimed that the *the law of mind* amounted to the tendency of nature to form habits. By viewing this fundamental *semiosis* — for a habit is the general form of an *interpretant* — as the innermost property of the universe, he worked out a cosmology in which the individual life and the individual consciousness appeared as evolutionary concretizations of a more general propensity of the universe. Restricting myself only to the examination of *living* (bio) *systems*, in this book, I have characterized this propensity among organisms as an evolutionary growth in semiotic freedom.

Thus, the realization of Peircean cosmology in the form of modern biosemiotics becomes a lever for the repositioning of the human being as a part of nature. Accordingly, Monod's idea of the human being as "a gypsy at the edge of time" (ibid., 188) may finally be dismantled — to be replaced by a conception of human beings as embedded in the general biosemiosis of living nature. Human mind is not, then, an alien element in the universe — but rather, an instantiation of evolutionary trends that penetrate the life sphere and that (I suspect) is deeply rooted in the general dynamics of the universe.

For the rest of this chapter, I wish to consider some of the consequences that may be drawn from this grand shift in the conception of *Man's place in nature* in a range of disciplinary contexts where, until now, the rationale and impetus has been based on the philosophy of *objectivity* so eloquently expressed by Monod. But before leaving the discussion of the way human beings belong to nature, I want to briefly draw attention to the German-American philosopher Hans Jonas who, in many respects, has developed a view of human belonging in nature that is related to that of biosemiotics (Jonas 1966 (2001)); Weber 2002; Frølund 2002).

Jonas, like Peirce, arrives at the conclusion that causality and teleology are not opposed to each other. On the contrary, teleology is needed for causality to make any sense. And *telos* is something we know about before we know

---

world that is deaf to his music, just as indifferent to his hopes as it is to his suffering or his crimes" (Monod 1971b, 172–73).

about anything else — for we know *telos* from our own bodily life. (Yet perhaps even my saying that we know this *from* our bodies reflects my own lingering Cartesianism — far truer to say that we know telos because we *are* our bodies or our bodies *are* us.) If my body is hit by something solid, I get hurt — and my attempts to avoid such situations nicely bring out the interrelation of materially efficient cause and *telos*. In much the same vein, Jonas's philosophy attempts to rehabilitate an anthropomorphic understanding of nature:

> Perhaps, rightly understood, man *is* after all the measure of all things — not indeed through the legislation of his reason, but through the exemplar of his psychophysical totality, which represents the maximum of concrete ontological completeness known to us: a completeness *from which*, reductively, the species of being may have to be determined by way of progressive ontological "subtraction" down to the minimum of bare elementary matter (instead of the complete being constructed from the basis by cumulative addition) (Jonas 1966, 23–24).

Importantly, Jonas is careful to underline that this anthropomorphic conception of nature in no way implies *vitalism*. Driesch's big error, according to Jonas, was to understand *final causality* as a *kind* of *efficient causality* that, mysteriously, worked backwards in time. This formulation, of course, got Driesch into big trouble, because such inversed causes could not be made accessible to empirical tests — as must be demanded if efficient causes are postulated. Driesch's misconceived vitalism, therefore, only helped to confirm prevailing prejudices against all and any kind of finalistic thinking. But final causes should not be confused with efficient causes since, as Jonas observes, the relation between these two kinds of causes is not one of identity (x *equals* y), but one where final causes make use of efficient causes as a means for executing their finality (x *via* y) (Frølund 2002). This, as we saw, was also the position of Peirce.[4]

How close Jonas gets to a biosemiotic understanding is perhaps especially evident in his critique of the *cybernetic model* of life that is based on a dyadic structure connecting perception and movement. Against this model, Jonas (1973) claims that the processes of life "are composed of the triad of perception, mobility and feeling. Feeling, more basic than the two other capacities, and rather linking them, is the animal translation of the basic tendency at work already from the undifferentiated, pre-animal stage on, in the continuous realization of metabolism" (cited from Weber 2002, 188). Thus, Weber notes that feeling "is the *interpretant* necessary to make up a biosemiotic entity. . . . Feeling that rises from the intrinsic teleology of organism which Jonas is calling here 'basic

---

4   This may be noted in Santaella-Braga's paraphrasing of Peirce in Chapter 3: "Final causation without efficient causation is helpless, but efficient without final is worse than helpless, 'by far, it is mere chaos; and chaos is not even so much as chaos, without final causation; it is blank nothing'" (*CP* 1:200; Santaella-Braga 1999, 502).

tendency' is the *tertium comparationis* that links the *causa*, perception, with its *effect,* mobility" (ibid.).

That Jonas in the above citation uses the word *metabolism* should be seen against the background of his conception of metabolism as a defining quality of life. Jonas sees metabolism as the formative determination whereby *substance* — in the sense of the chance accumulation of molecules — is ordered in the service of life's capacity for self-integration. Seen from the point of view of biochemistry, this identification of *the formative aspect* with *metabolism* may seem a little confusing, for metabolism is nothing but a transfer of energy from food (or sunlight) to a biologically useful form — usually ATP — and the utilization of this energy to run the ordinary business of the body (e.g., supporting the incessant maintenance work required by tissues and organs). Metabolism thus may well be seen as the precondition for self-organizing life processes — but only in the same way that a power station is a precondition for the continued reproduction and growth of societal organization. The point is that in both cases, what is being examined is brute energy production, without any *necessary* specification of purpose or direction. Understood biochemically, metabolism is connected to the dyadic — not the triadic — aspects of life. But this somewhat pedantic terminological objection should not detract from the essential core of a deep relatedness between Jonas's philosophy and a biosemiotic conception of *human-hood* as a legitimately understood *part* of nature.

Aesthetics and Ethics

The biosemiotic conception of the psychological life of humans as one modeled upon mental themes that, in at least some very primitive proto-form, are present already in animals, has implications both for our understanding of human aesthetic needs and for the understanding of human ethical dilemmas. Indeed — and perhaps most importantly of all — biosemiotics leads us to consider these two aspects of human life as inherently bound to one another.

The oddly under-esteemed work of Susanne Langer, on the bodily and semiotic roots of human symbolism and mind, offers an unavoidable source for insights concerning the biosemiotic analysis of these questions — a "treasure house of precise, provocative, and empirical exemplifications," as Robert Innis (2007) has put it. Langer (1942, 32–33) writes,

> I believe there is a primary need in man, which other creatures probably do not have, and which actuates all his apparently unzoölogical aims, his wistful fancies, his consciousness of value, his utterly impractical enthusiasms, and his awareness of a "Beyond" filled with holiness . . . *this basic need, which certainly is obvious only in man, is the need of symbolization.* The symbol-making function is one of man's

primary activities, like eating, looking, or moving about. . . . It is not the essential act of thought that is symbolization, but an act *essential to thought*, and prior to it. Symbolization is the essential act of mind; and mind takes in more than what is commonly called thought. Only certain products of the symbol-making brain can be used according to the canons of discursive reasoning. In every mind there is an enormous store of other symbolic material, which is put to different uses or perhaps even to no use at all — a mere result of spontaneous brain activity, a reserve fund of conceptions, a surplus mental wealth.

Langer, like Peirce and Dewey — and, as we saw, Deacon — is aware that we share with other life forms "the dual functioning, and yoking together, of index-icality and iconicity," and that "these two 'lower' strata of semiosis prepare the way for 'symbolization' proper, which makes the 'great shift' that Langer wants to trace from animal mentality to human mind" (Innis 2007). Comparing Langer's conception of art with that of John Dewey, Innis (2007) quotes Langer (1953, 397) from *Feeling and Form*: "In art, it is the impact of the whole, the immediate revelation of vital import, that acts as the psychological *lure to long contemplation*" (italics added) and goes on to observe that "Langer and Dewey both share a noninstrumental notion of the primary value of art: long contemplation culminates in a consummatory experience wherein the experience is savored for its own sake or, for Langer, for the sake of 'insight,' since an art work gives us not just heightened experience, but knowledge, although not in concepts."

Art is presumably the most uniquely human way to express one's aesthetic being — and for that reason, it has traditionally been the point of departure for most attempts at defining human aesthetic psychology. Seen from a biosemiotic perspective, however, the essential question that must be asked is if a key to an understanding of our human aesthetic being might eventually be discovered through an analysis of the daily biosemiotic challenges that our big-brained ancestors of millions of years ago had to cope with.

Susanne Langer's conception of aesthetics as based on a "general expressivity of life" seems important in this context: "The same feeling may be an ingredient in sorrow and in the joys of love. A work of art expressing such an ambiguously associated effect will be called 'cheerful' by one interpreter and 'wistful' or even 'sad' by another. But what it conveys is really just one nameless passage of *'felt life'* knowable through its incarnation in the art symbol — even if the beholder has never felt it in his own flesh" (Langer 1953, 374). She, in this way, sees art as springing from what might be called both the expressed *objectivization of feeling* and the simultaneous *subjectivization of nature*. Here, on the one side, art is a kind of inter-subjectivity guaranteed by the universal appeal of the aesthetic form as *living* form — and on the other side, such a guarantee is only valid in the first place because of the universality of subjectivity for any currently living creature.

That the aesthetic needs of humans are basically of flesh and blood implies that they must have their roots in the conditions of human life in general. These conditions are, among other things, characterized by a range of biological universals, such as breathing, pulse, movement schemas, digestion, and rhythm — as well as a long shared history of social universals such as threats, competition, embracement, and prayer. And behind these human universals, we must suppose the existence of a plethora of corporeal-psychological patterns that have become incorporated in line with the establishment of such universals.

Such corporeal-psychological interleavings open up the potential for an abundance of both feelings and expressive forms. For in the daily attempts to cope with the conditions of life, we all have to know how to interpret our own feelings and expressions as well as the feelings and expressions of others. And since this project is recursive and the possibilities of interpretation effectively infinite, there must necessarily arise a need for some kind of formalization. I shall suggest, then, that the biosemiotic core at the heart of the human aesthetic sense is rooted in an easing of the burden of interpretive inventiveness, or — in cases where this fails us — in the lowering of the level of social confusion that necessarily follows from the formalization, and thus the making sensible, of human expressions and feelings.

To work out a *natural history of aesthetics* along these lines obviously transcends the ambitions of the present work, and I shall pursue this question here no further. I shall, however, shortly ask the reader to consider how this biosemiotic understanding of *aesthetic origins* might change our conception of the interrelation between aesthetics and ethics. For, perhaps more than anything else, this whole discussion is important because it offers a new theoretical framework for considering the possible existence of some universal constituents in the way that human beings biosemiotically value their life-world.

It has been shown, for example, that most small children instantaneously distinguish between pictures depicting something alive from pictures depicting something that is not alive. Children also much more easily imitate actions that they can see a *purpose* to, than actions that seem random or purposeless — the existence of *purposiveness* is, of course, the most striking, if enigmatic, property of animate nature.

Similarly, the unprecedented quantity of pet animals that have become so much a part of family life in the modern world, testifies — like its mirror image, the aversion to the pervasive insect world of creeping, crawling things, as well as to slime, spit, blood, and excrement — to the immediate and strong appeal of *life itself* to our emotional life. For ultimately it is *life* that, depending on the circumstances, calls out of us our gentleness, fear, fascination, awe, confidence, brutality, loathing, and hate. And while the sea, a rainbow, the firmament, the mountains, and the rivers may all raise strong feelings within some people, these feelings rarely elicit the same degree of *spontaneous concern*. A stone in the road, a cloud in the sky, a drop of rain, or a snowflake may all in some situations seem cognitively interesting — but they do not, in general, appeal to our visceral and emotional life with the same surprising intensity as do the howling of nearby wolves or the peeping of just-hatched birds.

There is an *irreplaceability* to living things, claims the Danish philosopher Peter Kemp (1991), that we cannot avoid recognizing because it comes from a corporeal existence that we know in ourselves. Likewise, in the phenomenology of Merleau-Ponty, we exist corporeally neither as pure subject nor as pure object. When I experience myself and when I experience another, corporeality per se is a common denominator — we are identical by being *incarnated*. My experience of having any kind of *self* at all is therefore necessarily an experience of a kind of corporeality. And, as such, it cannot be separated from the experience of something being *outside of myself* (Merleau-Ponty 2002(1945)).

"I am always a stranger to myself and therefore open to others" is the way that the Danish phenomenologist Dan Zahavi (1999) puts it. The ability to feel empathy towards other people is, in this philosophy, part of the fundamental human condition, and it is this ability that makes us ethical beings. We cannot avoid recognizing that every other human person is vulnerable and existentially irreplaceable like ourselves. "This idea of the irreplaceable," Kemp writes, "derives from a growing consciousness of the fundamental importance for the ethical life of the experience of loss, and it may be stretched to count for everything that may irrevocably get lost, but which we do not want to disappear: the species of plants and animals, and the supporting natural cycles in general" (Kemp, Lebech, and Rendtorff 1997, 28).

Biosemiotics adds to these ideas the further standpoint that the psychological and corporeal reality of each individual is so tightly interconnected that the aesthetic and ethical needs of a human being can not any longer be coherently understood as *incorporeal* — which is to say that, such needs cannot any longer be conceived as being based only on *intellectual will power* but rather, as always arising from a conjoined corporeal/intellectual effort. To accept an ethics of vulnerability, justified by the fundamental irreplaceability of life, must be seen, in this light, to be as much a bodily need as it is an intellectual decision. Accordingly, the role played by intellectual judgment in this context is that of a *guide* rather than that of an *executive officer*.

This does not, of course, imply a reduction of ethics to brute ethology — for the whole point of this argument is that the semiotization of the body has already pushed the potential for genuinely free action back *into* corporeal nature. Psychic life is enabled by the corporeal and corporeal life is in a deep sense psychic (Chapter 7). The corporeality of the human being is therefore, for better or worse (for not only the human good but also the human evil has human empathy as its precondition), the key to our ability to empathize with *the other*. This is not because the other *has* a body, but because he or she *is* a body just like I *am* a body myself. Empathy is not just a mental effort, and in particular is it not a duty, for it is a corporeally felt necessity.[5] And the fact that we often choose to disregard this necessity is the cause for much harm to ourselves as well as to others. In a modern society (as perhaps in every human society), the biosemiotic integrity of people is perpetually challenged by the sheer multitude of social and mental demands for interaction. And because no person is capable of honoring them all, such demands latently threaten to corrupt our corporeal-psychological harmony. Ethics, in this view, is thus an expression of the culturally mediated experience of how to best protect the semiotized body self. We must empathize with the other for the sake of the biosemiotic body itself.

And here we can see how the aesthetic, as well as the ethical, demand on life is rooted in a biosemiosis that integrates body and mind. Our unquenchable longing for symbolizations of the *felt life* (to use Langer's eloquent expression) originates in the selfsame biosemiotic reality that also causes our empathy with the vulnerable irreplaceability of other human beings. Biosemiotics thus adds to the aesthetic and ethical discourse a possibility for the anchoring of such theories in the undeniable universality of the *lived life*.

Medical Bioethics

In our time, a threatening gap is widening between humankind's customary conceptions of the human condition and the new understandings of life that relentlessly intrude upon us under the pressure of the provocative possibilities being offered by an ever-advancing medical technology. Not only have the many technical interventions in the reproductive process created new human dilemmas, but even more fine-grained interventions such as gene therapy and organ transplantation are also critically challenging long-held moral and ethical traditions (Wheeler 2006).

The ethical problems connected to such new medical interventions involve both direct risk factors and the more general consideration of threats to human dignity and integrity. In the medical test called *amniocentesis*, for example, one of the risks concerns the occurrence of *false positives* — i.e., embryos that return a result of "probable defects" in test samples although the embryos are, in fact, perfectly healthy. The direct risk in such cases is that parents decide to abort an otherwise healthy embryo. Procedures such as amniocentesis, however, also lead to worries that are more indirectly connected to the *outcome*, but more directly

5   In this regard, see Favareau's (2001) biosemiotic analysis on the data coming out of the research into mirror neurons.

connected to the entire *purpose* of the test — i.e., the elimination of eventual embryos with genetic deficiencies or other traits that are potentially suboptimal (as might be determined by the culture at the time). Here it might be claimed that this purpose in itself goes against the Kantian imperative of never treating another person (here an embryo) only as a means.

While biosemiotics can hardly add anything important to the *mistake risk* discussion as such, it does throw new light upon the question of the ethical status of human life, personal integrity, and dignity. For, as we saw in Chapter 4, the very concept of an individual being (human, animal, plant, or fungal) is far from as obvious in a biosemiotic optic as it is when seen in the more traditional light, and this leads to a certain ambiguity in the conception of irreplaceability. For what exactly is it that is irreplaceable? Is it the germ cells? Is it the DNA? Is it the fertilized egg cell? Or is it the newborn child? And what should be meant quite generally by the beginning of a life — or the end of life for that matter?

Seen from the biosemiotic perspective, these questions may perhaps not be as decisive as is usually supposed. The semiotization of biological life implies a loosening of the narrow binding of life to certain privileged and distinct structures, in favor of a process conception that instead identifies the phenomenon of life with a continued code-dual semiosis. In prokaryotes (who make up by far the majority of all living creatures on Earth), life is indeed an unbroken chain of daughter cells following upon daughter cells in a continuous flow of cell growth and division. To ask when, exactly, any one of these given cells started its life is utterly abstract — since the cell, both as far as its substance (cytoplasm) is concerned, and as concerns its form (both its membrane and its genome), is connected to all ancestral organisms in the same unbroken line. In multicellular organisms such as plants or animals, one may of course speak of an *individual's* life that starts with the establishment of a single cell (the zygote in organisms that sexually reproduce) that multiplies through a web of cellular divisions, leading to the formation of an adult individual. But as soon as one turns to consider the life cycles found in fungi, it readily becomes clear that the concept of the *individual* that seems suitable for the life cycles of animals and humans cannot be generalized to life processes per se. Many fungal species, for example, spend the major part of their life cycle carrying two autonomous nuclei (one male and one female) in each cell. Conception has been frozen in place, so to say, at the moment right before these two nuclei would normally fuse in plants or animals. The fusion of the two nuclei takes place only at the ultimate *end* of this phase of dikaryotic life (*karyon* = cell nucleus) whereby, finally, a proper diploid cell is formed. This "normal" diploid cell, however, does not last for long but instead quickly divides itself again under the formation of haploid spores (sex cells with only one set of chromosomes). These haploid cells are then ready to fuse with haploid cells of the opposite sex whereby a new phase of *dikaryotic* life begins.

Thus, in these fungi, the diploid phase — that condition that we humans without further reflection take to be a proper *individual's* life — is reduced to only one short intermezzo in between two extended phases of tight common haploid life. Here the process of fertilization has been stretched, so to say, to occupy nearly the whole of the organism's life. Diploid life, such as we know it from ourselves, cannot therefore bio-logically be seen as being the privileged form of individual life. And one consequence of this is that *the* starting point *of an individual human life is not a biologically meaningful concept.* To the extent this concept has any meaning at all, it is a meaning that is derived elsewhere than from biology. For human life, just like all other forms of life, is from the biological perspective an unbroken chain of growth and cell divisions.

However, the fact that every human being nevertheless is very intimately connected to just one individual life therefore calls for an explanation that cannot be derived from the mere biological fact that human beings are living creatures. If we want to explain why human life is seen as so narrowly connected to our experience as individuals, we had better search for the explanation in the peculiarity of the human form of life, and not as the result of life processes in general. And here the overarching distinguishing mark between the human form of life and all others is the extraordinary semiotic agility possessed by this big-brained animal.

From these deliberations follows a conclusion of deep significance for ethical discussions in the medical area: *The individuality of a human life cannot be justified by its uniqueness as a particular genetic combination, but must be justified by its uniqueness as a particular semiotic creature.* Semiotic uniqueness is different in kind from genomic uniqueness because it depends on a potential that bears an inherent *indeterminacy* and *creativity* — a potential, furthermore, that we humans know much better how to evaluate and enjoy than is made possible through the severe determinacy of genomic specification. *Human* uniqueness, then, is not *molecular*, but *semiotic*. And from this it follows that the readily felt irreplaceability of a human being is rooted in its *semiotic individuation process.*

This conclusion leads us to the emotionally troublesome consequence that an objective answer can not be given to the question of *when* a human life starts and stops being irreplaceable. Considered as a genome, it is irreplaceable from the moment of fertilization — and the ethics of irreplaceability within the optic of genomic individuality logically leads to a rejection of any and all manipulation with embryos. But if instead we ask the question of when the semiotic individuation process of a fertilized egg cell is so far advanced that a human life is violated by its abortion, no such clear answer presents itself.

The biosemiotic understanding in reality forces us into an evaluation of a more-or-less situation. Humanhood thus becomes a more-or-less property! And thus — as with most other essential questions of life — the question of whether a technical intervention destroys a *human* life depends on a process of reflexive pros and cons, and thus each single instance of technical intervention must be evaluated through a concrete analysis. There simply is no *unitary procedure* — neither rational nor ethical — upon which such ethical judgments can otherwise be safely based.

Through this loosening of *individuality* from the genomic absolute, the inviolability of life becomes not theoretical and absolute, but specifically embodied and concrete. The stricture that you must not kill another human being does not then depend on a principle of human inviolability, but upon the fact that such an act will lead to the elimination of an invaluable semiotic richness — one that demands our deepest empathy. For, as we saw in the preceding section, it is precisely the existential ability to identify ourselves with other semiotically free systems that stands as the pivotal point of ethics within the biosemiotic perspective.

Thus, the statement that *the irreplaceability of a human being is connected with its semiotic individuation process* implies that an intervention at the early stages of embryonic development — where semiotic individuation has not yet proceeded very far — as for instance in early abortion or therapeutic cloning, do not violate the ethic of irreplaceability. Also, at the other end of life, this line of argumentation has ethical consequences. For it implies that patients whose brain processes have become irreparably dysfunctional, and who are therefore kept "alive" only by the mechanics of permanent treatment in a respirator, can no longer be conceived of as genuinely *human* individuals. For when *only* physiological and biochemical maintenance functions are left functional, semiotic individuation has been brought to a definitive stop — and the superficial similarity of such an artificially sustained body to a formerly alive person is nothing but a cosmetic cover for this once fully human person's current state of nonlife. Life-maintenance practices of this kind are — seen from the standpoint of biosemiotics — meaningless. What is being maintained here is not life, but merely chemistry.

And finally, a biosemiotic understanding of human life as an ongoing process of open-ended change has deep implications for our conceptions of identity and integrity. Semiotic individuation continues all through life, and the experience that many elderly persons have of being, in some strange sense, *another person* now than they were in their younger days is not wholly unfounded. We must reject the popular idea of identity as being connected to a particular biological entity or body, and learn to see identity rather as a temporary nodal point along a process that each of us is ceaselessly engaged in — a process of identity formation and change that does not end until death. It is, in other words, not in the heritable endowment of genetic singularity that identity is to be found, but rather in the lifelong attempt to adjust our personal development to our own unique needs and experiences — often in spite of the genetically caused defects that we might have to overcome. A person that is born blind is not, for that reason, left to a certain destiny but can, as we all know, counter his or her handicap in numerous semiotically unique and individuated ways. Integrity of identity is, I suppose, before anything else, a designation for the coherent personality that one tries ceaselessly all through life to work out, often against great odds.

Ecology and Environment

Much confusion surrounds the debates about human beings' responsibility to the environment. Are we to care for nature's ecological richness and coherence for the sake of its continued sustainability as our productive apparatus — or do we, in addition to this notion of self-interested *stewardship*, have some more profound moral obligation to protect nature's numerous life forms against the growing threat of their extinction? As is often the case in such situations, a polarization has occurred between fundamentalists and pragmatists. One faction has devoted itself to what they call *deep ecology* that ascribes an autonomous moral status to nature itself (Naess 1989), while the other side holds that the purpose of protecting nature is to maintain sustainable human societies (an attitude deep-ecologists have somewhat self-righteously nicknamed *shallow ecology*).

It probably won't surprise the reader to learn that the biosemiotic position opens up a possible intermediate theoretical standpoint — although deep ecologists conceivably might brush this off as shallow ecology in disguise. This position recognizes that human utility does not exhaust our ethical obligations towards natural systems, but it also refuses to ascribe moral status to all of nature's products and effects as such. Life cannot — neither in all abstractness, nor in the shape of individual living systems — be said to have a claim to our protection at any price. What *does* have such a claim is not life per se, but an emergent trait pertaining to life as an evolving process: semiotic diversity and elegance.

This position, it should be noted, is merely a logical extension of the ethical reflections discussed in the preceding sections — but now applied to nature at large. For if we accept *vulnerability* and *irreplaceability* as central ethical concerns that we can not disregard without brutalizing our own human integrity, then these two concepts can not a priori be delimited to the human sphere. Even an earth worm is vulnerable in the sense that it risks destruction if one treads on it. The same goes for bacteria in a test tube that may be killed by the addition of just a dash of strong acid. And, more obviously, this is true for complex creatures like bird young or mammals. Yet this designation does not apply to a stone on the road which cannot be either hurt or offended by being stepped on or kicked. True, a stone may lose some of its internal coherency by being crushed into little pieces, but it seems misguided to use the term *vulnerability* for this kind of material fragility.

Still, even if most people will be willing to accept *vulnerability* as a general characteristic of living systems, *irreplaceability* is yet another thing — a word that, I suspect, most of us would not be predisposed to use with such overarching generality. To call a mosquito larva irreplaceable is, for most language users, I suppose, to stretch the concept more than could be considered reasonable. In Danish, for example, we use the expression to denote a nice person that "she wouldn't hurt a fly," but here the whole point is, of course, to stress how highly unusual it is for a person to take pity on something as unworthy of our pity as a fly. Normally we just kill them. Their irreplaceability doesn't occur to us as any kind of pressing concern.

Yet seen from a genetics standpoint, each single fly may indeed *be* irreplaceable, since most often there will not be another fly in the world that carries a copy of the exact same genome. This objection, however, is not likely to carry much weight with people on the street. For genomic uniqueness here is not the thing that matters — though it *is* when it comes to human embryos. What matters, instead, is something very different — something that has to do, I believe, with the degree of semiotic freedom that we see expressed in another living creature (Hoffmeyer 1996b, 140–42).

From this, I shall suggest that the decisive factor in triggering empathic feelings towards organisms of other species is the degree of *semiotic individuation* that we perceive in them. For not only human beings — but, in fact, most living organisms — will engage in at least some modest degree of semiotic individuation. And the fact that especially birds and mammals appeal intensely to our capacity for empathy may be because these animals have extended personal life histories that, like us, influence their whole life and behavior. We can therefore not, for the sake of our own peace of mind, content ourselves with protecting nature merely for the sole purpose of sustaining business. Rather, we must protect other life forms because we also, in a deep sense, are not able to ignore their relatedness to ourselves. For in a universe that, as Monod said, is deaf to our music and indifferent to our pains and hopes, it remains strangely reassuring that at least some other kinds of life forms exist that do go through a process of semiotic individuation that brings them into the same boat we are in.

And here again it can be seen that the biosemiotic understanding leads to a holistic yet more-or-less approach to life, rather than to dyadic and mechanistic either-or thinking. Here, animals are not thought to have rights in the same way that humans have human rights — for such human rights are not derived, in the first instance, from any absolute principles, whether biological or philosophical, but reflect, as we saw, social and cultural experience in a certain historical conjunction (see Chapter 8, the section entitled, "The Problem of Freedom"). Neither can animals, as living creatures, automatically be considered to be moral subjects with an a priori ethical claim to human protection. We nevertheless must care for the living creatures on this Earth, because they all are expressions (albeit more or less pronounced) of the general tendency of the evolutionary process to create life forms with semiotically controlled life histories — and in this semiotic expressivity, we cannot but recognize the relatedness of these life forms to ourselves. It follows from this that our responsibility to other living creatures must be gradated according to a scale that reflects the sophistication and complexity of the semiotic individuation that individuals of a given species are biologically able to attain.

What I am *not* arguing here is that environmental protection should then mainly aim at conserving life conditions for panda bears, whales, rhinoceroses, and other animals with similar appeals to human imagination and empathy. That we may be emotionally disposed towards the plant and animal life on the planet does not preclude us from also developing a rational and ecologically informed relation to nature. On the contrary, the whole point is that ecological diversity is *in itself* an important component in the semiotic freedom that makes nature so appealing to us. The homogenous forests, the gigantic monocultures, or the impoverished plots of land left by rain forest clearings are not only poor in species diversity, but are also — and for the same reason — poor in semiotic diversity. A rich semiotic world goes hand in hand with a rich ecological world.

And it is no coincidence that the book that more than anything else contributed to raising a worldwide concern for the environmental question forty years ago was titled *Silent Spring*. Rachel Carson's best-seller from 1962 described how the spread of manmade pesticides to every corner of the planet threatened to extinguish a huge number of bird species — and with them also, the existence in our world of bird song (Carson 1962). This last relation may have been the crucial one, for it was not the de-semiotized image of nature drawn up by university ecologists talking of energy flows or biotopes that had the potential to mobilize whole populations to a defense of ecological diversity — rather, it was the fear of losing the semiotic diversity and pleasure of nature, as epitomized by bird song, that got people off their chairs and into action. An important message is to be drawn from this, and it is one that should not be brushed aside as naïve or sentimental. The insignificant fungi that takes care of the degradation of pine needles may, for all we know, have a greater importance for the well-being of a pine forest than do all of its birds and mammals — but if people did not value pine forests in and of themselves, the conservation of pine plantations would be left strictly to those with economic concerns. If however, in addition to the concerns for sustainability that supposedly will be met in a functional democratic society, we also want to take responsibility for the species diversity of natural systems for the sake of semiotic diversity itself, then it will be important for us to accustom our understanding of, and thinking about, natural systems in terms that will allow us to *experience* its richness. And this is something the biosemiotic approach is naturally tuned to do. *A semiotization of ecology, then, will not only bring the discipline in harmony with a general biosemiotic understanding of life processes, but will also revitalize ecology as a resource for humane environmental administration.*

I have earlier in this book (and especially in Chapter 6) discussed the insufficiency of the de-semiotized approach to understanding ecology, and have suggested a perspective in which the ecological niche concept could be extended and specified to cover the concept of a semiotic niche. I also suggested the term *semiosphere* as a more holistic and, in fact, *ecologically correct* counterpart to the rather physicalistic notion of the biosphere as usually defined. The point of the present section has been to argue that this reframing of ecological terminology and research horizons would not just be advantageous for the discipline of ecology itself, but would also support exactly those aspects of our ecological insights that call for human engagement in the fight for protection of the environment. It is through a re-semiotization of nature that the science of nature will come to stand as genuinely relevant for ordinary people — and, as a result, for nature. One promising line of research here is the field of zoosemiotics as introduced in 1963 by Sebeok and currently pursued by Dario Martinelli (2002)

Medical Science

Medical science may well be the area where the biosemiotic perspective might have the greatest significance. For a hospital is the place where the conflicting conceptions of the human being — as a person and as a biological piece of machinery, respectively — reaches its most acute expression through the clinical practice with patients. This conflict is rooted, of course, in the fact that the ruling theoretical paradigm governing clinical work has no way of making these two dimensions of human life meet. Understanding a human being (such as yourself for instance) unavoidably presupposes that one comes to terms with what it means to be an "I" (a first-person *singularis* phenomenon). But as we have seen, biology is in principle prevented from including "I" phenomena into its theory structure, and it does not even have the faintest theory to explain how such a thing could occur in the world at all.[6] Instead, biology exclusively

6 I here deliberately disregard a currently fashionable approach to the positing of biological foundations to human mental faculties called Evolutionary Psychology — an approach which, to the extent its smart clichés are scientific at all, is built upon a surprising lack of insight into the

deals with phenomena that may be described in the language of third-person phenomena, and thus — in this very point of departure — excludes this science from arriving at a theoretical understanding of the human biosystem as a first-person being.

In their clinical practice, physicians of necessity admit the assistance of psychologists and psychiatrists whose task is to take care of the patient as a first-person creature. And nurses are also primarily concerned with this dimension of the patient's needs, to the great (and underestimated) benefit of all concerned. But these "assistants" are rarely considered worthy of influencing the treatment of the patients as third-person cases. Since, however, (as we all know from personal experience) the first-person and third-person aspects of our being unavoidably belong to the same undissolvable unity, it is some wonder that the borders between these two aspects of a patient's need for caretaking have persisted as inviolable for so many years. But if, as one may well fear, the punishment for breaking with the current scientific paradigmatic consensus may be permanent exclusion from the social milieu of scientific medicine, then such a science has effectively sealed itself off with an immunity against the open expression of doubts concerning its own scientific fundament. The power structure of scientific societies may at times exhibit almost Mafia-like coercion to "loyalty."

As we saw in Chapter 7, however, there are courageous breakaway physicians who have insisted on dealing with the corporeal interrelations taking place between the psychological and the somatic dimensions of human health. But, as the reader may also recall from the discussion of the psychoneuroimmunological (PNI) research field, such work as currently framed lacks a robust theoretical foothold. And as I suggested earlier, this nascent field might greatly profit from the transcendence of its still too narrowly conceived biochemical paradigm, in order to attempt a more explicit integration of the biosemiotic approach into its theoretical resource base.

In fact, an explicitly biosemiotic understanding of health care was proposed — at least in preliminary form — by the late Thure von Uexküll (son of Jakob von Uexküll), who was at the time still professor emeritus in internal medicine at Ulm University (Uexküll 1999). Uexküll introduced the Peircean categories of Firstness, Secondness, and Thirdness as a means for allocating patients' symptoms into three *feeling layers*, each of which are manifestations of a phase

---

irreducible "I" aspect of human existence, the understanding of the human psychological world as a *home for experiences* in the true sense of these words. It is interesting to note in this regard, too, that in Scandinavian languages, the word for *experience* is *oplevelse* — which literally translates as "living made up." This aspect of experience as something occurring in the mind that is not the same as — but nevertheless is closely related to or even formally identical to — what actually occurs (or might have occurred), makes it intuitively obvious, I believe, that an experience (*oplevelse*) cannot be an illusion unless life as a whole is an illusion.

in the early life history of the patient. Firstness thus rules the earliest stages of development "where babies only experience qualities or differences of qualities in the form of moods and impressions that are lived through. As in a 'quasi-hypnoid state,' quantities are not yet attributed as *properties* to particular phenomena, but are experienced as 'pre-actual atmosphere' (*vorwirckliche Atmosphäre*, (Uexküll 1953)) in which only the sign connection exists: that of similarity (iconicity) between different grades of intensity of a mood or of an impression, or between moods and impressions" (Uexküll 1999, 652).

In this phase of a baby's life, taste and smell are the dominating sense qualities — and this is where the root forms are produced for the experience of qualities of things, experiences of similarity with earlier experiences, and, more generally, the correspondence between the organism and its environments. In this connection, Uexküll observes that the ability *to taste* is at the core of the Latin term for wisdom, *sapientia* (from *sapere* = taste) — which testifies to the importance that, at least in an earlier time, was ascribed to this iconic layer of the minded body. Such an iconic layer constitutes a semiotic core that, in many patients, is expressed in the subjective aspect of symptoms such as the intensity of pain.

On top of this iconic layer of the developing child is settled another layer — one that is concerned with specific acts. It is in the course of the establishment of this layer of *Secondness* that the subject-object distinction is conceived. This is an indexical layer, writes Uexküll, which may express itself through the attitude of the patient to his own body and in the body's relation to its surroundings. It is in this layer that the feeling of the *specificity* of a pain is grounded — for instance, the feeling that the pain in the region of the heart (and the feeling that this pain is the "same pain" that "grows worse" under certain conditions).

Finally, there is a layer of Thirdness through which symbolic and social conceptualizations become inscribed in the body. This layer has to do with the way that the patient *experiences* his symptoms — e.g., the recall of the signification that a given kind of pain had in the disease history of a relative or a sick friend.

It is Uexküll's point that the life history of the patient is expressed through these three layers "and the change of symptoms will depend on the weight that each of these layers has added throughout his life" (ibid., 653). The *symptoms* (which in medical terminology means "group of signs") of the patient are in the first place a *message* to the patient himself, leading to a resulting cascade of interpretants that perhaps — if they become conscious — may then be told to a doctor. The doctor's task then is to create a meta-interpretant (an *interpretation*) on the basis of his understanding of the patient's own interpretations — and in order to do this, it is important that the doctor succeeds in creating a *shared reality* with the patient. It is in the handling of this task that the relation between patient and physician will become decisive for a successful treatment. This is consequently mismanaged when the nurse is the only caretaking person to whom the patient is an "I" in the process.

And this is where the poverty of modern medical practice is most evidently illustrated. For while it is relatively easy to establish a shared code between the physician and the patient where the indexical layer is concerned (i.e., as concerns voluntary activities, surgical treatments or biotechnical interventions), it is usually much harder to build the same kind of common code when it comes to the symbolic or iconic layers. For the doctor and the patient do not ascribe the same status of *objective symptoms* to the interpretants that are formed under the influence of these ineliminable layers. To make the influences of these layers visible in clinical work will be a main task for a future biosemiotically based medical science.

And at this point a thread may be traced back to the physiologist Walther Cannon's seminal book from 1932, *The Wisdom of the Body*. Cannon showed that the body possesses a wisdom that is difficult for the physician to imitate, and he disclosed many of the relations that would contribute to the maintenance of bodily health — but the deeper mechanisms behind this wisdom have remained hidden to this day. Now, in the wake of decades of New-Age follies, one should take care not to anthropomorphize or exaggerate this "wisdom of the body" that, unfortunately, is too often defeated by life-threatening diseases. Uexküll's approach suggests the possibility that the utterly natural secret behind the body's wisdom rests in the establishment of a harmonic interplay between the iconic, indexical, and symbolic layers of the experiencing body. When we get sick, something has gone wrong in this optimally balanced interaction. Yet as long as physicians only take an interest in just one of three layers, the indexical layer, they will, of course, instead of wisdom, find only mechanics.

Cognitive Science

The term *cognitive science* initially referred to an attempt to produce a cross-disciplinary research area that could unite insights from formerly separate research fields, each concerned with their own special aspect of the cognitive processes — such as perception, categorization, language, memory, and thought. The term still has the ring of a belief in computer models as a means for simulating cognitive processes, a belief that in the 1980s and 1990s was yet sufficiently untested to support the cross-disciplinary optimism.

It is undeniable that a cross-disciplinary undertaking is absolutely necessary if genuinely new understandings are to be produced in this area, but I find it misguided to build this ambition upon a belief in the all-embracing power of computer modeling. In fact, in the light of biosemiotics, much — perhaps even most — of the work done in cognitive science seems more or less useless because it has been ignorant of (or, worse, has deliberately chosen to disregard) the literal corporeality of cognition and the "I" aspect that is part and parcel of such corporeality.

One major step forward was taken when the idea of cognition, as based upon a "symbol-processing brain" (e.g., Chomsky 1965; Newell and Simon 1975), was challenged by *dynamic systems theory* (Kelso 1995; Port and van Gelder 1995; van Gelder and Port 1995). For where the symbol-crunching paradigm saw cognition and language use as based on calculation processes inherent to the brain, the dynamic systems paradigm introduced real-time decision-making as an irreversible factor in the processes of cognition. In computer modeling, the problem of time is typically solved by the application of a series of distinct states, t_1, t_2, t_3 . . . . Time is converted to a series of discrete beats and the continuous intervals in between the beats are effectively disregarded. Thus, in the proper sense of the word, nothing really *happens* here — just logical operations clicking through a series of static states. At a minimum, it may be said that a strange disregard for *biological realism* is hiding in this computer-fixated way of thinking. Following biological realism, it must be recognized that life processes *happen* in continuous time, like a snail track slowly moving forward. Or, to allow myself the pleasure of my opponents' brand of dualism, biological reality should be seen as consisting of blood and slime — not of logical states.[7]

Reflections in this direction were the starting point for Robert Port's attempt to show that when people think, they do a whole lot of things other than merely manipulating symbols. There is nothing in the brain like the discrete beats of the computer, says Port. Rather, it seems there are many systems simultaneously operating that create oscillations at different rates. What happens seems to be organized dynamically rather than linearly — that is to say, via processes in time rather than via computer-like digital input/output transformations. Port (1998), among many other things, studied the rhythms of people's ways of talking in different languages and observed that the rhythms of human speech depend upon the same few but apparently universal rules. It is significant that Port's findings confirm that there are universal components in our ways of thinking and talking, but that, on the other hand these universals do not — as the Chomsky school has claimed so ardently — mirror any logical *context-free syntactic rules* that might lend themselves to simulations in the computer. Instead, they reflect dynamic traits in the biological functionality of humans as interacting organisms.

Yet, while dynamical systems theory thus puts cognition back into the biological organism (i.e., into the body) where it obviously belongs, the dynamicists

7   In Peirce's philosophy, time is continuous — i.e., it does not consist in distinct points on an axis of time.

themselves do not incorporate first-person perspectives into their understanding of cognition. Dynamicists therefore, no less than their adversaries in the Chomsky camp, appear to have cut themselves off from explaining why humans are not just *zombies* (to use the expression that has become standard in philosophical discussions of these matters).

The word *zombie* stems from Haitian folk-religion and refers to a person who, after having being killed by sorcery and buried, is made to come alive once again (by yet more sorcery, of course) and to continue a life where he, deprived of all will, unresistingly executes the slave work he is set to do.[8] In cognitive science, the term *zombie* is used to characterize a fictive robot-like creature that has been constructed on the basis of a totally mechanical understanding of the material and the computational capacity of a human being. Claus Emmeche (1998, 51) has described this sad creature: "Even if we could construct, on the basis of these principles, a robot that would be capable of answering 'intelligently' on questions of the environmental crisis or the meaning of Shakespeare's *Hamlet*, we could still not be assured that the robot would feel or *experience* anything at all; it would be like an automatic zombie (although the intuitions about this diverge)."

Emmeche's parenthetic remark may be central. Natural scientists mostly do not seem too worried about the implications of this unsettling intuition — which is, of course, the one that concerns biosemioticians so very much. Such scientists — and here we are, unfortunately, talking about many scientists who belong to the absolute elite within the diverse fields of natural science — do not apparently see any reason to believe that what happens in the experiential life of human consciousness could *not*, in principle, (and probably in practice) be derived from processes that are perfectly describable in grammatical third-person terms. We are here confronted with a difficulty that does not seem to have any easy solution. For how does one decide between disagreements about intuitions? (And make no mistake, the mechanist's intuition is no *less* an intuition than the biosemiotician's in this matter. One might even argue that, of the two, the mechanist's intuition is the one that is *more* at odds with empirical reality).

Given such an impasse, biosemiotics must take cognizance of the fact that dynamic systems theory does not exhibit any apparent curiosity about the evolutionary problem of deriving first-person experiential worlds from an ancestry that exhibit nothing but third-person phenomena. Apparently, dynamical systems theory assumes that purely dynamical processes — perfectly describable in third-person language — exhaust what we need to understand in order

8   My source for this explanation of the word *zombie*, the Danish National Encyclopedia (1998), adds — somewhat reassuringly — that the phenomenon has not been scientifically confirmed.

to describe and to explain human experiential and conscious life.[9] But again, as first-person phenomena ourselves, it is hard to see how this could be so — and thus the onus is on those who would argue otherwise to establish the validity of their position (which is contrary to all human experience).

Mainstream cognitive science in this regard puts itself very much in the same position as traditional medical science.[10] And it is my guess that the problems that medical science has experienced in its clinical practice because of its inhumanely reductive strategy will also eventually come to cripple a cognitive science too inanimately conceived. If and when such dead-end approaches come to be known as such, might this lesson in the de-semiotization of living systems count as support for the biosemiotic intuition? My personal expectation is that it will, and that when it does, the conceptual tools made available by biosemiotics will be turned to as a resource that will greatly stimulate cognitive science by suggesting eye-opening new paths of investigation.

9   Here I should warn the reader to keep the *qualia* concept and the *sign* concept separate as analytical concepts. Whether or not *qualia* covers a reality in primitive life forms is a fascinating subject, but it would be unfortunate to decide a solution to this problem in advance. The *semiosis* concept, on the other hand, might fruitfully be applied in the living world at large, as has been argued throughout this book. And my personal opinion is that it would be well-founded to consider semiosis as an elementary trait pertaining to our universe at large — in accordance with Peircean cosmology.

10   I am aware, of course, that this over-generalized categorization of cognitive science does not pertain to all the diverse fields that see themselves as belonging under this umbrella concept (e.g., see Roepstorff 2001; Roepstorff and Jack 2004). What I am speaking of here particularly are those attempts to examine and explain *cognition* as a de-semiotized, computational process that can be fully captured by third-person description.

Biosemiotic Technology

The Semiotization of Technology: A Historical Sketch

Classical technologies were concerned with upholding subsistence and rationalizing work load. The plow, for instance, was introduced when slash-and-burn agricultural practices were challenged by an increase in population density. Grass roots do not easily disappear and slash-and-burn practices were, contrary to what most of us may have learnt in school, a quite efficient way of upholding subsistence, because the grass cannot grow in the deep shadows of the forests so that no work had to be invested in removing it. But as population density increased, cultivators were forced to reduce their share of forest — which meant that the forest areas had to be cleared before they had fully regenerated, so that the grass was no longer properly wiped out. And since grass roots resist the burning practice, cultivators now had no choice but to start plowing the fields.[1]

Thousands of years later, water mills and wind mills were introduced to solve corresponding problems with growing populations. The mills immediately furnished a rotational power and the invention of the camshaft made it possible to transform this to a vertical power used to lift a fall hammer. Apart from crushing grain, such mill technology was also used for a variety of other production processes, depending on local circumstances — to weave cloth, tan leather, smelt and shape iron, saw wood, press olives, etc.. Productivity increased, dependence on human- and animal-muscle power gradually declined, and locations with good water-power resources became centers of economic and industrial activity.

Thus, in preindustrial peasant societies, nearly all energy consumption was of biological origin and was directly or indirectly provided on the basis of the

---

1 This not only meant more work for cultivators, but also meant the introduction of draught animals — and thus the requirement for additional cultivated land to feed the animals. Consequently, not only was the workload dramatically increased in the process, but the ownership of land, as well as new needs for collaboration and planning, introduced a social stratification unseen in egalitarian slash-and-burn farming groups — and, in fact, established the perfect conditions for a propertied class to eventually arise. In the schools of my childhood, this process of loss was portrayed as a significant progress for civilization under the emblem of *state formation*.

photosynthetic processes of plants. These were food, feed, fuel, tractive power, materials for construction, fibers, fertilizer, etc. But as long as only biological power sources were accessible, there remained very narrow limits on productive possibilities. Milling technology, in this respect, constituted a radical new departure, one fraught with radical historical consequences. With it, humans for the first time attained systematic access to the much larger, nonbiological energy resources beginning with natural energy flows such as wind and water — and later also, by drawing on artificial energy sources such as oil and coal.[2] In a fulling mill, one man manipulating the mill might replace forty fulling workers (Gimpel 1978, 22).

The industrial revolution may then be seen as a chain of technological inventions based, in every case, on the principle of freeing the productive process from the limitations put upon it by the materials of living nature, so that these materials were gradually replaced by materials from inanimate nature. Coal replaced wooden fuel; the locomotive replaced horses for transport; bricks, tile, and glass replaced trees as building materials; and a range of nonbiological chemicals such as chlorine and soda replaced the organic resources of an earlier time such as seaweed and sour milk (Wilkinson 1973; Hoffmeyer 1982; 1988). Against this background, it may be further claimed that the ecological basis for industrialization consisted in the transformation of biologically based energy and material economies into economies of non-biologically based energy and materials — and this transformation required an increasing flow of energy to be released on the planet's surface in order to sustain itself.

For centuries, industrialization — understood as the continuing effort to increase the efficiency of *energy transformations* of every kind — has remained the dominating trend in economic development. But towards the end of the twentieth century, a whole new set of principles started to get a foothold in the modes of production. At first these new principles were still directed at maximizing efficiency of work, but it soon appeared that they had inherent possibilities of quite a different kind — possibilities which have only begun to be seen more clearly in the last two decades. These new possibilities do not primarily aim at streamlining the work load. Rather, they aim at coordinating work efforts in all directions and across traditional barriers. As an overarching term for the focus of these new techniques we may suggest the term *managing of interfaces*, and for the main principle enabling it, *controlled semiosis*.

2   Fossil fuels, of course, are not artificial as such, having been created through geological processes transforming the organic sediments of ancient oceans. But the extraordinarily rapid release of these huge energy stores up to the surface of the earth through a very short span of time is indeed *artificial* (the work of human craftsmanship).

In everyday talk, these new techniques are most frequently designated by the abbreviation *IT* (or in full, *Information Technology*) — indicating that the public has picked up the engineering view that takes the basic property of such techniques to be their huge capacity for information processing. In a purely technical sense, this term is indeed correct, since an important property of such techniques is their unparalleled capacity to store, manipulate, and transport information. But this is only so in the technical sense of computer-switching-type *Shannon information*, as discussed in Chapter 3. And since the mathematical concept of information that is being used there has really nothing at all to do with the more primary definition of *information* as referring to *meaningful* messages or instructions, the designation Information Technology is potentially confusing. As a consequence of this unfortunate terminology, modern society is often said to be an *information society*, which then is contrasted with the term *industrial society* — indicating that the information society is a *postindustrial society*.

Here the ambiguity concerning the engineering way of understanding the term *information* and the everyday language conception becomes directly misleading. For if the technological principle at play behind the Industrial Revolution was indeed the more and more powerful mastering of energy transformations (i.e., *the ability to manipulate and command the physical arrangements of the natural world*), then the corresponding technological principle behind the present cultural transformation is not just the data-packet-transfer kind of information processing, but rather *controlled semiosis* (i.e., *the ability to mediate the significations of all the diverse ideas, expressions, thoughts, and events that take place in the world*).

This does not mean that the perspective of physical command will now become replaced by a perspective of mediation, for the need for sophisticated physical command of energy transformations will continue to play a decisive role for production in the future, just as agricultural production remained a vital component of the industrial societies. What is new is that we now, in addition to food production, and in addition to our brute command over physical processes, have also learned how to implement human interests, signification, in the world of machines and media.[3]

For indeed, the most pronounced feature of the new technologies is not just their capacity for the mechanization of information processing, but their use-

---

3 I must stress here that this analysis is not about values. Whether these changes are also good — in the sense that they will bring about a progressive trend in cultural evolution (or not) — is not of concern here. To be sure, I personally fear that the promising potentials of the technological changes described here will need accompanying changes in the form of radical cultural transformations if they are not to be harmful in many ways. Considering present political trends, however, it is not easy to be optimistic in this respect.

fulness as tools for semiotic activity of every sort. Thus their proper designation should be *semiotic technologies* instead of the ambiguous *information technologies*. Nineteenth-century industrialists were indifferent to the actual physical laws at work behind the steam engine. And in the same way, consumers of modern technology for the most part take no interest in the functional principles behind the capacity of these machines to process (Shannon) information. What does concern consumers, when they activate their cell phones or open a new program at the computer, is the ocean of significations upon which he or she is now going to surf. When the technologies are seen from the point of view of the *user*, instead of the engineer, the technologies are more than anything else *semiotic technologies*.

Summing up this brief sketch of technological history, we may say that the Stone Age was characterized by a technology that rarely, in any systematic way, used energy forms surpassing that of human muscular power. The technology and social organization of agricultural societies were designed to obtain the maximum yield from available biological energy sources (e.g. the field and its inherent photosynthesis). Industrial production transcended these limitations by developing means of energy transformations of almost every conceivable kind (including nuclear) — and accordingly based its huge productivity upon the consumption of (seemingly) infinite sources of artificial energy. Finally, the kind of society we are now entering appears to be one that will derive its enabling power more than anything else from the ability to produce the technological means for ever more sophisticated command over the semiotic dimension of the natural world.[4]

To the extent that this is true, we may further conclude that we are now finally prepared to complete the project that was initiated in the late Middle Ages with the introduction of windmills and water mills — i.e., to set societal production free of the constraining bonds given by the peculiarities of organic life. The first time around, under the Industrial Revolution, only the dimension of *energy* was set loose from the constraints of organismic life. Therefore, what we are now facing — and to some extent have already engaged in — is the setting free (and harnessing) of the semiotic dimension from its bindings in organic life. This is what I have called the development of *biosemiotic technologies*.

---

4   The parallel between energy consumption and semiotic control should not be neglected here, as the importance of the entropy law (the second law of thermodynamics) has been severely underestimated in the industrial period. Most people recognize the importance of access to energy, but few people know of the importance of reducing the entropy production that necessarily accompanies all kinds of energy consumption. Since this control is no longer assured by the natural semiotics of living systems, semiotic control on energy expenditure has become as crucially necessary as the access to energy per se.

The Biosemiotics of the Industrial Revolution

Throughout this book, I have been dealing with the analysis of the manifold semiotic controlling mechanisms that have been developed by living systems. Seen from a thermodynamic point of view, the overriding engineering task of evolution consists in the accomplishment of the efficient use of the available energy flows that organisms (or ecosystems) are canalizing through themselves as optimally as possible — in regard to the ongoing construction and maintenance of the biological systems and their manifestations, as well as in regard to the ongoing export into the environment of the entropy produced by such life processes.

And as noted above, I have argued that it is therefore the same two parameters — the magnitude of the flow of energy and the semiotic controls guiding the utilization of that energy — that have constituted the pivotal points in both the historical project of civilization, and in the evolution of life on Earth. Evolution has worked incessantly to fine-tune the semiotic control mechanisms of organisms to the nature of the surrounding environment's available energy flows, and evolution has therefore guaranteed an optimal fit between those two essential parameters of life in the organisms of this Earth.

I would like to propose that the proper term for this optimized organism-environment relationship is *semiotic fitness*. Semiotic fitness may — at least in a heuristic sense — be formalized as a relation between a component (S) referring to the efficiency of the semiotic control of life processes, and a component (E) that expresses the magnitude of the energy flow canalized through the system. If — for the sake of the argument — it was supposed that the size of the S component might indeed be measured, for instance as some function of the entropy efficiency, then on the basis of the above discussion, we would expect S in natural systems to somehow balance the energy flow (E) so that, for the sake of convenience, the ratio might be expressed as $S/E = 1$.[5] If we correspondingly measured the S- and E-values of, for instance, a given instance of agricultural production, we would arrive at a kind of measure for the sustainability of that agriculture. For the trick behind agriculture is to produce a simplified ecosystem (the field) through which a considerable amount of the locally available energy

5   It is of course not my intention to claim that a genuine numerical measure for semiotic fitness, S, might be constructed, as S by its nature is almost incalculably complex. And while the E component is measurable in normal energy units, the S component can not of course be reduced to units of the same sort. (We are not here intending to invent Maxwell demons!) But as a thought experiment, I think that the very concept of such a measure may throw an illustrative light on the main point that I am aiming at here — namely, that the disproportion between our semiotic and our energetic command of natural systems is the key to the lack of sustainability of industrial and agricultural production.

flow is canalized towards the benefit of one single species (i.e., the crop species). In pre-industrialized societies, this took an enormous amount of human labor, which in industrialized agriculture is replaced by a diversity of petrochemical products, such as fertilizer, herbicides, pesticides, and gasoline for motive power. Added to this, of course, are the indirect energy costs of draining and digging ditches and canals, and of distributing the agricultural input as well as the agricultural output. Thus, the energy-consuming (and entropy-producing) construction of roads, railroads, bridges, and transportation vehicles (including tankers transporting oil and materials around the globe) must also be factored into the thermodynamic *cost* here.

Clearly, already in preindustrial agriculture, the amount of exosemiotic control (understood as the controlling activity of the cultivator) on energy flows is lowered relative to the tight endosemiotic control exerted by the totality of organisms in wild ecosystems. On the other hand, this lowering of the S/E ratio is at least partially compensated by human skill and ingenuity, and ultimately through the semiotic control exercised by the input of human knowledge. Thus, in healthy traditional agriculture, the yield is still low and essentially constrained by biosemiotically controlled natural nutrient cycles, as well as by a whole range of nonbiotic limit factors (weather, seasons, rainfall, etc.). The industrial revolution resulted in our learning how to bypass all or most of these natural constraints through an explosive increase in the use of artificial energy. In an agricultural production system based on the petrochemical service industry, an enormous homogenization of nature can be obtained, yielding unequalled amounts of crop. Or to state this in different words, industrializing agriculture meant that we learned to circumvent nature's own semiotic controls on energy flows, replacing them with a few comparatively unsophisticated controls, such as market-driven time schedules for various operations such as sowing; irrigating; the spreading of fertilizer, herbicides, and pesticides; and harvesting.

I do not pretend here to offer a satisfactory comparison between the ecosemiotics of petrochemical agriculture and the ecosemiotics of more traditional kinds of agriculture, but I think that even in the absence of such a deep analysis, the overall picture is indisputable: the S/E ratio is dramatically lower today. And this dramatic decrease, I suggest, is the deep source of the modern environmental crisis.

In addition, when we talk about an industrialized production system, we cannot limit ourselves to analyzing just the agricultural sector, for agriculture is just one interdependently integrated component of the diversified production apparatuses of industrial society. From en ecological point of view, too, industrial agriculture may well be the worst single factor contributing to the deterioration of nature's capacity for sustainability — but it is by no means the only one. The overwhelming use of energy from chemicals ultimately derived from oil in almost every context of modern society inevitably ends up destabilizing, and often fatally overwhelming, the semiotic controls operative inside living systems at both the organismic as well as the ecological level (our own bodies included). Thus, into our estimate of the E component, we must include all kinds of energy flows staged by our manifold production systems.

In sum, we can see that the Industrial Revolution was, at best, only one half of a full revolution. For starting with the water wheel in the Middle Ages, and continuing through the mastery of coal, oil, and uranium, we gradually liberated the energy component of natural systems from their biosemiotic controls. However, we did not until very recently try to evolve techniques to compensate for this liberation of the power of energy by a corresponding mastering of the power inherent in the increasingly displaced biosemiotic controls. As a result, we have created an ecological impoverishment of the environment, reflecting the uninhibited dependence of our production systems on the brute force of artificial, non-biosemiotically controlled energy. And as a consequence, we are now unable to reestablish a sustainable production system, because we have neglected to take heed of the naturally occurring checks and balances of life on Earth as a result of the second law of thermodynamics.

The task ahead of us is to embark upon the second half of the industrial revolution. And this will consist in the development of a mastery of the biosemiotic controls that can match (and thus sophisticate) our present mechanical mastery of the gigantic energy flows that, in an overpopulated world, necessarily must destabilize nature's optimal balance points. Another way to say this is that *we need to develop a biosemiotic technology base for our production systems* — a technology base that can replace natural biosemiotic control mechanisms with biosemiotic control mechanisms artificially set to fulfill human and environmental needs.

Therefore, instead of speaking about the *industrial* revolution, we should now realize that the deeper principle at work behind this major, and yet sadly unfinished, historical transformation has always been something quite different — i.e., the substitution of a resource base relying on naturally controlled energy flows for a resource base relying on energy flows controlled by human ingenuity and skill. In fact, what we are witnessing is one long *process of the humanization of nature*. A process which slowly took off back in Neolithic times, then acquired an enormous momentum through the industrial revolution of the last three centuries, and is now reaching a final turn where the dangerous destabilizing of the energetic, semiotic, and entropic aspects of the natural world have to be continually restored artificially, and at an increasingly high level.

Clever Technologies

The American plant physiologist J. S. Boyer once calculated that the average harvest yield for a range of crops even in the most efficient U.S. agriculture was only 21.6 percent of a calculated maximum (Boyer 1982). Diseases, weeds, and pests would account for less than 10 percent out of this total loss of nearly 80 percent — whereas the rest would be due to suboptimal physico-chemical conditions: lack of water, cold, high salt levels, etc. As Boyer himself pointed out, the reason for this lack of efficiency is to be found in the current breeding strategies of the crops chosen to be grown.

Boyer found that breeders typically produce a small number of "miracle" varieties that give extremely high yields, provided they are grown under favorable conditions. Now, the bad thing about the natural world is that conditions in it are rarely reliably favorable — resulting in the fact that farmers have to buy services from the petrochemical industries in order to artificially create a situation of favorable conditions. However, as Boyer's numbers disclose, this strategy not only makes farmers dependent on expensive and ecologically destructive practices, it also apparently fails to deliver the true product, i.e., the favorable conditions promised.

The obvious solution to this problem is that instead of homogenizing the soils of the Earth so as to fit only a few new miracle crops, we should breed crops that fit the multitude of "unfavorable" conditions under which real-world farmers actually live. We need one particular kind of wheat for traditionally favorable conditions, another for salty soils, yet another for soils that are too wet, and so on. Such a strategy, however, was not feasible within the paradigm of traditional crossbreeding, where typically ten to fifteen years would be needed for the production of each new variety. But clever new breeding technologies based on modern biotechnology, and particularly on genetic manipulation, have the potential to support a diversified breeding strategy that is aiming at producing varieties fitted to local conditions.[6]

---

6   Whether or not political and economic conditions will permit the implementation of breeding for local needs is quite another question, of course. And since multinational petrochemical industries are main actors in the breeding market, one should perhaps not be too confident that breeding for the purpose of uncoupling agriculture from petrochemical services will soon become a major priority. My purpose here is simply to point out the potential capacity of biosemiotically influenced techniques to solve the environmental problems that have become endemic to high-energy society. For if it is not possible to make profits by solving these problems with biosemiotic technologies, I find it hard to see how they are going to be solved at all. While many people believe in a return to the frugal ways of life that characterized the societies of our grandparents, I personally find such suggestions to be nothing but escapist daydreams, considering how many billions of poor people are still in need of an elementary subsistence that can only be supplied at the cost of higher energy consumption in the areas where they live. Thus, it seems to me that the main

Using gene-technology-based breeding strategies to produce varieties for the world's variable local conditions is fundamentally a biosemiotic strategy — first, because it aims at naturally persuading the plant species to grow well under specified conditions, rather than commanding them to do so by the application of brute force (petrochemistry), and second, because the strategy will only succeed if actually developed in a collaborative effort with local farmers.

The strategy of breeding for local conditions is just one example out of a multitude of cases where the use of biosemiotic technology instead of petrochemical technology would move our production systems towards the goal of sustainability. To mention just one other example, let me point out the striking fact that fewer than thirty plant species cover 95 percent of our total need for living provisions, and that just three grasses — wheat, rice, and corn — alone furnish 75 percent of the total human need for food. Why, then, are so few of the world's estimated 300,000 species of plants in human use?

The explanation seems to be nothing more than historical contingency — and this promises that unknown potentials are hiding in many other plant species, not only for the production of crops for food and fodder, but also for the production of crops intended as raw materials for the fermentation industry, e.g. the making of biodegradable plastic. Indeed, prior to World War II, chemical engineering was based mainly on biological raw materials, but this has all been forgotten in the subsequent petrochemical rush. Returning to biomass that has been biosemiotically tuned to our needs as raw material will also fight the greenhouse effect, because the CO_2 intake of growing biomass balances out the CO_2 output from using that same biomass.

And agriculture is by no means the only sector in which biosemiotic technology holds great promise for the goal of sustainability. Substituting chemical technology with biosemiotic technology in many cases would be clearly favorable to the environment and probably less hazardous to health, too. I would suggest, therefore, that the formula $S/E = K$ (with K in naturally occurring ecological contexts attaining the average value of one) be considered as a guiding heuristic in our reflections on ecosemiotic strategies. The general effect of introducing biosemiotic technologies will be to decrease the necessary deployment of energy, due to a far better control of its effect on specific parameters. And this, of course, amounts to saying that S is increased at the same time that E is decreased in the system. Thus, the introduction of biosemiotic technology will contribute to our goal of approaching the situation where $S/E \geq 1$, i.e., a sustainable production system.

---

difficulty in this case is not the absence of technological possibilities, but the absence of the ability (or will) to find new and more appropriate political possibilities — a problem that this book, I am sad to say, takes no direct aim at solving.

Modern Biotechnology

The misguided belief in the autonomy of the gene as the controlling agency in the processes of life characterized much of the early work in biotechnology — but has now gradually been replaced by a growing preoccupation with the signal processes responsible for the integration of gene functions into the whole matrix of cellular and organismic life (such as, for instance, obeying boundary conditions as determined by the state of the organ or tissue in which the cells are situated, or with controls induced upon cells from neighboring cells as well as from more distant sources, such as in the psychoneuroimmunological system). The power of modern biotechnology, in fact, is based exactly on its ability to hitchhike upon the semiotic controls that are native to the cell or organism.

For an illustration of this phenomenon, let us here consider just one classical case. In their ongoing fight against pests, farmers long ago learned to use the bacterium *Bacillus thuringiensis* (Bt) as an ally. Like so many other bacteria, when starved, Bt forms a very resistant spore to keep it alive until better conditions arise. Inside the spores, the bacteria leaves a protein crystal, an endotoxin, that dissolves in the intestine of susceptible insects, should they be unfortunate enough to munch upon leaves sprayed with Bt spores. The crystals paralyze the cells in the gut, interfering with normal digestion and triggering the insect to stop feeding on host plants. Bt spores can then invade other insect tissue, multiplying in the insect's blood, until the insect dies.

Bacillus thuringiensis was first isolated in 1911 from sick Japanese silk worms and from 1958, the bacterium was brought to commercial use as a *biopesticide* — a biological weapon to fight pests. A suspension of Bt is sprayed over the crop when certain specific pest attacks begin. The toxin has no effect in mammals, but specifically targets the species of a particular group of insects called *lepidoptera,* i.e., butterflies and moths. Using this biopesticide, however, is not without problems. Often the crops may have to be sprayed repeatedly, because most of the time the lepidopteran larvae are hiding inside the stalks, where the bacterial poison cannot reach them. And although Bt has a rather high specificity, benign insect species are often also damaged by the treatments.

With the help of gene technology, the gene for the same toxic protein that is secreted by Bt has now been transferred directly into a range of ordinary crop species such as tobacco, cotton, and corn — and this results in a much narrower coupling of the toxic effect to the actual cause for the damage. For since the toxin is now excreted by the plants themselves, it will destroy the larvae no matter where they are located in the plant. Ideally, and perhaps in the future, the gene can be implemented in the genome in such a way that it would only be expressed when induced by the saliva from lepidopteran larvae munching upon the plant.

But many environmentalists have protested against the introduction of such genetically modified crop plants, and the reader may remember the story about how the beautiful monarch butterfly was severely threatened by the use of GM (genetically modified) crops. Accordingly, the Danish Society for Nature Conservation inserted full-page ads in major newspapers telling readers that "American experiments have shown that larvae of the Monarch butterfly are dying from feeding on pollen from corn plants genetically modified to produce insect toxins." But ecological concerns are not well managed by conservation groups blowing hot air and barking up the wrong old-growth-forest tree. For since the larvae of butterflies are themselves agricultural pests, it is obviously not possible to have rational farming without somehow hampering the life of butterflies. It is also unavoidable that problems of resistance will appear, but planting GM crops, in this respect, is no worse than spraying with Bt. And, as might have been expected, Monarch butterflies do, in fact, persist quite well in the vicinity of fields with GM crops. In fact, repeated Bt sprayings probably have far worse effects upon neighboring insect species than do GM crops with localized and specific effects.

Naturally, however, since the flow of money into this developing field of commerce is so incredibly huge, it will be important to watch over the whole business very carefully. But if the intent really is to care for the conservation of living nature without giving up on the need for rational farming, then some of the key concepts needed will be precision and gentleness — and the operationalization of such concerns is exactly what biosemiotic technology offers as its most important potential.

Proteomics

The study of *genomics* (i.e., the study of all the genetic material in a genome) has lately become supplemented by a new and far more complex field of investigation called *proteomics* — i.e., the study of all the proteins expressed by a genome, or in more practical terms, the study of the totality of protein-based activity that goes on in a cell.[7] And because of the incredible complexity involved, sophisticated computers are an absolutely necessary tool for this new kind of research — research that also includes the still-developing field of *bioinformatics*, which attempts to simulate cellular processes based on huge amounts of empirical data.

These new research strategies will no doubt drive even further the innovation of new biotechnological processes, as, for example, the development of

---

7 The tacit implication of this term is that a functional entity, a *proteome*, exists in cells just as does the physical entity of the genome. The idea seems a little far-fetched, since a structural and temporal definition equivalent to that of the genome cannot be given to this proteome.

new pharmaceuticals based on the specified interference of interaction patterns between cellular proteins is expected to profit from these new techniques. It will be interesting to follow the development of this field, and to see how far that it will get us. For from the theoretical perspective, the big question is whether or not the implied *systems approach* will manage to overcome the handicap introduced by the fact that the computer operates according to dyadic machine logic, and not according to triadic bio-logic.

And yet, the pivotal point where triadic logic may be expected to make a crucial difference is at the dynamic interfaces *between* different compartments of a biological system. These are the previously discussed *interfaces* that inevitably come into play as soon as semiotic control functions begin exerting influence across different levels in the hierarchical arrangement of interaction. An animal, for instance, has an *immediate* level of behavioral activity that is regulated through the intricate interactions between its brain and its endocrine system. Yet this endocrine system likewise depends upon the activity of a distributed gland tissue, each of whose single cells must respond accordingly to the holistic requirements of the moment. And inside each of these glandular cells, millions of highly integrated catalytic processes are continually taking place, some to maintain normal cellular metabolism, and some to assure the required production and secretion of specific signal molecules (e.g., hormones) to the surrounding cell media, as such secretion itself has been requested by signals instantiated by higher level controls.

So, in this case the activities of (at least) three highly interactive levels — the level of the single glandular cell, the level of the endocrine system, and the level of animal behavior — must be tuned to each other. This process of a reciprocal tuning between activities at different levels presupposes the presence of significant semiotic competence at each of the levels. The endocrine system produces a hormonal output that is supposed to optimize the ability of the animal to execute the activities that the brain tells it to initiate. But for this to work, the organism must be capable of regulating the activity of a variety of glandular tissues under the guidance of multiple signals from the nervous system, as well as from the immunological system, simultaneously.

Seen from a biosemiotic point of view, the activity of the endocrine system from moment to moment instantiates *interpretants* in the form of (evolutionarily established) weighted adjustments to the multitude of sign inputs by which the momentary situation of the organism (spatially as well as temporally) lets itself be felt in the body — while corresponding sets of sign relations take place on the level of the cell and on the level of the animal's behavior. It seems likely that any technological intervention in complex biosemiotic networks such as this one would stand a better chance of success if it was based upon an understanding of the triadic relations operative in the maintenance of the target system's own biochemical organization.

There are, however, no principled reasons to deny a priori that a dyadic systems-based technological solution could not function well, even in the absence of any knowledge of the triadic relations that are actually ruling the system. The degree of genuine semiotic freedom — both at the level of the single cell and at the level of the endocrine system — is, as we well know, very limited. Thus, a repertoire of possible responses could presumably be enumerated in a matrix that might catch as many as perhaps 99 percent of all situations, and this might be seen as satisfactory from a technological point of view. Another way to express this is to say that the semiotic buttressing of the organismic interaction patterns via evolution have been frozen to such an extent that they may well now be treated with a model based on purely dyadic relations. Such a model does not, of course, explain why the patterns of interactions have become frozen in exactly the way they have. But with a sufficient amount of empirical data and calculatory power, one may quite possibly make some qualified guesses about how such interaction patterns work.

In principle, this task is somewhat like the task that a visitor from Mars would face if asked to discover the dynamic organization of, say, the electricity supply in New York City without taking any notice of the lifestyle and the customs of the inhabitants. Perhaps it might indeed not be necessary for such a calculation to know anything about the circadian rhythms of the population, the activities of their family and work lives, their habits of entertainment consumption, or their methods of transport and social organization. The only thing needed would be a combination of enough raw empirical data and calculatory tools and then, sooner or later, the Martian and his machine could come up with a model that — to an extent — could predict regular changes in the pattern of electricity consumption in New York.

Or could it?

There is no need even to venture an answer to this question, for in any case, such a procedure is, on the face of it, both counterintuitive and clumsy. The more promising approach, of course, would be to invest the proper resources into a study of what really goes on within the system (the metropolis) each time power is applied around the city. In the absence of a theory of what electricity is good for, the task of calculating its dynamics based solely on raw (unprocessed) empirical data would be unnecessary difficult, and perhaps ultimately futile.

And yet proteomics essentially builds upon the conception of computer calculation as a substitute for a theoretical understanding, and although the technique most probably will lead to many important results, the idea that it will really help us understand what is going on within and between all these networked interactions seems dubious. Indeed, Nobel Prize laureate and biochemist Alfred Gilman has told *Scientific American*, "I could draw you a map of all the tens of thousands of components in a single-celled organism and put all the proper arrows connecting them (and even then) I or anybody else would look at that map and have absolutely no ability whatsoever to predict anything" (Gibbs 2001, 53).

But even if the supposition that using computer simulations is a superior way of understanding cellular life bears some practical fruits, it is, in a way, pulling the wool over our own eyes. For one should not let the occasional success of a dyadic-systems approach be sufficient reason for believing that organisms do in reality function in the same way as do systems of dyadic relations. Perhaps it might be useful, in this connection, to ponder what might have happened if the physicists at the time of Copernicus had had access to computers. In that case, the need for rejecting the fundamental belief in the Sun as moving around the Earth — rather than the opposite — might not have been felt as pressing. For a computer could surely come up with a satisfactory model of a geocentric universe based, as required, on infinitely many epicycles. And if such software was then installed on the computers to be used by the officers of ships, they may have equally managed to navigate just fine by this Advanced Geocentric Model — perhaps dubbing the project *astronomics*.

In all seriousness however, there is little doubt that proteomics promises to develop as an important technological tool. But since the essential principle pertaining to a genuinely biosemiotic technology is based on the possibilities of taking advantage of our knowledge about the biosemiotic controls that actually do govern the processes of life, a biosemiotically informed technology must take care not to forget these irreducibly triadic, naturally occurring (not merely computer-simulated) controls. For nature is not stupid — although science has more or less treated it as such, at least ever since Newton. Nature, in fact, is incredibly refined in its internal ways of regulation. So if we want high productivity based on a sustainable technology, we must make the effort to arrange our technological means in such a way that it works in accordance with, rather than against, the internal workings of nature. We must learn to *speak with nature* in its own language. And to do so, of course, is to implement a biosemiotic technology.

Postscript:
Short Historical Notes

Sebeok's Vision Comes True

A biosemiotic approach to life had been independently suggested several times throughout the twentieth century, yet — in all but one case — without getting any firm hold. Only in the last decade of the century did the different and sporadic attempts that had nourished the idea in its preliminary phases assemble to become a stronger current. Most decisive for this combining of the efforts was the unremitting diligence that the late Thomas Sebeok (1920–2001), American linguist and semiotician, invested in furthering the development and general recognition of the attempt to bridge biology and semiotics. This he did for more than four decades, and in the 1990s, the time finally was ripe for the biosemiotic project to take hold.

During most of these decades, Thomas Sebeok was the head of the world's leading center for semiotics, the *Center for Language and Semiotic Studies,* at Indiana University in Bloomington, and president of the *International Association for Semiotic Studies* — as well as editor-in-chief of the scientific journal of this association, *Semiotica.* Throughout his life, Sebeok nourished an intense interest in animal communication as well as in biology in general. It was Sebeok who, as early as 1963, launched the interdisciplinary field of zoosemiotics — and who, in 1968, suggested that "a full understanding of the dynamics of semiosis (may), in the last analysis, turn out to be no less than the definition of life" (Sebeok 1968; Sebeok 1985 (1976), 69).

Sebeok often mentioned that he had read a 1926 English translation of Jakob von Uexküll's book *Theoretiche biologie* in 1936, but found it uninteresting and nearly incomprehensible. Only many years later did he come upon an original German edition of the book, and there the text was both clear and meaningful. He concluded that the quality of translation in the English version had been miserable — and that this may have been a significant reason for the scant attention paid to Uexküll's work in the anglophone world at that time.[1] At this point,

---

1   However, the English version that Sebeok read in 1936 was based on the German first edition from 1920. In 1928, Uexküll's book appeared in a new revised edition, and this was the one Sebeok much later read in German (Sebeok 2001a).

Sebeok sought out Uexküll's entire oeuvre, and arranged for competent English translation of his 1940 *Bedeutungslehre*. The book appeared in 1982 in a special issue of *Semiotica* that included an introductory article by Jakob von Uexküll's son, the late Thure von Uexküll, then professor emeritus of internal medicine at the University of Ulm (Uexküll, T. v. 1982; Uexküll, J. v. 1982 (1940)).

It was through their collaboration on this project that Tom Sebeok and Thure von Uexküll became personally acquainted, and the friendship between these two men resulted in the first English-language international meeting devoted exclusively to *biosemiotics,* held in Glottertal, Germany in 1990. The meeting was organized by Jörg Hermann and took place in the beautiful Schwartswald landscape at the ReHa Klinik Glotterbad where Herman was, and still is, head of the medical section. This meeting resulted in the publication of a book assembling contributions from many of the people that were to go on to develop the field over the course of the next decade (Sebeok and Umiker-Sebeok 1992). This first meeting was followed in 1991 by a second meeting on biosemiotics, and I vividly remember the euphoric mood characterizing these early meetings which — at least as we attendees saw it — ushered in the birth of a frail but exciting new field of research that all of us were extremely enthusiastic about.

Biosemiotics has since then become a featured theme of many special sessions at scientific conferences on biology, semiotics, and philosophy, and a fast-growing number of publications on biosemiotically focused topics continues to appear each year. At the University of Copenhagen in 2001, the first of a series of annual international conferences entitled "Gatherings in Biosemiotics" was convened around the study of biosemiotics — with a special focus on its significance for theoretical biology.[2]

Needless to say, however, biosemiotics still plays only a marginal role, if any, in mainstream biological thinking. Perhaps this is partly due to the inherent resentment of science against the seemingly anthropomorphic and teleological aspects that — admittedly — inevitably accompany the idea of *natural semiosis* (as discussed in depth several times in the present volume, most notably in Chapters 3 and 4), as well as to the instrumentalistic preferences of empiricist experimentalists (as discussed in Chapter 9). (I shall not rehash the rebuttals to all those arguments here.)

My intention here, rather, is to offer a brief account of the early beginnings or root forms of the biosemiotic approach to the study of life, as can be found in twentieth-century biological and philosophical thinking. It is, of course, a

---

2   Gatherings Two took place in Tartu, Estonia in 2002; Gatherings Three in Copenhagen, Denmark in 2003; Gatherings Four in Prague, Czech Republic in 2004; Gatherings Five in Urbino, Italy in 2005; Gatherings Six in Salzburg, Austria in 2006; Gatherings Seven in Groningen, Netherlands in 2007, and Gaterings Eight is scheduled for Syros, Greece 2008.

task for a historian of science to unravel all the relevant threads. The following account should be seen only as an attempt to single out some of the contributions seen — through the biased eyes of this author — to have been important for the unfolding of the biosemiotic idea. For a more thorough analysis of the history of biosemiotics, the reader is referred to Don Favareau's extensive chapter in *Introduction to Biosemiotics: The New Biological Synthesis* (Favareau 2006), and Kalevi Kull's shorter account in *Biosemiotica* (Kull 1999a).

Beginnings

The two main pioneers in this story, as I see it, are the American scientist and philosopher Charles Sanders Peirce (1839–1914) and the Estonian-born German biologist Jakob von Uexküll (1864–1944). As far as we know, these two major figures were not acquainted with each other's work — and it is also somewhat dubious, I believe, to suppose that their respective intellectual styles would have been conducive to a reciprocal understanding. Their thinking was, however, undoubtedly connected by their broad-minded conceptions of the phenomenon of life, which was not current, let alone acceptable, in their own time (or today, for that matter).

Peirce's semiotics was taken up again after his death by Charles Morris (1901–79), while Uexküll's Umwelt theory continued as *behavioral biology* (ethology) through his pupil, Konrad Lorenz (1903–89). But the radical dimensions that lay hidden in Peirce's and Uexküll's ideas were left behind by their successors in the next generation. A main cause for this may be that *scientistic* conceptions hardly ever exerted a more hegemonic hold over scientific inquiry than in the first half of the twentieth century, before the innocence of science was lost due to the atomic bomb and other technological disasters such as those caused by thalidomide.

Nevertheless, Morris's and Lorenz's work, each in its own ways, has provided insights that, in the end, have helped combine the two separate streams of biosemiotics — biology and semiotics. Morris's foremost contribution here was his insistence that the science of signs must to built upon biological science, which in his context meant a science of behavior — and this has led many to call Morris a behaviorist. But in her biographical essay entitled "Charles Morris's Biosemiotics," Susan Petrilli emphasizes that Morris's use of this word *behavior* does not imply *behaviorism* in the proper sense of the philosophical-psychological trend of B.F. Skinner and the like. Rather, the idea is that a special science must be created to study behavior as a biologically anchored phenomenon (Petrilli 1999). According to Petrilli, this proposal does not imply the same reductionist naturalism that would so forcefully come to characterize later behavioral science under the headings of ethology, sociobiology, and (lately) evolutionary psychology (ibid., 67).

Admittedly, Morris's semiotics does part from Peirce's semiotics at several points, the most important of which is his tight coupling between semiosis and behavior that has, as its implication, the result that the concept of *interpreter* more or less fuses with the concept of *interpretant*. For Morris, semiosis was always connected to an organism (an interpreter) — and for that reason he could not accept Peirce's suggestion that *thought* (and thus semiosis) "is not necessarily connected with a brain. It appears in the work of bees, of crystals, and throughout the purely physical world, and one can no more deny that it is really there than that the colors, the shapes, etc., of objects are really there" (*CP* 4:551).

This is a point of view Morris rejects as "idealistic metaphysics" (Petrilli 1999), and it must be admitted that the quotation above puts one's tolerance for Peirce to a serious test. Peirce here, as Frederik Stjernfelt (2007, 43) has noticed, may be seen to have fallen into the trap of absolute idealism: "The Hegelian tendency to globalize notions like mind, teleology, even personality to the whole of cosmic evolution and the whole of the universe." Yet although it is tempting to agree with Morris against Peirce on this point, it should be remarked that the idea of bee intelligence, at any rate, has been taken up lately within the research field of *artificial life* under the designation *swarm intelligence*. Peirce thus has a strong point in rejecting the notion that thinking should be reserved for brains, which Morris surely took for granted.[3]

For Morris, semiosis is connected specifically to the behavior of animals, for semiosis presupposes an ability of the organism to respond to stimuli, and this again requires, he assumes, the presence of both sensory receptors, as well as effectors (e.g., muscles and glands). Therefore, for Morris, microorganisms — as well as individual cells within an organism's body — fall below the threshold where semiosis can take place. Although Petrilli may be right in claiming that Morris did not nourish reductionistic inclinations, it seems to me that Morris's strict localization of the *agency* of nature in the muscular processes of individual organisms implies an inherent reification of behavior that can not properly be reconciled with a processual (biosemiotic) conception of nature. For as I have maintained throughout this work, semiotic causation, and thus agency, is a process inherent in life from the earliest formation of living systems.

---

3   I personally would reject the notion that thought should appear in crystals or "throughout the purely physical world" since I don't believe that the necessary semiotic sophistication has evolved in such inanimate systems. But as soon as living creatures are found populating the world, I find it fruitful to imagine that thinking-like processes do occur within them — or at least between them, as with the bees — and at the meta-level of the evolution of their lineage.

Thomas Sebeok tells us that Morris, in the late 1930s, had been the first to urge him to "assiduously read whatever fragments of Peirce's semiotics were accessible at Chicago." Nevertheless, in the same paragraph, Sebeok (1986, 66–67) quotes John Dewey, who dubbed Morris's *behavioral semiotics* "a complete inversion of Peirce" — "a judgment," Sebeok adds, "with which I happen to concur."

Similarly, Konrad Lorenz is probably best known as the founder of *ethology* — a science that in 1973 became canonized through the awarding of the Nobel Prize in Physiology and Medicine to be shared among Lorenz and fellow ethologists Karl von Frisch and Nicolas Tinbergen. Although Lorenz has clearly stated the debt of ethology to the early work of Jakob von Uexküll,[4] it must be admitted that the recognition of ethology as a proper scientific discipline was obtained only because Lorenz turned his attention away from the Uexküllian Umwelt. The question of how animals *conceive* their surroundings was replaced in ethology by the question of animal *behavior*, which was then described to a large extent as the result of *inborn instincts*.

And so, instead of seeking a proximal explanation of behavior from the specific Umwelt of the animal, ethology devoted itself to the study of more distal explanations based on genetic dispositions and, ultimately, on natural selection. Thus, Tinbergen (1942, quoted in Lehrman 1953 (2001), 26) defines instinctive behavior as "highly stereotyped, coordinated movements, the neuromotor apparatus of which belongs, in its complete form, to the hereditary constitution of the animal," and Lorenz (1939, (quoted in Lehrman 1953 (2001), 26) talks about characteristic behavior as "hereditary, individually fixed, and thus open to evolutionary analysis."

The American psychologist Daniel S. Lehrman (from whom the above quotations are taken) objected to these conceptions as early as in 1953 since, as he pointed out, patterns of instinctive behavior — such as, for example, the ability of newly hatched bird young to peck seeds, or a female rat's typical nest-building behavior — only superficially *appear* to be unitary behaviors. When studied in more detail, such instinctive behaviors are found to actually consist rather of *conglomerates* of motoric activities, *each* with its own developmental dynamics (Lehrman 2001 (1953)). The strong focus of ethology on the conception of *instincts* as unitary packages of behavior thus became a sort of midwife for the concept of *genes for traits* that has distorted our understanding of the role of genes up to this day (see Chapter 5, and particularly the section on *C. elegans* and the human genome).

---

4 In 1971 he wrote that ethology "certainly owes more to his [Uexküll's] teaching than to any other school of behavior study" (Lorenz 1971, cited in Sebeok 2001a, 72; also see Stjernfelt 2001).

Against this, Lehrman observed that *inborn instincts* usually arise only through the real-time accomplished *interplay* between many independent factors that all must be in place for a successful behavior to obtain. For example, the communicative activity inside the embryo is decisive to the realization of such behaviors, as it assures the coordination of activities into one functional unit (e.g., reflexes and motor schemas). And in many cases, as for instance in the nest-building behavior of rats, the necessary coupling between the separate activities that, taken together, constitute the "instinct" can *only* occur in interaction with the external environment. The distinction between instinct and learning, then, is far from as clear-cut as has often been assumed.

In the same year that Lehrman launched this critique of ethology's concept of instinct, an epoch-making new discovery took place that signaled the establishment of a third — and wholly different — track leading to the biosemiotic conception of life. For in 1953, James Watson and Francis Crick published their famous double-helix model for the structure of the DNA molecule, and this discovery opened the doors to a deep understanding of the nature of the genes that, at the same time, made *exhaustive* explanations in terms of gene action very attractive to many scientists.

The Language Metaphor

With such fundamental concepts as *genetic code, messenger RNA, transcription* and *translation* it is obvious that the revolution of molecular biology, right from its beginnings, implied a powerful semiotic input to biology. But unfortunately, the communication sciences of that time were strongly attached to classical (dyadic and mechanistic) information theory, with the result that this ripe moment for the long-overdue semiotization of biology was castrated, so to say, at the outset by the ontological reduction of *semiosis* to *information transfer* (see Chapter 3, the section "The Central Dogma").

Nevertheless, molecular biology, the field in which I myself was trained, soon gave rise to a deep understanding of cellular communication processes that indirectly came to pave the way for the formulation of the modern project of *biosemiotics*. For very early on in the development of molecular biology, attempts were made to apply linguistic analysis to DNA's sequential syntax and coding function. Laura Shintani (1999) reports that, in 1968, French TV broadcast a debate with the title *"Vivre et parler"* ("To Live and To Speak") that featured the participation of two biologists, François Jacob and Phillipe L'Héritier; the anthropologist Claude Lévi-Strauss; and the linguist Roman Jakobsen.

Strongly impressed by Jacob's farsighted book *The Logic of Life: A History of Heredity* (1974), Jakobsen returned several times in the ensuing years to the relation between DNA and language, and he treated this idea quite extensively in several chapters of his *Main Trends in the Science of Language* (Jakobsen 1973). Jakobsen finds that the genetic texts and their structure involves the same principle of *double articulation* that is operative in all human language — and which in both cases is responsible for the ability to produce an infinite quantity of messages from a finite quantity of sign vehicles (phonemes and trinucleotide codons, respectively). In this connection, Shintani (1999, 109) offers us the following quotation from an undated manuscript of Jakobsen's (once scheduled for publication in the *New York Review of Books* which, for unknown reasons, never published the article):

> The two turning points in evolution are, first, the emergence of life and, nearly two billion years later,[5] the evidently conjoint emergence of thought and language. Thus, perhaps, the biological confrontation of these two widely separated evolutionary gains — life and language — from the perspective of the earlier turning point, may be supplemented by attempts toward a retrospective, *ergo* initially "linguistic" interpretation, the more so as François Jacob himself has convincingly endowed the linguistic model with "an exceptional value for the molecular analysis of heredity" (Paris journal *Critique,* March 1974).[6]

Jakobsen did not go on to develop these ideas any further, but the conception of DNA as a linguistic-like system has since then inspired many theoreticians in linguistics as well as in biology (Emmeche and Hoffmeyer 1991; Ji 2002). In the context of this book, we have seen this conception in Pattee's distinction between the dynamic and linguistic aspects of life processes (Chapter 4).

Gregory Bateson, too, had concerned himself with language as a model for life, but his conception of this model stretched beyond the level of the DNA, not limiting itself to the merely syntactical level, but drawing in also the pragmatic dimensions of language: "Both grammar and biological structure are products of communicational organizational process. The anatomy of the plant is a complex transform of genotypic instructions, and the 'language' of the genes, like

---

5   Today this span of time normally is given as nearly four billion years.

6   I cannot in this context resist the temptation to quote my own first book (from 1975, not translated to English): "Over the entire course of evolution, only one event may perhaps be comparable in kind to the evolution of the human species, and this is the formation of the first cell out of lifeless nature. The point of similarity between these two radical departures is this: The cell transcended the dictates of physical law by making chemistry biological. Humans transcended the control of chemistry by making biology mental. The human being has by force of its 'free will' evaded the dictates of the chemical code in exactly the same way that the cell evaded the dictates of the second law of thermodynamics — i.e., by the help of its biological organization that (locally) allowed it to decrease the surrounding entropy" (Hoffmeyer 1975, 104–5).

any other languages, must of necessity have contextual structure. . . . The tissues of the plant could not 'read' the genotypic instructions . . . unless [both] cell and tissue exist, at that given moment, in a contextual structure" (Bateson 1972, 154). Through this and similar observations, Bateson brings us from the purely linguistic to a proper biosemiotic frame of understanding, although he never used the term *biosemiotics*. Seen from a biosemiotic point of view, the whole problem with the linguistic metaphor for the genetic system is precisely that it serves all too well to privilege the digitally coded communication system of the organism that is based on the sequential structure of the DNA molecules — at the expense of communicative activities based upon analog codes that, in this context, are often seen as just a biochemical substrate without autonomous communicative function. Indeed, I would argue that Bateson's most important contribution to the later development of the biosemiotic approach to the study of life was exactly his unfailing emphasis on the interplay between analog and digital codings in the organization of living systems.

Thus, while Bateson's work has only been touched upon sporadically in this book, it has indeed been a decisive source of inspiration for modern biosemiotics and, not the least, for this author (Hoffmeyer 2008a). Most importantly, Bateson's famous definition of the smallest unit of information being *a difference that makes a difference* has been an early eye-opener for all of us in the field — for this definition stems from Bateson's insight that everything that may be subjected to our understanding must rely on temporal or spatial differences in order to be so apprehended. Receptors and sense organs can only react upon differences or changes, and everything that may be known must therefore in the final analysis be built upon the detection and analysis of differences. But a difference is par excellence a subjective phenomenon, one that does not exist or occur as a *difference* in the real world unless someone makes it so by responding to it as one — unless, in other words, the difference *makes a difference* to someone (or to some receptive system).

Another central preoccupation in Bateson's work that has been influential in biosemiotics is that of *the pattern that connects*. This notion expresses the basic problem that fascinates him in the study of the living world in its broadest sense — i.e., as comprising both the biological and the social. Here the idea comes expressed as the question he asks himself in the introduction to *Mind and Nature*: "What pattern connects the crab to the lobster and the orchid to the primrose and all the four of them to me? And me to you? And all the six of us to the amoeba in one direction and to the back-ward schizophrenic in another?" (Bateson 1979, 8). Behind this attempt to trace *the pattern that connects*, we find Bateson's concept of *abduction* as an unavoidable methodological principle in the sciences of communication — and biology, in his conception, belongs to the sciences of communication.

Abduction — a concept that he explicitly acknowledges to have inherited from Peirce — was, as Bateson understood it, a way of constructing hypotheses by use of an associative transfer of abstract relations from one area to another, under the presumption that the two areas are governed by similar rules. As such, abduction has close relations to another central element in Bateson's thinking, the principle of *double description*, a method that was inspired by the concept of logical types which he borrowed from the work of Russell and Whitehead (1910–13). Bateson claimed that the product of double description belongs to a higher logical type than do the phenomena that were abductively compared. The similarities reached by abduction are here seen as cases on which to build an inductive inference that brings us to a higher logical type.

In a recent analysis of this aspect of Bateson's thinking, Julie Hui, Ty Cashman, and Terrence Deacon (2008) suggest that even Bateson's seemingly far-fetched claim of a unifying pattern connecting the crab-lobster pair to the orchid-primrose pair may be seen to hold true when analyzed in terms of molecular genetics: "The general developmental logic of duplication and differentiation is effectively a unifying pattern underlying the generation of biological form in organisms with respect to environmental constraints."

Peter Harries-Jones (1995, 177) notes,

> [For Bateson,] abduction was like qualitative modeling — a means of undertaking formal comparisons through contrasts, ratios, divergences of form, and convergences. His uses of the techniques of abduction were remarkably similar to those of identifying "resemblances" in a comparison of "language games" which Wittgenstein had originally proposed. . . . The notion of resemblances derived through comparison of differences is vividly captured in Wittgenstein's image of the thread among resemblances in language games: or, as Wittgenstein puts it, in overlapping and criss-crossing fibers of a woven pattern.

Seen from the biosemiotic standpoint, the reason why abductive inferences lead to fruitful results — even though such kinds of inferences are not logically "clean" — is that nature itself possess semiogenic creativity, or the tendency of things to take habits, as Peirce called it. What this means is that natural systems tend to arrive at available solution models that, due to the inherent universals of life processes, tend to exhibit important common features. The feathers of extinct dinosaurs that in their origins were means of thermoregulation, were later in evolution — abductively, so to say — to become the means of flight in birds, through the process that Gould and Vrba (1982) call *exaptation*. Sciences based on the ideal of induction may thus well abhor abduction, but — as Bateson reminds us — nature itself does not. Accordingly, he felt that naturalists afraid of tautologies will overlook "many important orders of phenomena" (cited in Harries-Jones 1995, 179).

Among the many scientific approaches that have been inspired by Bateson, I must mention here *second-order cybernetics* — a field that Søren Brier, since the 1990s, has led in an effort to combine cybernetics with biosemiotics to form the new discipline of *cybersemiotics* (Brier 1995; 1999). Also, Maturana's and Varela's *autopoetic analysis* — which in many ways is related to the biosemiotic approach — should be mentioned here, although space considerations deny me the opportunity for a comparative analysis.

Code-Tappping Animals

As mentioned, it was Thomas Sebeok who had the broadminded intellect and indefatigable energy to assemble all the threads that would serve as the foundation for the modern biosemiotic project, both intellectually and socially.[7] Yet Sebeok's own major interest was the semiotic relations between humans and animals (Sebeok 1963; 1968; 1972; 1977; 1979; Sebeok and Umiker-Sebeok 1980). In this connection, Sebeok called attention to Heini Hediger as one of the biosemiotic pioneers. Hediger was professor in ethology at the University of Zurich and the head of Zurich's zoological garden — and his contribution to the development of biosemiotics mainly stems from his deep grasp of the semiotic dynamics involved in the taming and training of wild animals (Sebeok 2001b). Hediger, as Sebeok (2001b, 19) puts it, "remained perennially intrigued by the theoretical, as well as the applied, aspects and consequences of animal taming. He famously defined *taming* as the reduction or even elimination of the 'flight-distance.' . . . In this frame of reference, he, increasingly after 1934, came to perceive, perhaps more than any other scientist in our time, an important connection with understanding possibilities for communication between man and animals that continued to concern him to the end of his life."

The *flight-distance* referred to here denotes the distance between the animal and other animals (or humans), that can not be trespassed without provoking the animal to flee or attack. Sebeok termed the space around the animal that was delimited by the flight-distance *the Hediger bubble* (see Chapter 6), in obvious reference to the Umwelt notion of Jakob von Uexküll. Unlike most other ethologists, Hediger felt comfortable with the Uexküllian Umwelt concept, and he considered Sebeok's zoosemiotics to be a legitimate extension of his work. And like Sebeok, Hediger was extremely interested in the multiple reports of animals that are alleged to be capable of executing the most surprising operations.

---

7   Sebeok's far-reaching disciplinary and social network was legendary. I have hardly mentioned a figure here from the early history of biosemiotics that Sebeok didn't know personally (apart from the long-deceased "fathers," Charles Sanders Peirce and Jakob von Uexküll).

The most well-known of these is Clever Hans — the famous German horse that, it was claimed by its owner, could "do arithmetic" by its tapping out with one hoof the correct answer to a simple arithmetic question.[8] The psychologist Oscar Pfungst took an interest in the horse and subjected its behavior to a closer examination, which disclosed that Clever Hans, firstly, could not perform its "math" unless it could see the trainer during the performance, and, secondly, that he could not answer correctly when the questioner did not himself know the correct answer. The experiments showed that the horse would just keep tapping with its foreleg until the trainer by the slightest of body movements, unwittingly, indicated that it should stop.

Commenting on this paradigmatic story, Hediger (1974, 27–28) remarks, "The apparent performance of these 'code-tapping' animals is only explainable by the continually repressed fact, that the animal — be it horse, monkey, or planarian — is generally more capable of interpreting the signals emanating from humans than is converse the case. In other words, the animal is frequently the considerably better observer of the two, or is more sensitive than man; it can evaluate signals that remain hidden to man" (cited in Sebeok 2001b, 23). The Clever Hans phenomenon repeatedly reappears in new disguises and was, for example, involved again in many of the early attempts at proving apes' capacity for linguistic expression. Another recent example is the sorry story about how severely mentally retarded children were thought to exhibit surprising abilities for reading when guided by a teacher with a so-called spelling plate — an example that, by the way, once again indicates the skills of pre-linguistic children in reading bodily cues that the adults are not themselves conscious of having offered.

Another central figure in the final establishment of modern biosemiotics is Jakob von Uexküll's son, Thure von Uexküll, whose contribution to the semiotics of medical care we discussed in Chapter 10. In addition to his pioneering efforts to establish a semiotic track within psychosomatic medicine, Thure von Uexküll has himself made decisive theoretical contributions to biosemiotics and has helped in bringing his father's work into harmony with a modern Peirce-inspired biosemiotics (Uexküll 1982; 1986; 1999; Uexküll, Geigges, and Hermann 1993).

The Tartu Group

In choosing to put the Glottertal meetings as the dividing line between what we might call the initial period in the establishment of the modern interdiscipline of biosemiotics, and biosemiotics as it is presently practiced, we shall not pursue

---

8   In Chapter 6, this case was discussed as an example of semethic interaction.

the history beyond these meetings.[9] It must be mentioned, however, that among the participants in the Glottertal meetings was the Estonian ecologist Kalevi Kull, who had arrived at the biosemiotic synthesis through paths that we have not yet portrayed. Kull is a professor at Tartu University — the same university that also hosted the Russian semiotician Yuri Lotman — and Lotman nourished a strong interest in living systems, an interest that was reinforced when his son, Alexei Lotman, chose to study biology.

Kull (1999c, 117) recounts that in 1978, "the Tartu group of theoretical biology, together with similar groups from Moscow and St. Petersburg, organized a conference entitled 'Biology and Linguistics,' and this conference was held in Tartu on February 1–2 of that year. One of the key lecturers was Yuri Lotman, and many of his colleagues also participated." During the ensuing years a fruitful exchange was upheld between the Estonian and Russian researchers, and this collaboration prepared the way for an independent development of a *biosemiotic* understanding. Lotman is generally considered to be the originator of the term *semiosphere*[10] — a term that Lotman claims to have coined through his inspiration by the Russian geologist and biogeographer V. I. Vernadsky. Kull (1999c, 120–21) quotes Lotman as reporting,

> I'm reading Vernadsky with much interest and find in him many ideas of my own. ... I am amazed by one of his statements. Once in our seminar in Moscow, I was brave enough to declare my belief that "text" can exist (i.e., can be socially recognized as a text) if it is preceded by another text, and that any developed culture should be preceded by another developed culture. And now I find Vernadsky's deeply argued idea arising from his great experience of investigations in cosmic geology that life can arise only from the living, i.e., that it is preceded by life. ... Only the antecedence of a *semiotic sphere* makes a message a message. Only the existence of mind explains the existence of mind.

The Copenhagen Group

It might perhaps be appropriate to end this with the personal story of how Claus Emmeche and I also independently arrived at the concept of *biosemiotics* near the end of the 1980s.

Impressed by the radical new agendas of technological development opening up in the 1970s in biochemistry (my own discipline), I had, for an extended period in the late 1970s and early 1980s, spent a considerable amount of time studying the significance of ecological factors for technological development,

9 But again, see Favareau (2006) and the volume *Introduction to Biosemiotics* (Barbieri 2006) for more on the subsequent history of biosemiotics from that time till now.

10 But see note 4 in Chapter 1.

as seen in a macrohistorical perspective (Hoffmeyer 1982; 1987; 1988). (I presume that these kinds of studies have come to be called *ecological history*, but this concept as such didn't really exist at the time.) One result of this work was an understanding of human history as essentially interwoven into ecologically-based thresholds that could be transcended only by the introduction of radically new technological principles — principles that so changed the living conditions of people that decisive changes were bound to occur subsequently at even deeper explanatory levels (e.g., how people would see and understand their world), and that such changes would subsequently hold great consequences for philosophy, religion, and science. It followed from this understanding that the rapid introduction of *information technologies* in the 1970s should be expected to set off rather radical changes even inside single sciences such as biology and biochemistry.

Simultaneously, in 1985 Claus Emmeche started working on the thesis for his PhD degree with the aim of throwing light upon the concept of *information* as this concept was at that time being used within the life sciences. Through our frequent discussions of the set of problems connected to the application of the information concept in biology, the realization eventually became clear to us that what biologists used to call *information* was indeed, properly speaking, a lot closer to what semioticians call *signs* than it was to the denuded, quantized conception of *information* that was being used in the exact sciences (physics and IT mathematics) at the time. We therefore set out to examine the eventual gains that might be had by applying a semiotic understanding to *informational biology*. The dominating tradition in semiotics in Denmark at the time was still strongly based on French structural linguistics, and accordingly, in Emmeche's and my first publications in biosemiotics (which, at the time, we called investigations into the *semiotics of nature*) the Saussurean inspiration is still evident.

Gradually, however, it dawned upon us that the Peircean tradition offered a more obvious starting point for this *semiotics of nature* project that, at the time, we still naively imagined we had invented ourselves. The Danish physicist Peder Voetmann Christiansen was very influential in opening our eyes to the potential of Peircean semiotics. Christiansen had for years taken an intense interest in Peirce's semiotics and especially in the possibility he saw for building a nonparadoxical interpretation of quantum mechanics based upon Peircean semiotic philosophy. Christiansen was at the time the leading figure in a cross-disciplinary study group in Copenhagen among philosophers, scientists, and other people with a keen interest in taking a scientific approach to "the big questions." And through the many meetings of this little group, he came to exert a decisive influence upon the development of a Peirce-inspired semiotics of nature in Denmark. If today we may, with a certain right, talk about a new *Copenhagen school*[11] in semiotics, it is now not so much Hjelmslev that we should thank, as it is Christiansen.

But of course, and as always, it is forebears such as Niels Bohr, Søren Kierkegaard, and Hans Christian Andersen that constitute an inescapable — although largely unconscious — ballast for all attempts of Danish thinkers to reach a theoretical grasp of the world.

11   This designation was suggested by Frederik Stjernfelt (2002, 337), who himself has contributed to the school — which, by the way, has more the character of a local milieu for discussions than of any close-knit movement or doctrine.

Literature

Ader, R., and N. Cohen (1975). "Behaviorally Conditioned Immunosuppression." *Psychosomatic Medicine* 37: 333–40.

——— (1993). "Psychoneuroimmunology: Conditioning and Stress." *Annual Review of Psychology* 44: 53–85.

Alberch, P. (1982). "Developmental Constraints in Evolutionary Processes." *Evolution and Development.* J. T. Bonner (ed.) Berlin, Springer, 313–32.

Albertsen, L. (1990). "Hvor Kommer Ordet Omverden fra?" *OMverden* 1(2): 39.

Allen, C., M. Bekoff, and G. Lauder, Eds. (1998). *Nature's Purposes: Analysis of Function and Design in Biology.* Cambridge, MA, MIT Press.

Allen, G. E. (1975). *Life Science in the Twentieth Century.* New York, John Wiley.

Altmann, S. (1967). "The Structure of Primate Social Communication." *Social Communication among Primates.* S. Altmann, Ed. Chicago, University of Chicago Press.

Ambros, V. (2004). "The Functions of Animal MicroRNAs." *Nature* 431: 350–55.

Ammundsen, R., and G. V. Lauder (1994). "Function without Purpose: The Uses of Causal Role Function in Evolutionary Biology." *Biology and Philosophy* 9: 443–69.

Anderson, M. (1998). "Apoptosis, the Eradication of Old Ideas, and the Persistence of Weeds." *Semiotica* 120(3/4): 231–41.

———, et al. (1984). "A Semiotic Perspective on the Sciences: Steps toward a New Paradigm." *Semiotica* 52: 7–47.

Andrade, L. E. (1999). "Natural Selection and Maxwell's Demons: A Semiotic Approach to Evolutionary Biology." *Semiotica* 127(1/4, Special Issue on Biosemiotics): 133–49.

Avery, L., C. I. Bargmann, and H. R. Horovitz (1993). "The *Caenorhabditis Elegans* unc–31 Gene Affects Multiple System-Controlled Functions." *Genetics* 134: 455–64.

Bains, P. (2001). "Umwelten." *Semiotica* 134(1/4): 137–66.

——— (2007). *The Primacy of Semiosis: An Ontology of Relations*. Toronto, Toronto University Press.

Baldwin, J. M. (1896). "A New Factor in Evolution." *The American Naturalist* 30: 442–43.

——— (1902). *Development and Evolution*. New York, Macmillan.

Barbieri, M. (2001). *The Organic Codes: The Birth of Semantic Biology*. Ancona, Italy, peQuod.

——— (2003). *The Organic Codes: An Introduction to Semantic Biology*. Cambridge, Cambridge University Press.

———, Ed. (2006). *Introduction to Biosemiotics: The New Biological Synthesis*. Dordrecht, Netherlands, Springer.

Bateson, G. (1963). "The Role of Somatic Change in Evolution." *Evolution* 17, Reprinted in Bateson 1972, 346–63.

——— (1968). "Redundancy and Coding." *Animal Communication: Techniques of Study and Results of Research*. T. A. Sebeok, Ed. Bloomington, Indiana University Press: Reprinted in G. Bateson 1972, 411–25.

——— (1972). *Steps to an Ecology of Mind*. New York, Ballantine Books.

——— (1979). *Mind and Nature: A Necessary Unity*. New York, Bentam Books.

Behe, M. J. (1996). *Darwin's Black Box: The Biochemical Challenge to Evolution*. New York, The Free Press.

Ben-Jacob, E., et al. (2004). "Bacterial Linguistic Communication and Social Intelligence." *Trends in Microbiology* 12(8): 366–72.

Berg, R. L. (1960). "Evolutionary Significance of Correlation Pleiades." *Evolution* 14: 171–80.

Berque, A. (2004). "Milieu et Identité Humaine." *Annales de Géographie* 638–39: 385–99.

Bickerton, D. (1990). *Language and Species*. Chicago, University of Chicago Press.

Bickhard, M. H., and D. T. Campbell (1999). "Emergence." *Downward Causation*. P. B. Andersen, et al., Eds., Aarhus, Denmark, Aarhus University Press: 322–48.

Boyer, J. S. (1982) "Plant Production and Environment." *Science* 214: 443–48

Brenner, S. (1974). "The Genetics of *Caenorhabditis Elegans*." *Genetics* 77: 71–94.

Brier, S. (1995). "Cyber-Semiotics: On Autopoiesis, Code-Duality, and Sign Games in Biosemiotics." *Cybernetics and Human Knowing* 3(1): 3-14.

——— (1999). "Biosemiotics and the Foundation of Cybersemiotics." *Semiotica* 127(1/4): 169–98.

——— (2000). "Biosemiotics as a Possible Bridge between Embodiment in Cognitive Semantics and the Motivation Concept of Animal Cognition in Ethology." *Cybernetics and Human Knowing* 7(1): 57–76.

Brooks, D., and E. O. Wiley (1986–88). *Evolution as Entropy: Toward a Unified Theory of Biology*. Chicago, University of Chicago Press.

Brooks, R. (1991). "Intelligence Without Representation." *Artificial Intelligence Journal* (47): 139–59.

Brown, J. R. (2001). *Who Rules in Science: An Opinionated Guide to the Wars*. Cambridge, MA, Harvard University Press.

Bruin, J., and M. Dicke (2001). "Chemical Information Transfer between Wounded and Unwounded Plants: Backing Up the Future." *Biochemical Systematics and Ecology* 29: 1103–13.

Bruner, J. (1968). *Processes in Cognitive Growth: Infancy*, Worcester, MA, Clark University Press.

——— (1990). *Acts of Meaning*. Cambridge, MA, Harvard University Press.

Bruni, L. E. (2002). "Does 'Quorum Sensing' Imply a New Type of Biological Information?" *Sign Systems Studies* 30(1): 221–43.

——— (2003). A Sign-Theoretic Approach to Biotechnology. *Dissertation. Institute of Molecular Biology, University of Copenhagen*. Copenhagen.

——— (2007). "Cellular Semiotics and Signal Transduction." *Cellular Semiotics and Signal Transduction*. M. Barbieri, Ed., Springer 365–407.

Bruno, G. (1584 (2000)). *De la causa, principia e uno*. Danish translation from Ialian by Ole Jorn: "Om årsagen, princippet og enheden" Copenhagen: Reitzel.

Burkhardt, R. D. (1977). *The Spirit of System: Lamarck and Evolutionary Biology*. Cambridge, MA, Harvard University Press.

Buss, L. (1987). *The Evolution of Individuality*. Princeton, NJ, Princeton University Press.

Campbell, D. T. (1974). "Downward Causation." *Studies in the Philosophy of Biology*. F. I. Ayala and T. Dobzhansky, Eds., Berkeley, University of California Press: 179–86.

Cannon, W. B. (1932). *The Wisdom of the Body*. New York, Norton.

Carson, R. (1962). *Silent Spring*. New York, Houghton Mifflin.

Chalmers, D. (1996). *The Conscious Mind*. Oxford, Oxford University Press.

Chandler, J., and G. Van de Vijver, Eds. (2000). *Closure: Emergent Organizations and Their Dynamics*. New York, New York Academy of Sciences.

Chen, S. H., et al. (1987). "Apolipoprotein B-48 is the Product of a Messenger RNA with an Organ-Specific In-Frame Stop Codon." *Science* 238: 363–66.

Cheng, M.-F. (1992). "For Whom Does the Female Dove Coo? A Case for the Role of Vocal Self-Stimulation." *Animal Behaviour* 43: 1035–44.

Chomsky, N. (1965). *Aspects of the Theory of Syntax*. Cambridge, MA, MIT Press.

——— (1972). *Language and Mind*. New York, Harcourt Brace Jovanovich.

Christiansen, P. V. (1999). "Macro and Micro-Levels in Physics." *Downward Causation: Minds, Bodies and Matter*. P. B. Andersen, et al., Eds., Aarhus, Denmark, Aarhus University Press: 51–62.

——— (2002). "Habit Formation as Symmetry Breaking in the Early Universe." *Sign Systems Studies* 30(1): 347–60.

Churchland, P. M. (1991). "Folk Psychology and the Explanation of Human Behavior." *The Future of Folk Psychology*. J. D. Greenwood, Ed. Cambridge, Cambridge University Press.

Churchland, P. (1984). *Matter and Consciousness*. Cambridge, MA, MIT Press.

Churchland, P. S. (1986). *Neurophilosophy: Toward a Unified Theory of Mind-Brain*. Cambridge, MA, MIT Press.

Clark, A. (1997). *Being There: Putting Brain, Body, and World Together Again*. Cambridge, MA, MIT Press.

——— (2002). "Is Seeing All It Seems? Action, Reason and the Grand Illusion." *Journal of Consciousness Studies* 9(5/6): 181–202.

Coleman, W. (1977). *Biology in the Nineteenth Century: Problems of Form, Function, and Transformation*. Cambridge, Cambridge University Press.

Collier, J. (2000). "Autonomy and Process Closure as the Basis for Functionality." *Closure: Emergent Organizations and Their Dynamics.* J. Chandler and G. Van de Vijver, Eds., New York, Annals of the New York Academy of Sciences 901: 280–90.

Cooper, B. (1981). *Michel Foucault: An Introduction to His Thoughts.* New York, Edwin Mellan.

Coulombre, A. J. (1965). "The Eye." *Organogenesis.* R. L. DeHaan and H. Ursprung, Eds. New York, Holt, Rinehart and Winston, 219–51.

Crick, F. (1988). *What Mad Pursuit.* New York, Basic Books.

Cummins, R. (1975). "Functional Analysis." *Journal of Philosophy* 72: 741–65.

Damasio, A. (1994). *Descartes' Error: Emotion, Reason, and the Human Brain.* New York, Putnam Books.

Danesi, M. (2001). "Layering Theory and Human Abstract Thinking." *Cybernetics and Human Knowing* 8(3): 5–24.

Darwin, C. (1981 (1871)). *The Descent of Man, and Selection in Relation to Sex.* Princeton, NJ, Princeton University Press. (Free Online Books: http://www.darwin-literature.com/The_Descent_Of_Man/5.html)

Dawkins, R. (1976). *The Selfish Gene.* Oxford, Oxford University Press.

——— (1977). *Det selviske gen.* Copenhagen, Fremad.

——— (1982). *The Extended Phenotype: The Long Reach of the Gene.* Oxford, Oxford University Press.

——— (1989). *The Selfish Gene,* New Edition. Oxford, Oxford University Press.

Deacon, T. (1997). *The Symbolic Species.* New York, Norton.

——— (2002). "Problemet med Memer." *Kritik* (155/156): 120–26.

——— (2003). "Multilevel Selection in a Complex Adaptive System: The Problem of Language Origins." *Evolution and Learning: The Baldwin Effect Reconsidered.* B. Weber and D. Depew, Eds., Cambridge, MA, MIT Press, 81–106.

de Duve, C. (1991). *Blueprint for a Cell: The Nature and Origin of Life.* Burlington, NC, Neil Patterson.

Deely, J. (1990). *Basics of Semiotics.* Bloomington, Indiana University Press.

——— (1994). "How Does Semiosis Effect Renvoi?" *The American Journal of Semiotics* 11(1/2): 11–61.

——— (2001). *Four Ages of Understanding: The First Postmodern Survey of Philosophy from Ancient Times to the Turn of the Twenty-First Century.* Toronto, Toronto University Press.

Dembski, W. A. (1999). *Intelligent Design: The Bridge between Science and Theology.* Downers Grow, IL, InterVarisity Press.

Dennett, D. C. (1987). *The Intentional Stance.* Cambridge, MA, MIT Press.

——— (1991). *Consciousness Explained.* London, Allan Lane, Penguin.

——— (1995). *Darwin's Dangerous Idea.* New York, Simon and Schuster.

Depew, D. (2003). "Baldwin and His Many Effects." *Evolution and Learning: The Baldwin Effect Reconsidered.* B. Weber and D. Depew, Eds., Cambridge, MA, MIT Press: 3–31.

———, and B. Weber (1995). *Darwinism Evolving: Systems Dynamics and the Genealogy of Natural Selection.* Cambridge, MA, MIT Press.

Dewey, J. (1948). *Reconstruction in Philosophy* (Enlarged edition with a new forty page introduction by the author). Boston, Beacon.

Dinesen, A. M. and F. Stjernfelt (1994). "Om Semiotik og Pragmatisme." *Charles Sanders Peirce: Semiotik og Pragmatisme.* A. M. Dinesen and F. Stjernfelt, Eds., Copenhagen, Gyldendal, 7–24.

Donald, M. (1991). *Origin of the Modern Mind: Three Stages in the Evolution of Culture and Cognition.* Cambridge, MA, Harvard University Press.

Driesch, H. (1908). *Philosophie des Organischen.* Leipzig, Wilhelm Engelmann.

Eco, U. (1976). *A Theory of Semiotics.* Bloomington, Indiana University Press.

Eder, J., and H. Rembold (1992). "Biosemiotics — a Paradigm of Biology: Biological Signaling on the Verge of Deterministic Chaos." *Naturwissenshaften* 79(2): 60–67.

Einstein, A., and L. Infeld (1938). *The Evolution of Physics.* New York, Simon and Schuster.

El-Hani, Charbel Nino, and Antonio Marcos Pereira (2000). "Higher-Level Descriptions: Why Should We Preserve Them?" in *Higher-Level Descriptions: Why Should We Preserve Them?* P. B. Andersen, C. Emmeche, N. O. Finnemann, and P. V. Christiansen, Eds., Århus, Århus University Press: 118 –42.

El-Hani, C.N., J, Queiroz, and C. Emmeche (2006). "A Semiotic Analysis of the Genetic Information System." *Semiotica* 160 (1/4): 1–68.

Emmeche, C. (1992). "Modeling Life: A Note on the Semiotics of Emergence and Computation in Artificial and Natural Living Systems." *Biosemiotics: The Semiotic Web 1991*. T. A. Sebeok and J. Umiker-Sebeok, Eds., Berlin, Mouton de Gruyter: 77–99.

——— (1994). *The Garden in the Machine: The Emerging Science of Artificial Life*. Princeton, NJ, Princeton University Press.

——— (1997). "Den biosemiotiske tanke." *Anvendt Semiotik*. K. G. Jørgensen, Ed. Copenhagen, Gyldendal: 62–94.

——— (1998). "Kognitionsforskning." *Den Store Danske Encyklopædi*. J. Lund, Ed. Copenhagen. 11: 49–51.

——— (1999a). "The Biosemiotics of Emergent Properties in a Pluralist Ontology." *Semiosis, Evolution, Energy: Towards a Reconceptualization of the Sign*. E. Taborsky, Ed. Aachen, Germany, Shaker Verlag: 89–108.

——— (1999b). "The Sarkar Challenge to Biosemiotics: Is There Any Information in a Cell?" *Semiotica* 127(1/4): 273–93.

——— (2001). "Does a Robot have an Umwelt? Reflections on the Qualitative Biosemiotics of Jakob von Uexküll." *Semiotica*, Special Issue (K. Kull ed.), *Jakob von Uexküll: A Paradigm for Biology and Semiotics* 134(1/4): 653-693.

——— (2002). "Taking the Semiotic Turn, or How Significant Philosophy of Biology Should Be Done." *SATS — Nordic Journal of Philosophy* 3(1): 155–62.

——— (2004). "Organicism and Qualitative Aspects of Self-Organization." *Revue Internationale de Philosophie* (Special issue on Self-Organization) 228, 205–218.

——— and J. Hoffmeyer (1991). "From Language to Nature: The Semiotic Metaphor in Biology." *Semiotica* 84(1/2): 1–42.

———, Simo Køppe, et al. (2000). "Levels, Emergence, and Three Versions of Downward Causation." *Levels, Emergence, and Three Versions of Downward Causation*. P. B. Andersen, C. Emmeche, N. O. Finnemann, and P. V. Christiansen, Eds. Aarhus, Aarhus University Press: 13 - 34

———, K. Kull, and F. Stjernfelt, Eds. (2002). *Reading Hoffmeyer, Rethinking Biology* (Tartu Semiotics Library). Tartu, Estonia, University of Tartu.

Etxeberria, A. (1998). "Embodiment of Natural and Artificial Agents." *Evolutionary Systems: Biological and Epistemological Perspectives on Selection and Self-Organization*. G. Van de Vijver, S. Salthe, and M. Delpos, Eds., Dordrecht, Netherlands, Kluwer Academic Publishers: 397–412.

Favareau, D. (2001). "Beyond Self and Other: On the Neurosemiotic Emergence of Intersubjectivity." *Sign Systems Studies* 30 (1), 57–100.

——— (2002). "Constructing Representema: On the Neurosemiotics of Self and Vision." *SEED* 2 (4), 3–24.

——— (2006). "The Evolutionary History of Biosemiotics." *Introduction to Biosemiotics: The New Biological Synthesis*. M. Barbieri, Ed. Dordrecht, Netherlands: Springer: pp. 1–67.

——— (2008). "Collapsing the Wave Function of Meaning: The Epistemological Matrix of Talk-in-Interaction." *A Legacy for Living Systems: Bateson as Precursor to Biosemiotics*. J. Hoffmeyer, Ed., Dordrecht, Netherlands, Springer: 169-211.

Ferry, L. (1992). *Le Nouvel Ordre Écologique*. Paris, Editions Grasset.

Fodor, J. (1975). *The Language of Thought*. Cambridge, MA, Harvard University Press.

Fogel, A. (1993). "Two Principles of Communication, Co-Regulation and Framing." *New Perspectives in Early Communicative Development*. J. Nadel and L. Camaioni, Eds., London, Routledge: 9–22.

Foster, M. L. (1978). "The Symbolic Structure of Primordial Language." *Human Evolution: Biosocial Perspectives*. S. L. Washburn and E. R. McCown, Eds., Menlo Park, CA, Benjamin/Cummings: 77–121.

Foucault, M. (1970). *The Order of Things: An Archaeology of the Human Sciences*. London, Tavistock.

Frølund, S. (2002). "Teleology and the 'Natural History of Signification': The Implications of Hans Jonas's Bioontology for Biosemiotics." Unpublished manuscript.

Fyrand, O. (1997). *Det gådefulde sprog: Om hudens kommunikation*. Copenhagen, Gyldendal.

Gallant, J. A., and D. Lindsley (1998). "Ribosomes Can Slide Over and Beyond 'Hungry' Codons, Resuming Protein Chain Elongation Many Nucleotides Downstream." *Proceedings of the National Academy of Sciences of the United States of America* 95: 13771–76.

Garstang, W. (1922). "The Theory of Recapitulation: A Critical Restatement of the Biogenetic Law." *Zoological Journal of the Linnean Society*. 35: 81–101.

Ghez, C., et al. (1995). "Contributions of Vision and Proprioception to Accuracy in Limb Movements." *The Cognitive Neurosciences*. M. S. Gazzaniga, Ed. Cambridge, MA, MIT Press: 549–64.

Gibbs, W. (2001). "Cybernetic Cells." *Scientific American* 265(2): 52--57.

Gibson, J. J. (1979). *The Ecological Approach to Visual Perception*. Boston, Houghton Mifflin.

Gilbert, S. F. (1991a). *Developmental Biology* (Third Edition) Sunderland, MA, Sinauer.

——— (1991b). "The Role of Embryonic Induction in Creating Self." *Organism and the Origins of Self*. A. I. Tauber, Ed. Dordrecht, Netherlands, Kluwer: 341–60.

Gimpel, J. (1978). *Den Industrielle Revolution*. Copenhagen, Gyldendal.

Goodwin, B. (1989). "Evolution and the Generative Order." *Theoretical Biology: Epigenetic and Evolutionary Order from Complex Systems*. B. Goodwin and P. Saunders, Eds., Edinburgh, Edinburgh University Press.

——— (1994). *How the Leopard Changed Its Spots*. New York, Charles Scribner's.

Goodwin, C. (2003). "The Semiotic Body in Its Environment". *Discourses of the Body*. J. Coupland and R. Gwyn, Eds., New York, Palgrave Macmillan, 19–42.

Gordon, D. (1995). "The Development of Organization in an Ant Colony." *American Scientist* 83: 50–57.

——— (1999). *Ants at Work: How an Insect Society is Organized*. New York, The Free Press.

Gottlieb, G. (1981). "Roles of Early Experience in Species-Specific Perceptual Development." *Development of Perception*. R. N. Aslin, J. R. Alberts and M. P. Petersen, Eds., New York, Academic Press: 5–44.

Gould, S. J. (1989). *Wonderful Life: The Burgess Shale and the Nature of History*. New York, Norton.

——— (1996). "Triumph of the Root-Heads." *Natural History* 105: 10–17.

——— and S. Vrba (1982). "Exaptation — a Missing Term in the Science of Form." *Paleobiology* 8: 4–15.

Grimes, J. (1996). "On the Failure to Detect Changes in the Scene across Saccades." *Perception (Vancouver Studies in Cognitive Science*, Vol. 5. K. Akins, Ed., New York, Oxford University Press: 89–109.

Griffiths, P. E., and R. D. Gray (1994). "Developmental Systems and Evolutionary Explanations." *Journal of Philosophy* 91: 277–304.

Guilford, J. (1967). *The Nature of Human Intelligence.* New York, McGraw-Hill.

Haken, H. (1984). *The Science of Structure: Synergetics.* New York, Van Nostrand Reinhold.

Halvorson, H., and R. Clifton (2002). "No Place for Particles in Relativistic Quantum Theories?" *Philosophy of Science* 69: 1–28.

Hamburger, J. (1988). *The Heritage of Experimental Embryology.* Oxford, Oxford University Press.

Harries-Jones, P. (1995). *Ecological Understanding and Gregory Bateson.* Toronto, Toronto University Press.

Hartshorne, C. (1970). *Creative Synthesis and Philosophic Method*, Chicago, Open Court.

Havel, I. M., and A. Markos, Eds. (2002). *Is There a Purpose in Nature? How to Navigate Between the Scylla of Mechanism and the Charybdis of Teleology.* Prague, Vesmir.

Hediger, H. (1974). "Communication between Man and Animal." *Image Roche* 62: 27–40.

Hendriks-Jansen, H. (1996). *Catching Ourselves in the Act: Situated Activity, Interactive Emergence, and Human Thought.* Cambridge, MA, MIT Press.

Hinton, G. E., and S. J. Nowlan (1996 (1987)). "How Learning Can Guide Evolution." *Adaptive Individuals in Evolving Populations: Models and Algorithms.* R. Belew and M. Mitchell, Eds., Reading, MA, Addison-Wesley: 447–53.

Hoffman, M. (1992). "The Enemy of My Enemy is My Friend." *American Scientist.* 80: 536–37.

Hoffmeyer, J. (1975). *Dansen om Guldkornet: En bog om biologi og samfund.* Copenhagen, Gyldendal.

——— (1982). *Samfundets naturhistorie.* Copenhagen, Rosinante.

——— (1987). "The Constraints of Nature on Free Will." *Free Will and Determinism*. V. Mortensen and R. C. Sorensen, Eds., Aarhus, Denmark, Aarhus University Press: 188–200.

——— (1988). "The Historical Logic of Domestication." *The Triumph of Biotechnologies: The Domestication of the Human Animal, Acte du Cours de l'Inter-University Centre*, Dubrovnik, March 1986. G. Thill and P. Kemp, Eds., Namur, Belgium, Presse Universitaires de Namur: 107–15.

——— (1992). "Some Semiotic Aspects of the Psycho-Physical Relation: The Endo-Exosemiotic Boundary." *Biosemiotics: The Semiotic Web 1991*. T. A. Sebeok and J. Umiker-Sebeok, Eds., Berlin, Mouton de Gruyter: 101–23.

——— (1993). *En snegl på vejen. Om betydningens naturhistorie*. Copenhagen, Rosinante/Munksgaard.

——— (1994a). "The Global Semiosphere." *Semiotics Around the World: Proceedings of the Fifth Congress of the International Association for Semiotic Studies*, Berkeley, Berlin, Mouton de Gruyter: 933–36.

——— (1994b). "The Swarming Body." *Fifth Congress of The International Association for Semiotic Studies*, Berkeley, Berlin, Mouton Gruyter: 37–40.

——— (1995). "The Swarming Cyberspace of the Body." *Cybernetics and Human Knowing* 3(1): 16–25. .

——— (1996a). "Evolutionary Intentionality." *Third European Conference on Systems Science*, Rome, Edzioni Kappa: 99-703.

——— (1996b). *Signs of Meaning in the Universe*. Bloomington, Indiana University Press.

——— (1997a). "Biosemiotics: Towards a New Synthesis in Biology." *European Journal for Semiotic Studies* 9(2): 355–76.

——— (1997b). "Semiotic Emergence." *Revue de la Pensée d'aujourd'hui* 25–7 (6): 105–17 (in Japanese; English version available from http://www.molbio.ku.dk/MolBioPages/abk/PersonalPages/Jesper/Publications.html).

——— (1998a). "Semiosis and Biohistory: A Reply." *Semiotica* 120(3/4): 455–82.

——— (1998b). "Surfaces inside Surfaces. On the Origin of Agency and Life." *Cybernetics and Human Knowing* 5(1): 33–42.

——— (1998c). "The Unfolding Semiosphere." *Evolutionary Systems: Biological and Epistemological Perspectives on Selection and Self-Organization*. G. Van de Vijver, S. Salthe, and M. Delpos, Eds., Dordrecht, Netherlands, Kluwer: 281–94.

——— (1999a). "Order out of Indeterminacy." *Semiotica* 127(Biosemiotica II, Special Issue on Biosemiotics): 321–43.

——— (1999b). "The Vague Boundaries of Life." *Semiosis, Evolution, Energy: Towards a Reconceptualization of the Sign*. E. Taborsky, Ed., Aachen, Germany, Shaker Verlag: 151–70.

——— (2000a). "The Biology of Signification." *Perspectives in Biology and Medicine* 43(2): 252–68.

——— (2000b). "Code-Duality and the Epistemic Cut." *Closure, Emergent Organizations and Their Dynamics*. J. L. R. Chandler and G. Van de Vijver, Eds., New York, Annals of the New York Academy of Sciences 901: 175–86.

——— (2001a). "Life and Reference."*Biosystems*, Special Issue (Guest edited by Luis Mateus Rocha), *The Physics and Evolution of Symbols and Codes: Reflections on the Work of Howard Pattee* 60(1/3): 123–30.

——— (2001b). "S/E ≥ 1. A Semiotic Understanding of Bioengineering." *Sign System Studies* 29(1): 277–91.

——— (2001c). "Seeing Virtuality in Nature." *Semiotica,* Special Issue (K. Kull ed.), *Jakob von Uexküll: A Paradigm for Biology and Semiotics* 134(1/4): 381–98.

——— (2002a). "The Central Dogma: A Joke That Became Real." *Semiotica* 138(1): 1–13.

——— (2002b). "Obituary: Thomas A. Sebeok." *Sign Systems Studies* 30(1): 383–86.

——— (2004). *Uexküllian Planmässigkeit*. Proceedings from the Conference "Signs and the Design of Life — Uexküll's Significance Today." *Sign Systems Studies* 32 (1/2), 73–97.

——— (2007). "Semiotic Scaffolding of Living Systems". *Introduction to Biosemiotics*. M. Barbieri, Ed., Dordrecht, Netherlands, Springer:149–66.

——— (2008a). "From Thing to Relation: On Bateson's Bioanthropology." In *A Legacy for Living Systems: Bateson as a Precursor to Biosemiotics*. J. Hoffmeyer, Ed., Dordrecht, Netherlands, Springer. 27-44.

———, Ed. (2008b). *A Legacy for Living Systems: Gregory Bateson as Precursor to Biosemiotics*. Dordrecht, Netherlands, Springer.

——— and C. Emmeche (1991). "Code-Duality and the Semiotics of Nature." *On Semiotic Modeling*. M. Anderson and F. Merrell, Eds., New York, Mouton de Gruyter: 117–66.

——— (2005 (1991)). "Code-Duality and the Semiotics of Nature." *Journal of Biosemiotics* 1 (1), 27–64.

———, and K. Kull (2003). "Baldwin and Biosemiotics: What Intelligence is For." *Evolution and Learning: The Baldwin Effect Reconsidered*. B. Weber and D. Depew, Eds., Cambridge, MA, MIT Press: 253 –72.

Hogan, N., et al. (1987). "Controlling Multi-Joint Behavior." *Exercise and Sport Science Review* 15: 153–90.

Holley, A. J. (1993). "Do Brown Hares Signal to Foxes?" *Ethology* 94: 21–30.

Hollick, J. B., J. E. Dorweiler, and V. L. Chandler (1997). "Paramutation and Related Allelic Interactions." *Trends in Genetics* 13: 302–8.

Hui, J., T. Cashman, and T. Deacon. (2008). "Bateson's Method: Double Description. What Is It? How Does It Work? What Do We Learn?" *A Legacy for Living Systems: Gregory Bateson as Precursor to Biosemiotics*. J. Hoffmeyer, Ed. Dordrecht, Netherlands, Springer: 77–92.

Hull, D. (1980). "Individuality and Selection." *Annual Reviews of Ecology, Evolution, and Systematics* 11: 311–32.

Hutchinson, G. E. (1957). "Concluding Remarks." *Cold Spring Harbor Symposia on Quantitative Biology* 22: 415–27.

Innis, R. (2007). "Placing Langer's Philosophical Project." *Summer Institute in American Philosophy*, Boulder, July 14 .

Jablonka, E. (2002). "Information: Its Interpretation, Its Inheritance, and Its Sharing." *Philosophy of Science* 69: 578–605.

———, and M. Lamb (1995). *Epigenetic Inheritance and Evolution: The Lamarckian Dimension*. Oxford, Oxford University Press.

Jacob, F. (1974). *The Logic of Living Systems: A History of Heredity*. London, Allen Lane.

——— (1982). *The Possible and the Actual*. New York, Pantheon.

Jakobsen, R. (1973). *Main Trends in the Science of Language*. London, George Allen & Unwin Ltd.

Ji, S. (2002). "Microsemiotics of DNA." *Semiotica* 134: 1–18.

Johnson, M. (1987). *The Body in the Mind*. Chicago, Chicago University Press.

Johnstone, R. A. (1997). "The Evolution of Animal Signals." *Behavioral Ecology: An Evolutionary Approach*. J. R. Krebs and N. B. Davies, Eds., Oxford, Blackwell: 155–78.

Jonas, H. (1966, (2001)). *The Phenomenon of Life*. New York, Harper & Row.

——— (1973). *Organismus und freiheit. Ansätze zu einer philosophischen Biologie*. Göttingen, Germany, Vandenhoeck und Ruprecht.

Jordan, M., T. Flash, and Y. Arnon (1994). "A Model of the Learning of Arm Trajectories from Spatial Deviations." *Journal of Cognitive Neuroscience* 6(4): 359–76.

Juarrero, A. (1998). "Causality as Constraint." *Evolutionary Systems: Biological and Epistemological Perspectives in Selection and Self-Organization*. G. Van de Vijver, Ed., Dordrecht, Netherlands, Kluwer: 233–42.

——— (1999). *Dynamics in Action: Intentional Behavior as a Complex System*. Cambridge, MA, MIT Press.

Kampis, G. (1998). "Evolution as Its Own Cause and Effect." *Evolutionary Systems: Biological and Epistemological Perspectives on Selection and Self-Organization*. G. Van de Vijver, S. Salthe, and M. Delpos, Eds., Dordrecht, Netherlands, Kluwer: 255–65.

Karatay, V., and Y. Denizhan (2002). "Semiotics of the 'Window.'" *Sign Systems Studies* 30 (1), 259–70.

Kauffman, S. A. (1993). *Origins of Order: Self-Organization and Selection in Evolution*. Oxford, Oxford University Press.

——— (1995). *At Home in the Universe*. Oxford, Oxford University Press.

——— (2000). *Investigations*. Oxford, Oxford University Press.

Kawade, Y. (1992). "A Molecular Semiotic View of Biology: Interferon and 'Homeokine' as Symbol." *Rivista di Biologia / Biology Forum* 85: 71–78.

——— (1996). "Molecular Biosemiotics: Molecules Carry out Semiosis in Living Systems." *Semiotica* 111(3/4): 195–215.

Keller, Evelyn Fox (1995). *Refiguring Life: Metaphors of Twentieth-century Biology*. New York, Wellek Library Lectures, Columbia University Press.

Kelso, J. A. S. (1995). *Dynamic Patterns: The Self-Organization of Brain and Behavior*. Cambridge, MA, MIT Press.

Kemp, P. (1991). *Det uerstattelige. En teknologi-etik*. Copenhagen, Spektrum.

———, M. Lebech, and J. Rendtorff (1997). *Den bioetiske vending. En grundbog i bioetik*. Copenhagen, Spektrum.

Kierkegaard, S. (1944). *The Sickness unto Death*. Oxford, Oxford University Press. First published in 1849, in Danish.

Kilstrup, M. (1998). "Biokemi og Semiotik." *Anvendt Semiotik*. K. G. Jørgensen, Ed., Copenhagen, Gyldendal: 95–120.

——— (2000). lecture notes, Copenhagen, PhD course, Institute of Molecular Biology.

Kim, J. (1990). "Supervenience as a Philosophical Concept." *Metaphilosophy* 21(1/2): 1–27.

Krampen, M. (1992). "Phytosemiotics Revisited." *Biosemiotics: The Semiotic Web 1991*. T. A. Sebeok and J. Umiker-Sebeok, Eds., Berlin, Mouton de Gruyter: 213–20.

Kull K. (1992). "Evolution and Semiotics." *Biosemiotics: Semiotic Web 1991*. T. A. Sebeok and J. Umiker-Sebeok, Eds. Berlin, Mouton de Gruyter, 221–33.

——— (1993). "Semiotic Paradigm in Theoretical Biology." *Lectures in Theoretical Biology: The Second Stage*. K. Kull and T. Tiivel, Eds., Tallin, Estonian Academy of Sciences: 52–62.

——— (1998). "Semiotic Ecology: Different Natures in the Semiosphere." *Sign Systems Studies* 26: 344–369.

——— (1999a). "Biosemiotics in the Twentieth Century: A View from Biology." *Semiotica* 127 (Biosemiotica 2): 385–414.

——— (1999b). "Outlines for a Post-Darwinian Biology." *Folia Baeriana* 7: 129–142.

——— (1999c). "Towards Biosemiotics with Yuri Lotman." *Semiotica* 127: 115–31.

——— (1999d). "Umwelt and Evolution: From Uexküll to Post-Darwinism." *Semiosis, Evolution, Energy: Towards a Reconceptualization of the Sign*. E. Taborsky, Ed. Aachen, Germany, Shaker Verlag: 53–70.

——— (2000). "Organisms Can Be Proud to Have Been Their Own Designers." *Cybernetics and Human Knowing* 7(1): 45–55.

Lakoff, G. (1987). *Woman, Fire, and Dangerous Things: What Categories Reveal about the Mind*. Chicago, University of Chicago Press.

———, and M. Johnson (1999). *Philosophy in the Flesh*. New York, Basic Books.

Lamarck, J. B. (1809). *Philosophie Zoologique, ou Exposition des Considération Relatives a l'Histoire Naturelle des Animaux*. Paris, Dentu.

Langer, S. (1942). *Philosophy in a New Key: A Study in the Symbolism of Reason, Rite, and Art*. New York, The New American Library.

——— (1953). *Feeling and Form*. New York, Scribner's.

——— (1967–82). *Mind: An Essay on Human Feeling*, vol. 1–3. Baltimore, Johns Hopkins University Press.

Langton, C., Ed. (1989). *Artificial Life: The Proceedings of an Interdisciplinary Workshop on the Synthesis and Simulation of Living Systems Held September 1987 in Los Alamos*. Redwood City, CA, Addison-Wesley.

LeDoux, J. (1996). *The Emotional Brain: The Mysterious Underpinnings of Emotional Life*. New York, Simon & Schuster.

Lehrman, D. S. (1953 (2001)). "A Critique of Konrad Lorenz's Theory of Instinctive Behavior." *Cycles of Contingency: Developmental Systems and Evolution*. S. Oyama, P. E. Griffiths, and R. D. Gray, Eds., Cambridge, MA, MIT Press. 25–40.

——— (1970). "Semantic and Conceptual Issues in the Nature-Nurture Problem." *Development and the Evolution of Behavior*. D. S. Lehrman, Ed., San Francisco, Freeman: 17–52.

Lestel, D. (1995). *Paroles des singe. L'impossible dialogue homme-primate*. Paris, Éditions la Découverte.

Lewontin, R. C. (1983). «Gene, Organism, and Environment.» *Evolution from Molecules to Men*. D. S. Bendall, Ed., Cambridge, Cambridge University Press: 273–85.

——— (1992). "The Dream of the Human Genome." *The New York Review* May 28: 31–40.

Lieberman, P. (1975). "On the Evolution of Language: A Unified View." *Primate Functional Morphology and Evolution*. R. H. Tuttle, Ed., The Hague, Netherlands, Morton, 501-40.

Lloyd Morgan, C. (1896). "Of Modification and Variation." *Science* 4(99): 733–39.

Longa, V. (2006). "A Misconception about the Baldwin Effect: Implications for Language Evolution". *Folia Linguistica* 40 (3/4): 305–18.

Lorenz, K. (1939). "Vergleichende Verhaltensforschung." *Zoologsche Anzeitung* 12 (Suppl. band): 69–112.

——— (1971). *Studies in Animal and Human Behaviour*. Cambridge, MA, Harvard University Press.

Lotman, Y. M. (1990). *Universe of the Mind: A Semiotic Theory of Culture*. London, I. B. Taurus and Co.

Lund, J., Ed. (2001). *Den Store Danske Encyclopædi*. Copenhagen, Gyldendal.

Luria, S. E., S. J. Gould, and S. Singer (1981). *A View of Life*. Menlo Park, CA, Benjamin/Cummings.

Margulis, L. (1970). *Origin of Eukaryotic Cells: Evidence and Research Implications for a Theory of the Origin and Evolution of Microbial, Plant, and Animal Cells on the Precambrian Earth*. New Haven, CN/London, Yale University Press.

——— (1981). *Symbiosis in Cell Evolution: Life and Its Environment on Earth*. San Francisco, Freeman.

———, and R. Fester, Eds. (1991). *Symbiosis as a Source of Evolutionary Innovation: Speciation and Morphogenesis*. Cambridge, MA, MIT Press.

Margulis, L., and D. Sagan (2003). *Acquiring Genomes: A Theory of the Origins of Species*. New York, Basic Books.

Markos, A. (2002). *Readers of the Book of Life: Contextualizing Developmental and Evolutionary Biology*. Oxford, Oxford University Press.

Martinelli, D. (2002). *How Musical Is a Whale? Towards a Theory of Zoomusicology*. Helsinki Acta Semiotica Fennica.

Mathews, C., K. E. van Holde, and K. G. Ahern (1999). *Biochemistry* (Third Edition). San Francisco, Addison-Wesley.

Matsuno, K. (1989). *Protobiology: Physical Basis of Biology*. Boca Raton, FL, CRC Press.

——— (1996). "Internalist Stance and the Physics of Information." *Biosystems* 38, 111–18.

———, and S. Salthe (1995). "Global Idealism/Local Materialism." *Biology and Philosophy* 10: 309–37.

Maturana, H., and F. Varela (1980). *Autopoiesis and Cognition: The Realization of the Living*. Dordrecht, Netherlands, Reidel.

McFall-Ngai, J., and E. G. Ruby (1998). "Sepiolids and Vibrios: When First They Meet." *BioScience* 48(4): 257-265

McFarland, D. (1987). *Oxford Companion to Animal Behaviour*. Oxford, Oxford University Press.

McShea, D. W. (1991). "Complexity and Evolution: What Everybody Knows." *Biology and Philosophy* 6: 303–21.

Merleau-Ponty, M. (1995 (1968)). *La Nature: Notes, Cours Du Collège de France*. Paris, Editions du Seuil.

——— (2002 (1945)). *Phenomenology of Perception*, Transl. by Colin Smith from *Phénoménologie de la Perception*. London, Routledge.

Merrell, F. (1992). "As Signs Grow, So Life Goes." *Biosemiotics: The Semiotic Web 1991*. T. A. Sebeok and J. Umiker-Sebeok, Eds., Berlin, Mouton de Gruyter: 251–81.

Millikan, R. G. (1989). "In Defense of Proper Function." *Philosophy of Science* 56: 288–302.

——— (1998). "In Defense of Proper Function." *Nature's Purposes: Analysis of Function and Design in Biology*. C. Allen, M. Bekoff, and G. Lauder, Eds., Cambridge, MA, MIT Press: 295–312.

Milner, D., and M. Goodale (1995). *The Visual Brain in Action*. Oxford, Oxford University Press.

——— (1998). "The Visual Brain in Action (Precis)." *Psyche* 4(12), http://psyche.cs.monash.edu.au/v4/psyche-4-12-milner.html

Moeller, A. (1993). "Fungus Infecting Domestic Flies Manipulates Sexual Behavior of Its Host." *Behavioral Ecology and Sociobiology* 33(6): 403–07.

Monod, J. (1970). *Le Hazard at la Nécessité. Essai sur la Philosophie Naturelle de la Biologie Moderne*. Paris, Seuil. (English edition, *Chance and Necessity*, was published in 1971; see below.)

——— (1971a). *Tilfældigheden og nødvendigheden*. Copenhagen, Fremad.

——— (1971b). *Chance and Necessity: An Essay on the Natural Philosophy of Modern Biology*. New York: Knopf.

Morgan, T. H., et al. (1915). *The Mechanism of Mendelian Heredity*. New York, Henry Holt.

Morowitz, H. (1992). *Beginnings of Cellular Life: Metabolism Recapitulates Biogenesis*. New Haven, CN, Yale University Press.

Moss, L. (2001). "Deconstructing the Gene and Reconstructing Molecular Developmental Systems." *Cycles of Contingency: Developmental Systems and Evolution*. S. Oyama, P. E. Griffiths, and R. D. Gray, Eds., Cambridge, MA, MIT Press: 85–97.

Muller, H. J. (1964). "The Relation of Recombination to Mutational Advance." *Mutation Research* 1: 2–9.

Muskvitin, J. (1987). "Vitalisme." *Naturens Historiefortællere*. N. Bonde, J. Hoffmeyer and H. Stangerup, Eds., Copenhagen, Gads Forlag. 2: 322–34.

Naess, A. (1989). *Ecology, Community and Lifestyle: An Outline of an Ecosophy*. Cambridge, Cambridge University Press.

Nagel, T. (1986). *The View from Nowhere*. Oxford, Oxford University Press.

Neander, K. (1991). "Functions as Selected Effects: The Conceptual Analyst's Defense." *Philosophy of Science* 58: 168–84.

Neuman, Y. (2005). "'Meaning-Making' in Language and Biology." *Perspectives in Biology and Medicine* 48: 320–27.

Neumann-Held, E. M. (1998). "The Gene is Dead — Long Live the Gene: Conceptualizing the Gene the Constructionist Way." *Developmental Systems, Competition and Cooperation in Sociobiology and Economics*. P. Koslowsky, Ed., Berlin, Springer: 105–37.

Newell, A., and H. Simon (1975). "Computer Science as Empirical Enquiry." *Communication of the Association for Computing Machinery* 6: 113–26.

Noë, A. (2002). "Is the Visual World a Grand Illusion?" *Journal of Consciousness Studies* 9(5/6): 1–12.

Nöth, W. (2000). *Handbuch der Semiotik. 2., vollständig neu bearbeitede und erweiterte Auflage*. Stuttgart, Verlag J. B. Metzler.

——— (2001). "Biosemiotica." *Cybernetics and Human Knowing* 8(1/2): 157–60.

Nunn, C.(2007). Review in *Journal of Consciousness Studies* 14, 127–29.

Odling-Smee, F. J. (1988). Niche Constructing Phenotypes. *The Role of Behavior in Evolution*. H. C. Plotkin, Ed. Cambridge, MA, MIT Press: 72–132.

——— (2001). "Niche Construction, Ecological Inheritance, and Cycles of Contingency in Evolution." *Cycles of Contingency: Developmental Systems and Evolution*. S. Oyama, P. E. Griffiths, and R. D. Gray, Eds., Cambridge, MA, MIT Press: 117–26.

———, K. N. Laland, and M. W. Feldman (1996). "Niche Construction." *American Naturalist* 147(4): 641–48.

Odling-Smee, F. J., and B. Patten (1994). "The Genotype-Phenotype-Envirotype Complex: Ecological and Genetic Inheritance in Evolution." *Manuscript*.

Olmstead, J. M. D. (1938). *Claude Bernard, Physiologist*. New York, Harper & Brothers.

O'Regan, J. K., and A. Noë (2001). "A Sensorimotor Account of Vision and Visual Consciousness." *Behavioral and Brain Science* 24(5): 939–1031.

Oyama, Susan (1985). *The Ontogeny of Information*. Cambridge, Cambridge University Press.

——— (2003). "On Having a Hammer". *Evolution and Learning. The Baldwin Effect Reconsidered*, B. Weber and D. Depew, Eds., Cambridge, MA, MIT Press, 169–91.

———, et al., Eds. (2001). *Cycles of Contingency. Developmental Sytems and Evolution*. Cambridge, MA, MIT Press.

Panksepp, J. (2001). "The Neuro-Evolutionary Cusp between Emotions and Cognition." *Evolution and Cognition* 7(2): 141–63.

Pape, H. (1993). "Final Causality in Peirce's Semiotics and the Classification of the Sciences." *Transactions of the Charles S. Peirce Society* 29(4): 581–607.

Paterson, H. E. H. (1993). *Evolution and the Recognition Concept of Species: Collected Writings*. S. F. McEvey, Ed., Baltimore, Johns Hopkins University Press.

Pattee, H. (1972). "Laws and Constraints, Symbols, and Languages." *Towards a Theoretical Biology*. C. H. Waddington, Ed., Edinburgh, University of Edinburgh Press. 4: 248–58.

——— (1977). "Dynamic and Linguistic Modes of Complex Systems." *International Journal for General Systems* 3: 259–66.

——— (1997). "The Physics of Symbols and the Evolution of Semiotic Controls." In: Coombs M. , editor. Proc Workshop on Control Mechanisms for Complex Systems. Addison-Wesley; 1997 http://www.ssie.binghamton.edu/pattee/semiotic.html.

Peirce, C. S. (1908). A Letter to Lady Welby dated Dec.23. *Semiotics and Significs: The Correspondence Between Charles S. Peirce and Victoria Lady Welby.* Charles S. Hardwick & J. Cook (eds.) 1977. Bloomington, Indiana University Press.

——— (1931–35, 1958). *Collected Papers of Charles Sanders Peirce, vols. 1 - 6,* C. Hartstone and P. Weiss, Eds., *vols. 7–8* A. W. Burks, Ed., Cambridge, MA, Harvard University Press.

——— (1955). Philosophical Writings of Peirce. J. Buchler, Ed. New York, Dover.

Petrilli, S. (1999). "Charles Morris's Biosemiotics." *Semiotica* 127(1/4): 67–102.

Piaget, J. (1953). *The Origin of Intelligence in Children*, London, Routledge & Kegan Paul.

——— (1968). *La Naissance de l'iIntelligence chez l'enfant,* 6th ed. Neuchatel, Switzerland, Delachaux et Niestlé.

——— (1976). *The Grasp of Consciousness: Action and Concept in the Young Child*, Cambridge, MA, Harvard University Press.

Pinker, S. (1994). *The Language Instinct: The New Science of Language and Mind.* London, Penguin.

Polanyi, M. (1958). *Personal Knowledge.* London, Routledge.

——— (1968). "Life's Irreducible Structure." *Science* 160: 1308–12.

Polit, A., and E. Bizzi (1978). "Processes Controlling Arm Movements in Monkeys." *Science* 201: 1235–37.

Popper, K. (1990). *A World of Propensities.* Bristol, Thoemmes Antiquarian Books.

Port, R. (1998). "The Dynamical Approach to Cognition: Inferences from Language." *Advanced Studies in Semiotics: Biosemiotics and Cognitive Semiotics*, Farias P. and J. Queiros, Eds. Sao Paulo, Sao Paulo Catholic University: 93–122

———, and T. van Gelder, Eds. (1995). *Mind as Motion: Explorations in the Dynamics of Cognition.* Cambridge, MA, MIT Press.

Preer, J. R., Jr. (1996). "Tracy Morton Sonneborn." *National Academy of Sciences, Biographical Memoirs* 69: 268–92.

Prigogine, I., and I. Stengers (1984). *Order out of Chaos*. London, Heinemann.

——— (1985). *Den nye pagt mellem mennesket og universet*. Aarhus, Denmark, Ask.

Ramachandran, V. S., and S. Blakeslee (1998). *Phantoms in the Brain: Human Nature and the Architecture of the Mind*. New York, William Morrow & Co.

Rayner, A. D. M. (1997). *Degrees of Freedom: Living in Dynamic Boundaries*. London, Imperial College Press.

Regan, T. (1983). *The Case for Animal Rights*. Berkeley, University of California Press.

Resnick, M. (1994). *Turtles, Termites, and Traffic Jams: Explorations in Massively Parallel Microworlds*. Cambridge, MA, MIT Press.

Riddle, D. (1988). "The Dauer Larva." *The Nematode Caenorhabditis elegans*. W. Wood, Ed. Cold Spring Harbor, NY, Cold Spring Harbor Laboratory Press: 393–412.

Riedl, R. (1997). "From Four Forces back to Four Causes." *Evolution and Cognition* 3(2): 148–58.

Robbins, B., and A. Ross (1996). "Mystery Science Theater." *Lingua Franca* July/August, http://physics.nyu.edu/faculty/sokal/mstsokal.html

Rocha, L. (1998). "Syntactic Autonomy." *Proceedings of the Joint Conference on the Science and Technology of Intelligent Systems*, National Institute of Standards and Technology, Gaithersburg, MD, http://informatics. indiana.edu/rocha/sa.html#N_1_

——— (2001). "Evolution with Material Symbol Systems." *BioSystems*, Special Issue, (Guest edited by Luis Mateus Rocha): *The Physics and Evolution of Symbols and Codes: Reflections on the Work of Howard Pattee*, 60(1/3): 95–121.

Roepstorff, A. (2001). "Brains in Scanners. An Umwelt of Cognitive Neuroscience." *Semiotica* 134(1/4): 747–65.

———, and A. I. Jack (2004). "Trust or Interaction? Editorial Introduction." *Journal of Consciousness* 11(7/8): v–xxii.

Rosen, R. (1991). *Life Itself: A Comprehensive Inquiry into the Nature, Origin and Fabrication of Life*. New York, Columbia University Press.

Rothschild, F. S. (1962). "Laws of Symbolic Mediation in the Dynamics of Self and Personality." *Annals of the New York Academy of Sciences* 96: 774–84.

Rousseau, J.-J. (1910 (1755)). *Discours sur l'Origine et les fondements del'Inégalité parmi les Hommes*. Amsterdam, M. M. Rey, 1755.

Ruse, M. (1979). *The Darwinian Revolution*. Chicago, University of Chicago Press.

Russell, B., and A. N. Whitehead (1910–13). *Principia Mathematica vol. I –III*. Cambridge, Cambridge University Press.

Russell, M. J. (1971). "Human Olfactory Communication." *Nature* 260: 520-522.

Sacks, O. (1986). *A Leg to Stand on*. London, Pan.

Salthe, S. (1993). *Development and Evolution: Complexity and Change in Biology*. Cambridge, MA, MIT Press.

——— (1999). "A Semiotic Attempt to Corral Creativity via Generativity." *Semiotica* 127(1/4): 481–95.

Santaella-Braga, L. (1999). "A New Causality for the Understanding of the Living: Biosemiotics." *Semiotica* 127(1/4): 497–519.

Sarkar, S. (1996). "Biological Information: A Skeptical Look at Some Central Dogmas of Molecular Biology." *The Philosophy and History of Molecular Biology: New Perspectives*. S. Sarkar, Ed., Dordrecht, Netherlands, Kluwer: 187–231.

——— (1997). "Decoding 'coding': Information and DNA." *European Journal for Semiotic Studies* 9(2): 227–32.

Savage-Rumbaugh, S., and R. Lewin (1994). *Kanzi: The Ape at the Brink of Human Mind*. New York, John Wiley.

Schaffner, K. F. (1998). "Genes, Behavior, and Developmental Emergentism: One Process, Indivisible?" *Philosophy of Science* 65: 209–52.

Schegloff, E. (2007). *Sequence Organization in Interaction Analysis*. Cambridge, Cambridge University Press.

Schleidt, M., B. Hold, and C. Attili (1981). "A Cross-Cultural Study on the Attitude Towards Personal Odours." *Journal of Chemical Ecology* 7(1): 19–31.

Schrödinger, E. (1944). *What is Life? The Physical Aspect of the Living Cell*. Cambridge, Cambridge University Press.

Scriver, C. R., and P. J. Waters (1999). "Monogenic Traits Are Not Simple." *Trends in Genetics* 15(7): 267–72.

Searle, J. R. (1992). *The Rediscovery of Mind*. Cambridge, MA, MIT Press.

Sebeok, T. A. (1963). "Communication in Animals and Men." *Language* 39: 448–66.

——— (1968). "Is a Comparative Semiotics Possible?" *Echange et Communications: Mélanges offerts a Claude Lévi-Strauss à l'occasion de son 60ème anniversaire.* J. Pouillon and P. Maranda, Eds., The Hague, Netherlands, Mouton. 614–27.

——— (1972). *Perspectives in Zoosemiotics.* The Hague, Netherlands, Mouton.

———, Ed. (1977). *How Animals Communicate.* Bloomington, Indiana University Press.

——— (1979). *The Sign and Its Masters,* Austin, University of Texas Press.

——— (1985 (1976)). *Contributions to the Doctrine of Signs.* Bloomington, Indiana University Press.

——— (1986.) *I Think I Am A Verb: More Contributions to the Doctrine of Signs.* New York, Plenum Press

——— (1987). "Toward a Natural History of Language." *Semiotica* 65(3/4): 343–58.

——— (1988). "'Animal' in Biological and Semiotic Perspective." *What is an Animal?* T. Ingold, Ed., London, Unwin Hyman: 63–76.

——— (1999). "The Music of the Spheres." *Semiotica* 28(3/4): 527-33.

——— (2001a). "Biosemiotics: Its, Roots, Proliferation, and Prospects." *Semiotica* 134 (1/4): 61–78.

——— (2001b). *The Swiss Pioneer in Nonverbal Communication Studies Heini Hediger (1908–1992).* New York, Legas.

———, and J. Umiker-Sebeok (1980). *Speaking of Apes: A Critical Anthology of Two-Way Communication with Man.* New York, Plenum Books.

———, Eds. (1992). *Biosemiotics: The Semiotic Web 1991.* Berlin, Mouton de Gruyter.

Sengupta, P., et al. (1993). "The Cellular and Genetic Basis of Olfactory Responses in Caenorhabditis elegans." *The Molecular Basis of Smell and Taste Transduction* (Ciba Foundation Symposium 179). D. Chadwick, J. Marsh and J. Goode, Eds., Chichester, UK, Wiley: 235–50.

Sercarz, E. E., et al., Eds. (1988). *The Semiotics of Cellular Communication in the Immune System.* Berlin, Springer.

Seyfarth, R. M., and D. L. Cheney (1992). "Meaning and Mind in Monkeys." *Scientific American* 267: 122–29.

Shannon, C., and W. Weaver (1949). *The Mathematical Theory of Communication*. Urbana, University of Illinois Press.

Shapiro, J. A. (1999). "Genome System Architecture and Natural Genetic Engineering in Evolution." *Molecular Strategies in Biological Evolution*. L. H. Caporale, Ed. New York, Annals of the New York Academy of Sciences 870: 23–35.

Sharkey, N., and T. Ziemke (2001a). "Life, Mind and Robots: The Ins and Outs of Embodiment." *Symbolic and Neural Net Hybrids*. S. Wermter and R. Sun, Eds., Cambridge, MA, MIT Press.

——— (2001b). "Mechanistic versus Phenomenal Embodiment: Can Robot Embodiment Lead to Strong AI?" *Cognitive Systems Research* 2 (4): 251–62.

Sharov, A. A. (1992). "Biosemiotics: A Functional-Evolutionary Approach to the Analysis of the Sense of Information." *Biosemiotics: The Semiotic Web 1991*. T. A. Sebeok and J. Umiker-Sebeok, Eds., Berlin, Mouton de Gruyter: 345–74.

Shavit, Y., et al. (1986). "Involvement of Brain Opiate Receptors in the Immunosuppresive Effect of Morphine." *Proceedings of the National Academy of Sciences of USA* 83: 7114–17.

Sheets-Johnstone, M. (1990). *The Roots of Thinking*. Philadelphia, Temple University Press.

——— (1998). "Consciousness: A Natural History." *Journal of Consciousness Studies* 5(3): 260–94.

——— (1999). "Sensory-Kinetic Understandings of Language." *Evolution of Communication* 3(2): 149–83.

Shintani, L. (1999). "Roman Jakobsen and Biology: 'A System of Systems.'" *Semiotica* 127(1/4): 103–13.

Short, T. L. (2002). "Darwin's Concept of Final Cause: Neither New nor Trivial." *Biology and Philosophy* 17: 323–40.

Simpson, G. G. (1949). *The Meaning of Evolution; A Study of the History of Life and Its Significance for Man*. New Haven, CN, Yale University Press.

——— (1953). "The Baldwin Effect." *Evolution* 7: 110–17.

Skyrms, B. (2002). "Signals, Evolution and the Explanatory Power of Transient Information." *Philosophy of Science* 69(3): 407–28.

Sober, E. (1993). *Philosophy of Biology*. Boulder CO, Westview Press.

Soler, M., and J. J. Soler (1999). "Innate versus Learned Recognition of Conspecifics in Great Spotted Cuckoos *Clamator glandarius*." *Animal Cognition* 2: 97–102.

Solomon, E. P., L. R. Berg, and D. W. Martin (1999). *Biology*. Philadelphia, PA, Saunders College Publishing

Sonea, S. (1991). "The Global Organism." *The Semiotic Web 1990*. T. A. Sebeok and J. Umiker-Sebeok, Eds., Berlin, Mouton de Gruyter.

——— (1992). "Half of the Living World Was Unable to Communicate for About One Billion Years. *Biosemiotics: The Semiotic web 1991*. T. Sebeok and J. Umiker-Sebeok, Eds., Berlin, Mouton de Gruyter: 375–92.

Stepanov, Y. (1971). *Semiotika*. Moscow, Nauka.

Sterelny, K., and P. E. Griffiths (1999). *Sex and Death: An Introduction to Philosophy of Biology*. Chicago, University of Chicago Press.

Sternberg, R. v. (2000). Genomes, Form, Epigenetic Influences, and Morphological Attractors: "The Case for Teleomorphic Recursivity." *Closure: Emergent Organizations and Their Dynamics*. J. Chandler and G. Van de Vijver, Eds., New York, New York Academy of Sciences: 224–36.

Stjernfelt, F. (1992). "Categorial Perception as a General Prerequisite to the Formation of Signs? On the Biological Range of a Deep Semiotic Problem in Hjelmslev's as Well as Peirce's Semiotics." *Biosemiotics: The Semiotic Web 1991*. T. A. Sebeok and J. Umiker-Sebeok, Eds., Berlin, Mouton de Gruyter: 427–54.

——— (1999). "Formal Ontology." *Semiotica* 127(1/4): 537–66.

——— (2000). *Biology, Abstraction, Schemata*. Rationality and Irrationality: Proceedings of the Kirchberg Wittgenstein Conference 2000, Vienna.

——— (2001). "A Natural Symphony? To what Extent is Uexküll's *Bedutungslehre* Actual for the Semiotics of Our Time?" *Semiotica*, Special Issue (K. Kull ed.), *Jakob von Uexküll: A Paradigm for Biology and Semiotics* 134(1/4): 79–102.

——— (2006). *Diagrammatology: An Investigation on the Borderlines of Phenomenology, Ontology, and Semiotics*. Aalborg, Denmark, Aalborg Universitetscenter.

——— (2007). *Diagrammatology. An Investigation on the Borderlines of Phenomenology, Ontology, and Semiotics*. Dordrecht, Netherlands, Springer Verlag, Synthese Library.

Strohman, R. (2001). "The Human Genome Project in Crisis: Where is the Program for Life?" *California Monthly* 4, www.biotech-info.net/StrohmanMarch09.pdf

Swenson, R. (1999). "Epistemic Ordering and Development of Space-Time: Intentionality as a Universal Entailment." *Semiotica* 127 (1/4): 567–98.

———, and M. T. Turvey (1991). "Thermodynamic Reasons for Perception-Action Cycles." *Ecological Psychology* 3(4): 317–48.

Symonds, J. A. (1893 (reprint of 3rd ed.1920)). *Studies of the Greek Poets*. London, A. & C. Black.

Taborsky, E. (2001a). "The Internal and the External Semiosic Properties of Reality." Retrieved from the World Wide Web at: http://www.library.utoronto.ca/see/SEED/Vol1-1/Taborsky-Journal1.html

——— (2001b). "What is a Sign?" *Journal of Literary Semantics,* 30 (2): 12.

——— (2002). "Energy and Evolutionary Semiosis." *Sign Systems Studies* 30(1): 361–81.

Taylor, P. (2005). *Unruly Complexity: Ecology, Interpretation, Engagement*. Chicago, University of Chicago Press.

Thelen, E., and L. Smith (1994). *A Dynamic Systems Approach to the Development of Cognition and Action*. Cambridge, MA, MIT Press.

Thibault, P. J. (1998). "Code." *Encyclopedia of Semiotics*. P. Bouissac, Ed. Oxford, Oxford University Press: 125–29.

Tinbergen, N. (1942). "An Objectivistic Study of the Innate Behavior of Animals." *Biotheca Biotheoretica* 1: 39–98.

Tomkins, G. M. (1975). "The Metabolic Code." *Science* 189: 760–63.

Turlings, T. C. J., J. H. Tumlinson, and W. J. Lewis (1990). "Exploitation of Herbivore-Induced Plant Odors by Host-Seeking Parasitic Wasps." *Science* 250: 1251–53.

Uexküll, J. v. (1973 (1928)). *Theoretische Biologie*. Frankfurt, Suhrkamp.

——— (1982 (1940)). "The Theory of Meaning." *Semiotica* 42(1): 25–87.

Uexküll, T. v. (1953). "Ein Beitrag zur Pathologie und Klinik der Bereistellungsregulationen." *Verhandlungen des Deutschen Gesselshaft Innere Medizin* 59: 104–7.

——— (1982). "Introduction: Meaning and Science in Jakob von Uexküll's Concept of Biology." *Semiotica* 42(1): 1–24.

——— (1986). From Index to Icon. A Semiotic Attempt at Interpreting Piaget's Developmental Theory. *Iconicity: Essays on the Nature of Culture. Festschrift for Thomas A. Sebeok on his 65th birthday.* P. Bouissac, M. Herzfeld and R. Posner, Eds., Tübingen, Germany, Stauffenberg Verlag: 119–40.

——— (1999). "The Relationship between Semiotics and Mechanical Models: Of Explanation in the Life Sciences." *Semiotica* 127(1/4): 647–55.

———, W. Geigges, and J. M. Hermann (1993). "Endosemiosis." *Semiotica* 96(1/2): 5–52.

Ulanowicz, R. E. (1997). *Ecology, the Ascendent Perspective.* New York, Columbia University Press.

Ulvestad, E. (2007). *Defending Life: The Nature of Host-Parasite Relations.* Dordrecht, Netherlands, Springer.

Van de Vijver, G. (1996). "Internalism versus Externalism: A Matter of Choice?" (in Japanese). *Contemporary Philosophy* 24(11): 93–101.

———, S. Salthe, and M. Delpos, Eds. (1998). *Evolutionary Systems: Biological and Epistemological Perspectives on Selection and Self-Organization.* Dordrecht, Netherlands, Kluwer.

van Gelder, T., and R. Port (1995). "It's About Time: Overview of the Dynamical Approach to Cognition." *Mind or Motion: Explorations in the Dynamics of Cognition.* R. Port and T. van Gelder, Eds., Cambridge, MA, MIT Press: 1–43.

Varela, F. (1991). "Organism: A Meshwork of Selfless Selves." *Organism and the Origins of Self.* A. I. Tauber, Ed. Dordrecht, Netherlands, Kluwer: 79–107.

Vehkavaara, T. (2003). "Natural Interests: Interactive Representation, and the Emergences of Objects and Umwelt." *Sign System Studies* 30 (2): 547–87. Vernadsky, V. I. (1926). *Biosfera.* Leningrad, Nauka (French version: Paris 1929).

——— (1945). "The Biosphere and the Noösphere." *American Scientist* 33: 1–12.

von Neumann, J. (1955). *The Mathematical Foundations of Quantum Mechanics*. Princeton, NJ, Princeton University Press.

Vygotsky, L. S. (1986). *Thought and Language*. Cambridge, MA, MIT Press.

Waddington, C. H. (1956). "Genetic Assimilation of the Bithorax Phenotype." *Evolution* 10(1): 1–13.

——— (1957). *The Strategy of the Genes*. London, Allen & Unwin.

———, Ed. (1968–72). *Towards a Theoretical Biology, Vol. 1–4*. Chicago, Aldine.

Weber, A. (2002). "Origins of Meaning in the Biological Philosophy of Susanne Langer and Hans Jonas." *Sign Systems Studies* 30(1): 183–200.

Weber, B. (1998a). "Emergence of Life and Biological Selection from the Perspective of Complex Systems Dynamics." *Evolutionary Systems: Biological and Epistemological Perspectives on Selection and Self-Organization*. G. Van de Vijver, S. Salthe, and M. Delpos, Eds., Dordrecht, Netherlands, Kluwer: 59–66.

——— (1998b). "Origins of Order in Dynamical Models." *Biology and Philosophy* 13: 133–44.

———, et al. (1989). "Evolution in Thermodynamic Perspective: An Ecological Approach." *Biology and Philosophy* 4: 373–405.

West, M. L. (1971). *Early Greek Philosophy*. Oxford, Clarendon Press.

Wheeler, W. (2006). *The Whole Creature: Complexity, Biosemiotics and the Evolution of Culture*. London, Lawrence & Wishart.

Whitehead, A. N. (1929 (1978)). *Process and Reality: An Essay in Cosmology*. New York, The Free Press.

Wiener, N. (1962). *Cybernetics: or Control and Communication in the Animal and the Machine*. Cambridge MA, MIT Press.

Wilden, A. (1980). *System and Structure*. New York, Tavistock.

Wiles, J., et al. (2005). "Transient Phenomena in Learning and Evolution: Genetic Assimilation and Genetic Redistribution." *Artificial Life*, 11 (1/2), 1–13.

Wilkinson, R. G. (1973). *Poverty and Progress: An Ecological Model of Economic Development*. London, Methuen.

Wilson, E. O. (1975). *Sociobiology: The New Synthesis*. London, Belknap Press.

Wolpert, L., et al. (2002). *Principles of Development*. Oxford, Oxford University Press.

Yarbus, A. (1967). *Eye Movements and Vision*. New York, Plenum Press.

Yates, E. F. (1985). "Semiotics as Bridge Between Information (Biology) and Dynamics (Physics)." *Recherches Sémiotiques/ Semiotic Inquiry* 5: 347–60.

———, Ed. (1987). *Self-Organizing Systems: The Emergence of Order*. New York, Plenum.

——— (1992). "On the Emergence of Chemical Languages." *Biosemiotics: The Semiotic Web 1991*. T. A. Sebeok and J. Umiker-Sebeok, Eds., Berlin, Mouton de Gruyter: 471–86.

——— (1998). "Biosphere as Semiosphere." *Semiotica* 120 (3/4), Special Issue, *Semiotics in the Biosphere: Reviews and Rejoinder*: 439–53.

———, and P. N. Kugler (1984). "Signs, Singularities and Significance: A Physical Model for Semiotics." *Semiotica* 52(1/2): 49–77.

Zahavi, D. (1999). *Self-Awareness and Alterity: A Phenomenological Investigation*. Evanston, IL, Northwestern University Press.

Zeki, S. (1993). *A Vision of the Brain*. Oxford, UK, Blackwell.

——— (1999). *Inner Vision: An Exploration of Art and the Brain*. Oxford, Oxford University Press.

Ziemke, T., and N. Sharkey (2001). "A Stroll through the Worlds of Robots and Animals: Applying Jakob von Uexküll's Theory of Meaning to Adaptive Robots and Artificial Life." *Semiotica* 134 (1–4), 701–46.

Index